Deep Learning for Biomedical Image Reconstruction

Discover the power of deep neural networks for image reconstruction with this state-of-the-art review of modern theories and applications. The background theory of deep learning is introduced step by step, and by incorporating modeling fundamentals this book explains how to implement deep learning in a variety of modalities, including X-ray, CT, MRI, and others. Real-world examples demonstrate an interdisciplinary approach to medical image reconstruction processes, featuring numerous imaging applications. Recent clinical studies and innovative research activity in generative models and mathematical theory will inspire the reader towards new frontiers. This book is ideal for graduate students in electrical or biomedical engineering or medical physics.

Jong Chul Ye is a Professor in the Graduate School of AI at Korea Advanced Institute of Science and Technology (KAIST), Korea. He is currently an associate editor for *IEEE Transactions on Medical Imaging*, and a Senior Editor of *IEEE Signal Processing Magazine*. He is an IEEE Fellow, and was the Chair of IEEE SPS Computational Imaging TC and IEEE EMBS Distinguished Lecturer. He is the author of *Geometry of Deep Learning: A Signal Processing Perspective* (Springer 2022).

Yonina C. Eldar is a Professor in the Department of Mathematics and Computer Science, Weizmann Institute of Science, Rehovot, Israel, where she heads the Center for Biomedical Engineering. She is also a Visiting Professor at MIT, a Visiting Scientist at the Broad Institute, and an Adjunct Professor at Duke University. She is a member of the Israel Academy of Sciences and Humanities, an IEEE Fellow and a EURASIP Fellow, and the recipient of the Technical Achievement Award of the IEEE Signal Processing Society. She is the author of *Sampling Theory* (Cambridge, 2015), and co-editor of *Convex Optimization in Signal Processing and Communications* (Cambridge, 2009), *Compressed Sensing* (Cambridge, 2012), *Information-Theoretic Methods in Data Science* (Cambridge 2021), and *Machine Learning in Wireless Communications* (Cambridge, 2022).

Michael Unser is a Professor in the Institute of Electrical and Micro Engineering, EPFL, Switzerland, where he also heads the Center for Imaging. He is a Fellow of the IEEE, an elected member of the Swiss Academy of Engineering Sciences, and a EURASIP Fellow. He was the recipient of the 2008 Technical Achievement Award of the IEEE Signal Processing Society and the 2020 Academic Career Achievement Award from the IEEE Engineering in Medicine and Biology Society. He is co-author of *An Introduction to Sparse Stochastic Processes* (Cambridge 2014).

"This book is a comprehensive collection authored by luminaries in the imaging field. It is a must read for newcomers to image reconstruction."

Jeffrey A. Fessler, *University of Michigan*

"This is a timely textbook on medical imaging reconstruction with deep learning. The chapters related to MRI would be very useful for preparing students to work on fast imaging and multiple contrasts. The chapter on phase unwrapping is highly relevant to recent MRI advancements, including quantitative susceptibility mapping. I would recommend this book to all my students."

Yi Wang, *Cornell University*

"*Deep Learning for Biomedical Image Reconstruction* is an outstanding guide that immerses readers in the captivating realm of deep learning and its profound impact on medical imaging. Written by esteemed experts in the field, this comprehensive resource offers a meticulous introduction to the theory and implementation of deep learning across diverse modalities, including X-ray, CT, MRI, and more. The book's interdisciplinary approach, complemented by real-world illustrations and groundbreaking research insights, makes it an invaluable companion for graduate students in electrical or biomedical engineering, medical imaging, or physics. The authors' commendable expertise and contributions in deep learning, signal processing, and image reconstruction ensure that readers gain a profound comprehension of the subject matter. This book stands as an indispensable reference for anyone seeking to explore the forefront of medical imaging advancements and the transformative potential of deep learning in revolutionizing healthcare."

Xiaofeng Yang, *Emory University*

Deep Learning for Biomedical Image Reconstruction

Edited by

JONG CHUL YE
Korea Advanced Institute of Science and Technology (KAIST)

YONINA C. ELDAR
Weizmann Institute of Science, Israel

MICHAEL UNSER
École Polytechnique Fédérale de Lausanne, Switzerland

Shaftesbury Road, Cambridge CB2 8EA, United Kingdom

One Liberty Plaza, 20th Floor, New York, NY 10006, USA

477 Williamstown Road, Port Melbourne, VIC 3207, Australia

314–321, 3rd Floor, Plot 3, Splendor Forum, Jasola District Centre, New Delhi – 110025, India

103 Penang Road, #05–06/07, Visioncrest Commercial, Singapore 238467

Cambridge University Press is part of Cambridge University Press & Assessment, a department of the University of Cambridge.

We share the University's mission to contribute to society through the pursuit of education, learning, and research at the highest international levels of excellence.

www.cambridge.org
Information on this title: www.cambridge.org/9781316517512

DOI: 10.1017/9781009042529

© Cambridge University Press & Assessment 2023

This publication is in copyright. Subject to statutory exception and to the provisions of relevant collective licensing agreements, no reproduction of any part may take place without the written permission of Cambridge University Press & Assessment.

First published 2023

Printed in the United Kingdom by CPI Group Ltd, Croydon CR0 4YY

A catalogue record for this publication is available from the British Library.

A Cataloging-in-Publication data record for this book is available from the Library of Congress.

ISBN 978-1-316-51751-2 Hardback

Cambridge University Press & Assessment has no responsibility for the persistence or accuracy of URLs for external or third-party Internet websites referred to in this publication and does not guarantee that any content on such websites is, or will remain, accurate or appropriate.

To Andy, Ella, and Joo – JCY

To my husband Shalomi and children Yonatan, Moriah, Tal, Noa, and Roei for their boundless love and for filling my life with endless happiness - YE

To Lucia, Matthias, and Monica – MU

Contents

List of Contributors		*page* xiv
Preface		xix

Part I Theory of Deep Learning for Image Reconstruction

1 Formalizing Deep Neural Networks — 3
Michael Unser
  1.1 Introduction — 3
  1.2 Primary Components of Neural Networks — 3
      1.2.1 Vectorial Representation of a Deep Neural Network — 4
  1.3 Training — 5
      1.3.1 The Backpropagation Algorithm — 6
  1.4 Categorical Loss for Classification — 9
  1.5 Good Practice for Training DNNs — 11

2 Geometry of Deep Learning — 13
Jong Chul Ye and Sangmin Lee
  2.1 Introduction — 13
  2.2 Limitations of Classical Machine Learning — 13
      2.2.1 Kernel Machines — 13
      2.2.2 Shallow Neural Networks — 14
      2.2.3 Frame Representation — 15
  2.3 Understanding Deep Neural Networks — 17
      2.3.1 Multi-Layer Perceptron — 17
      2.3.2 Convolutional Neural Networks — 19
  2.4 Generalization Capability of Deep Neural Networks — 21
      2.4.1 Deep Double Descent — 21
      2.4.2 High-Dimensional Partition Geometry for Generalization — 24
  2.5 Summary — 25

3 Model-Based Reconstruction with Learning: From Unsupervised to Supervised and Beyond — 28
Zhishen Huang, Siqi Ye, Michael T. McCann, and Saiprasad Ravishankar
  3.1 Introduction — 28
  3.2 Classical Model-Based Image Reconstruction — 28

	3.3	MBIR with Unsupervised Learning	31
		3.3.1 Synthesis Dictionary Learning	32
		3.3.2 Sparsifying Transform Learning	33
		3.3.3 Autoencoder	36
		3.3.4 Generative Adversarial Network (GAN)-Based Methods	37
		3.3.5 Deep Image Prior	38
	3.4	MBIR with Supervised Learning	39
		3.4.1 Plug-and-Play	39
		3.4.2 Unrolling	41
	3.5	Case Study: Combined Supervised–Unsupervised (SUPER) Learning	41
		3.5.1 SUPER Reconstruction	42
		3.5.2 SUPER Training	44
		3.5.3 Mathematical Underpinnings of SUPER	44
	3.6	Discussion and Future Directions	45
4	**Deep Algorithm Unrolling for Biomedical Imaging**		**53**
	Yuelong Li, Or Bar-Shira, Vishal Monga, and Yonina C. Eldar		
	4.1	Introduction	53
	4.2	Development of Algorithm Unrolling	56
		4.2.1 Iterative Shrinkage and Thresholding Algorithm	56
		4.2.2 LISTA: Learned Iterative Shrinkage and Thresholding Algorithm	57
		4.2.3 Towards a Theoretical Understanding of Algorithm Unrolling	58
		4.2.4 Unrolling Generic Iterative Algorithms	60
	4.3	Deep Algorithm Unrolling for Biomedical Imaging	61
		4.3.1 Applications of Unrolling in Computed Tomography	62
		4.3.2 Unrolling in Super-Resolution Microscopy	66
		4.3.3 Applications of Unrolling in Ultrasound	68
		4.3.4 Applications of Unrolling in Magnetic Resonance Imaging	72
		4.3.5 Unrolling Techniques across Multiple Biomedical Imaging Modalities	76
	4.4	Perspectives and Recent Trends	80
		4.4.1 Why is Unrolling So Effective for Biomedical Imaging?	80
		4.4.2 Emerging Unrolling Trends for Biomedical Imaging	81
	4.5	Conclusions	82

Part II Deep-Learning Architecture for Various Imaging Architectures

5	**Deep Learning for CT Image Reconstruction**		**89**
	Haimiao Zhang, Bin Dong, Ge Wang, and Baodong Liu		
	5.1	General Background	89
	5.2	Major Problems and Deep Solutions	90
		5.2.1 Low-Dose CT Denoising	91
		5.2.2 CT Image Super-Resolution	93
		5.2.3 Limited Angle, Sparse-View, Interior CT	93

		5.2.4	Spectral CT	97
		5.2.5	Metal-Artifacts Reduction	98
		5.2.6	Motion-Artifacts Reduction	98
	5.3	Deep-Learning-Empowered Dedicated Systems		99
		5.3.1	C-Arm CT	99
		5.3.2	Dental CT	99
		5.3.3	Cabin CT	100
		5.3.4	CT for Covid-19	100
		5.3.5	Computed Laminography	100
	5.4	Data Synthesis and Transfer		101
	5.5	Important Topics		102

6 Deep Learning in CT Reconstruction: Bringing the Measured Data to Tasks 114
Guang-Hong Chen, Chengzhu Zhang, Yinsheng Li, Yoseob Han, and Jong Chul Ye

	6.1	Introduction		114
	6.2	CT Imaging Physics and Reconstruction-Problem Formulations		115
		6.2.1	CT Image Reconstruction Problem Formulation: A Deterministic Approach	115
		6.2.2	CT Reconstruction Problem Formulation (II): Statistical Learning Framework	117
		6.2.3	CT Reconstruction Problem Formulation (III): Deep-Learning Framework	121
		6.2.4	CT Reconstruction Problem Formulation (IV): Combining Deep Learning with either Analytical or Statistical IR Framework	122
	6.3	Deep Learning in CT Reconstruction: From Sinogram to Image Directly		123
		6.3.1	iCT-Net for Image Reconstruction with Diagnostic Purpose	125
		6.3.2	ScoutCT-Net: Reconstruction of 3D Tomographic Patient Models from Two-Scout-View of Projections	130
	6.4	Deep Learning in CT Reconstruction: Hybrid Deep Learning with DBP		138
		6.4.1	Region-of-Interest (ROI) Tomography	138
		6.4.2	Cone-Beam Artifact Removal	142
	6.5	Deep Learning in CT Reconstruction: Synergy of Deep Learning and Statistical IR		150
		6.5.1	DL-PICSS Reconstruction Pipeline	151
		6.5.2	Results: Reconstruction Accuracy Quantification and Generalizability Tests	152
	6.6	Synergy of FBP, Deep Learning, and Statistical IR		156
	6.7	Summary		160

7 Overview of the Deep-Learning Reconstruction of Accelerated MRI 166
Patricia Johnson and Florian Knoll

	7.1	Overview of Image Reconstruction for Accelerated MR Imaging	166

	7.2	Parallel Imaging	167
	7.3	Compressed Sensing	168
	7.4	Machine Learning	169
	7.5	Experimental Results	171
	7.5.1	Data Acquisition and Experimental Design	171
	7.5.2	Parallel Imaging	172
	7.5.3	Compressed Sensing	172
	7.5.4	Machine Learning	173
	7.6	Summary	174

8 Model-Based Deep-Learning Algorithms for Inverse Problems — 177
Mathews Jacob, Hemant K. Aggarwal, and Qing Zou

8.1	Introduction	177
8.2	Model-Based Approaches that Rely on Shallow Learning	178
8.2.1	Image Formation and Forward Model	178
8.2.2	Model-Based Algorithms	178
8.2.3	Challenges with Traditional Model-Based Algorithms	179
8.3	Direct Inversion-Based Deep-Learning Algorithms	180
8.4	Model-Based Deep-Learning Image Recovery Using Plug-and-Play Methods	182
8.4.1	Denoising Networks	182
8.4.2	Autoencoders	182
8.4.3	Generative Adversarial Networks (GANs)	183
8.4.4	Benefits and Challenges of Plug-and-Play Methods	183
8.5	Model-Based Deep-Learning Algorithms with Unrolling and End-to-End Optimization	184
8.5.1	Benefits and Challenges	186
8.6	Model-Based Deep-Learning Reconstruction Without Pre-Learning	187
8.6.1	Single-Image Recovery using DIP	187
8.6.2	Inverse Problems Involving Multiple Images in a Manifold	188
8.7	Model-Based Deep-Learning Image Reconstruction: General Challenges, Current Solutions, and Opportunities	189
8.7.1	Lack of Distortion-Free Training Data	189
8.7.2	Vulnerability to Input Perturbations and Model Misfit	191
8.7.3	Joint Design of System Matrix and Image Recovery	193
8.8	Summary	194

9 k-Space Deep Learning for MR Reconstruction and Artifact Removal — 200
Mehmet Akcakaya, Gyutaek Oh, and Jong Chul Ye

9.1	Introduction	200
9.2	Scan-Specific k-Space Learning	202
9.2.1	Scan-Specific Neural Networks for k-Space Interpolation	202
9.3	k-Space Deep Learning using Training Data	208
9.3.1	Data-Driven k-Space Deep Learning for k-Space Interpolation	208

	9.3.2 MR Motion-Artifact Removal	213
9.4	Summary and Outlook	218

10 Deep Learning for Ultrasound Beamforming 223
Ruud J. G. van Sloun, Jong Chul Ye, and Yonina C. Eldar

- 10.1 Introduction and Relevance — 223
- 10.2 Ultrasound Scanning in a Nutshell — 224
  - 10.2.1 Focused Transmits/Line Scanning — 224
  - 10.2.2 Synthetic Aperture — 225
  - 10.2.3 Plane Wave Ultrafast — 225
- 10.3 Digital Ultrasound Beamforming — 226
  - 10.3.1 Digital Beamforming Model and Framework — 226
  - 10.3.2 Delay-and-Sum — 227
  - 10.3.3 Adaptive Beamforming — 229
- 10.4 Deep-Learning Opportunities — 231
  - 10.4.1 Opportunity 1: Improving Image Quality — 231
  - 10.4.2 Opportunity 2: Enabling Fast and Robust Compressed Sensing — 232
  - 10.4.3 Opportunity 3: Beyond MMSE with Task-Adaptive Beamforming — 234
  - 10.4.4 A Brief Overview of the State-of-the-Art — 234
  - 10.4.5 Public Datasets and Open Source Code — 235
- 10.5 Deep-Learning Architectures for Ultrasound Beamforming — 235
  - 10.5.1 Overview and Common Architectural Choices — 235
  - 10.5.2 DNN Directly on Channel Data — 236
  - 10.5.3 DNN for Beam-Summing — 236
  - 10.5.4 DNN as an Adaptive Processor — 238
  - 10.5.5 DNN for Fourier-Domain Beam-Summing — 239
  - 10.5.6 Post-Filtering after Beam-Summing — 240
- 10.6 Training Strategies and Data — 241
  - 10.6.1 Training Data — 241
  - 10.6.2 Loss Functions and Optimization — 242
- 10.7 New Research Opportunities — 245
  - 10.7.1 Multi-Functional Deep Beamformer — 245
  - 10.7.2 Unsupervised Learning — 246

11 Ultrasound Image Artifact Removal using Deep Neural Networks 252
Jaeyoung Huh, Shujaat Khan, and Jong Chul Ye

- 11.1 Introduction — 252
- 11.2 Ultrasound Artifacts — 253
  - 11.2.1 Time–Space Mismatch — 253
  - 11.2.2 Speed-of-Sound Mismatch — 254
  - 11.2.3 Attenuation — 255
  - 11.2.4 Fundamental Limitations of US Physics — 256
- 11.3 Deep Learning for US Artifact Removal — 258
  - 11.3.1 Deconvolution — 258

	11.3.2 Despeckle	262
	11.3.3 Reverberation	264
	11.3.4 Phase Aberration	267
	11.3.5 Side Lobes	268
11.4	Summary and Outlook	272

Part III Generative Models for Biomedical Imaging

12	**Image Synthesis in Multi-Contrast MRI with Generative Adversarial Networks**	279
	Tolga Çukur, Mahmut Yurt, Salman Ul Hassan Dar, Hyungjin Chung, and Jong Chul Ye	
	12.1 Introduction	279
	12.2 Physics for MR Contrast	280
	12.3 Brief Review of Generative Adversarial Networks (GANs)	281
	12.4 MR Contrast Conversion using GAN	283
	12.4.1 Unconditional GANs	283
	12.4.2 Conditional GANs	284
	12.5 Collaborative GAN for MR Contrast Conversion	287
	12.5.1 Collaborative GAN	288
	12.5.2 MR Contrast Synthesis using CollaGAN	292
	12.6 Summary and Outlook	293
13	**Regularizing Deep-Neural-Network Paradigm for the Reconstruction of Dynamic Magnetic Resonance Images**	299
	Jaejun Yoo and Michael Unser	
	13.1 Introduction	299
	13.1.1 Challenges of dMRI	299
	13.1.2 Image-Formation Model of dMRI	300
	13.2 Reconstruction Approaches in dMRI	301
	13.2.1 Sparsity and Low-Rank-Based Methods	301
	13.2.2 Manifold Learning	302
	13.2.3 Supervised Deep Learning	303
	13.3 Regularizing Deep-Neural-Network Paradigm	303
	13.3.1 Deep Image Prior (DIP)	303
	13.3.2 Time-Dependent Deep Image Prior (TD-DIP)	304
	13.4 Results	306
	13.5 Discussion	306
	13.6 Conclusion	308
14	**Regularizing Neural Network for Phase Unwrapping**	312
	Thanh-an Pham, Fangshu Yang, and Michael Unser	
	14.1 Problem Formulation	312
	14.1.1 Classical Methods	313
	14.1.2 Deep-Learning-Based Approaches	314

	14.2 Phase Unwrapping with Deep Image Prior	314
	14.2.1 Problem Formulation	314
	14.2.2 Architecture and Optimization Strategy	315
	14.3 Experiments	317
	14.3.1 Simulated Data	317
	14.3.2 Phase Images of Organoids	320
	14.4 Discussion	322
15	**CryoGAN: A Deep Generative Adversarial Approach to Single-Particle Cryo-EM**	325
	Michael T. McCann, Laurène Donati, Harshit Gupta, and Michael Unser	
	15.1 The Reconstruction Problem in Single-Particle Cryo-EM	325
	15.1.1 The Quest for Protein Structures	325
	15.1.2 Single-Particle Cryo-EM	325
	15.1.3 The Image-Formation Model	326
	15.1.4 A Challenging Reconstruction Procedure	327
	15.2 Reconstruction Approaches in Single-Particle Cryo-EM	327
	15.2.1 Projection Matching	328
	15.2.2 Maximum-Likelihood Estimation	328
	15.2.3 Method of Moments	329
	15.3 The CryoGAN Framework	329
	15.3.1 Distribution Matching	329
	15.3.2 GANs for Distribution Matching	330
	15.3.3 Mathematical Framework for CryoGAN	330
	15.4 The CryoGAN Algorithm	333
	15.4.1 The Cryo-EM Physics Simulator	333
	15.4.2 The CryoGAN Discriminator Network	335
	15.4.3 Reconstruction from a Realistic Synthetic Dataset	335
	15.4.4 Next Steps	336
	15.5 Conclusion	338

List of Contributors

Hemant Kumar Aggarwal
Staff Scientist, GE Healthcare, Bangalore, India

Mehmet Akcakaya
Department of Electrical and Computer Engineering, University of Minnesota, MN, USA

Or Bar-Shira
Weizmann Institute of Science, Rehovot, Israel

Guang-Hong Chen
Department of Medical Physics, University of Wisconsin in Madison, WI, USA

Hyungjin Chung
Department of Bio and Brain Engineering, KAIST, Daejeon, Korea

Tolga Çukur
Department of Electrical-Electronics Engineering, Bilkent University, Ankara, Turkey

Salman U. H. Dar
Department of Electrical-Electronics Engineering, Bilkent University, Ankara, Turkey

Laurène Donati
Biomedical Imaging Group, École Polytechnique Fédérale de Lausanne (EPFL), Lausanne, Switzerland
Currently with the EPFL Center for Imaging, Lausanne, Switzerland

Bin Dong
Beijing International Center for Mathematical Research (BICMR), Peking University, Beijing, China

Yonina C. Eldar
Weizmann Institute of Science, Rehovot, Israel

List of Contributors

Harshit Gupta
Biomedical Imaging Group, École Polytechnique Fédérale de Lausanne, Lausanne, Switzerland
Currently with the Stanford Linear Accelerator Center (SLAC), California, USA

Yoseob Han
Harvard Medical School and Massachusetts General Hospital, MA, USA
Currently with the Department of Electronic Engineering, Soongsil University, Seoul, Korea

Zhishen Huang
Department of Computational Mathematics, Science and Engineering, Michigan State University, East Lansing, MI, USA

Jaeyoung Huh
Department of Bio and Brain Engineering, KAIST, Daejeon, Korea

Mathews Jacob
Department of Electrical and Computer Engineering, University of Iowa, IA, USA

Patricia Johnson
Center for Biomedical Imaging, Department of Radiology, NYU Grossman School of Medicine, New York, USA

Shujaat Khan
Digital Technology & Innovation, Siemens Medical Solutions USA

Florian Knoll
Center for Biomedical Imaging, Department of Radiology, NYU Grossman School of Medicine, New York, USA, and
Friedrich-Alexander University, Erlangen, Nuremberg, Germany

Sangmin Lee
Department of Mathematical Sciences, KAIST, Daejeon, Korea

Yinsheng Li
Department of Medical Physics, University of Wisconsin in Madison, WI, USA

Yuelong Li
Amazon Inc., Santa Clara, USA

Baodong Liu
School of Nuclear Science and Technology, UCAS, Beijing, China
Institute of High Energy Physics, CAS, Beijing, China

List of Contributors

Michael T. McCann
Department of Computational Mathematics, Science and Engineering, Michigan State University, East Lansing, MI, USA
Currently with the Applied Mathematics and Plasma Physics Group (T-5), Los Alamos National Laboratory, Los Alamos, NM, USA

Vishal Monga
Pennsylvania State University, State College, USA

Gyutaek Oh
Department of Bio and Brain Engineering, KAIST, Daejeon, Korea

Thanh-an Pham
Biomedical Imaging Group, École Polytechnique Fédérale de Lausanne (EPFL), Lausanne, Switzerland
Currently with the 3D Optical Systems Group, Department of Mechanical Engineering, Massachusetts Institute of Technology, Cambridge, MA, USA

Saiprasad Ravishankar
Department of Computational Mathematics, Science and Engineering, and Department of Biomedical Engineering, Michigan State University, East Lansing, MI, USA

Michael Unser
Biomedical Imaging Group, École Polytechnique Fédérale de Lausanne (EPFL), Lausanne, Switzerland

Ruud J. G. van Sloun
Department of Electrical Engineering, Eindhoven University of Technology, Eindhoven, The Netherlands

Ge Wang
Biomedical Imaging Center, Rensselaer Polytechnic Institute, New York, USA

Fangshu Yang
Harbin Institute of Technology at Weihai, Weihai, China

Jong Chul Ye
Graduate School of AI, KAIST, Daejeon, Korea

Siqi Ye
University of Michigan - Shanghai Jiao Tong University Joint Institute, Shanghai Jiao Tong University, Shanghai, China

Currently with Department of Radiation Oncology, Stanford University, Stanford, CA, USA

Jaejun Yoo
Graduate School of AI, UNIST, Ulsan, Republic of Korea

Mahmut Yurt
Department of Electrical-Electronics Engineering, Bilkent University, Ankara, Turkey

Chengzhu Zhang
Department of Medical Physics, University of Wisconsin in Madison, WI, USA

Haimiao Zhang
Institute of Applied Mathematics, Beijing Information Science and Technology University, Beijing, China

Qing Zou
Advanced Imaging Research Center, The University of Texas Southwestern Medical Centre, TX, USA

Preface

The incredible achievements of deep learning in recent years have led to remarkable transformations within the field of data science. As compared with methods that rely on "feature engineering" to extract handcrafted features to be fed to basic classifiers, one key advantage of deep neural networks is that they autonomously identify the features and create suitable classifiers in a data-driven manner. They strive to automatically uncover and incorporate model information by optimizing network parameters learned from real-world training samples, substituting for the traditional design of tractable models and priors based on the human analysis of physical processes. This has made deep learning a highly adaptable tool in data science, leading to numerous technological advancements.

Throughout the course of the past decade, the revolution in deep learning has been driven by the availability of extensive training datasets, by the access to powerful computational resources, and by advances in neural-network research such as the development of effective architectures and efficient training algorithms. These factors have contributed to the unparalleled success of deep learning across countless applications in computer vision, pattern recognition, and speech processing. In computer vision, they have led to a considerable improvement in the accuracy of image recognition. Groundbreaking performances have been showcased by AlexNet, even surpassing humans on classification tasks over the ImageNet dataset. In signal processing, learning-based approaches offer intriguing algorithmic alternatives to traditional model-based analytical methods. Some recent, more balanced, approaches also consider the merge of conventional iterative methods and deep-learning methods, possibly in the form of unfolded networks.

As early as 2016, the highly publicized successes of deep learning involved stunning feats such as the defeat by AlphaGo of the world champion Lee Sedol in Go matches. That same year, several algorithms based on deep learning resulted in breakthrough reconstructions of biomedical images, for instance for low-dose CT reconstruction (the 2016 Low-Dose X-Ray CT Grand Challenge organized by the American Association of Physicists in Medicine), for MR reconstruction from sparse samples (the 2016 International Society of Magnetic Resonance in Medicine Annual Meeting), and for sparse-view CT reconstruction (FBPNet). Since then, deep learning has achieved tremendous progress. It is now fair to say that the incorporation of machine-learning techniques has firmly reshaped the research landscape in the reconstruction of biomedical images. As of today, deep-learning-based image reconstruction has not only become a primary research focus, it has also reached some degree of maturity and is already an essential tool for clinical applications, as testified by the recent FDA approval of a deep-learning commercial product for low-dose CT.

However, despite the tremendous success of deep learning in image reconstruction, its development has remained primarily empirical and, to the best of our knowledge, there is currently a scarcity of organized materials that would provide a comprehensive overview of these areas and cover, at the same time, the theory and specific applications for various medical-imaging modalities. In view of the significance of this subject for the biomedical-imaging community, we believe it is important to coordinate global efforts and so create a monograph that offers a coherent review of, and outlook for, this field.

Roadmap to the Book

The chapters are organized in three parts, which are briefly described below.

Part I: Theory of Deep Learning for Image Reconstruction

The reconstruction of medical images is a typically ill-posed inverse problem, in the sense that the forward operator does not admit a stable inverse. Practically, this means that many possible candidate reconstructions can coexist, each one offering perfect consistency with the measurements.

To tackle ill-posed inverse problems, mild assumptions or regularization can be imposed upon the solution space to reflect our prior knowledge of the underlying class of images. For instance, it is widely acknowledged that natural images admit a compact representation in a wavelet basis and/or exhibit a sparse gradient, which justifies a preference for reconstructions whose total variation is reasonably small. Accordingly, one traditionally formulates the reconstruction problem as the minimization of a cost functional that balances two parts: (i) a data term, which enforces consistency with the measurements; and (ii) a regularization term, which favors "desirable" solutions. This strategy is the basis for compressed sensing, which relies on sparsity-promoting regularization to reconstruct medical images (with diagnostic quality) from noisy and/or incomplete measurement data. However, since sparsity-promoting regularization is not designed specifically for medical imaging, there is ample room for the improvement afforded by data-driven regularization – in other words, by machine learning.

It turns out that the training of a neural network uses the same kind of optimization strategy as does variational image reconstruction. This is the reason why we have included an introductory chapter, Chapter 1. It presents a self-contained treatment of neural networks within the vector formalism commonly used in imaging.

In the recent theory of deep convolutional framelets, Ye and coworkers have demonstrated that a deep neural network can be interpreted as a framelet representation. The main idea, which is explained in Chapter 2, is that the ReLU activations offer an adaptive selection mechanism by which the most relevant dictionary elements of some underlying frame learned from training data are selected to best encode the underlying input signal. This chapter also discusses the double-descent phenomenon and provides new insights on the amazing generalization capability of deep neural networks.

Recent studies have revealed a close relationship between deep-learning approaches and sparse representations. This link is explained and further developed in Chapter 3. It is illustrated with specific model-based approaches for deep learning in both supervised and unsupervised settings.

By expanding on the model-based approaches, neural network architectures can also be put in correspondence with the unrolled version of a classical iterative algorithm for the recovery of sparse signals. The twist is that each unrolled segment can be trained in data-driven fashion and optimized for best performance. Chapter 4 contains a comprehensive examination of these unrolling techniques, which enhances our understanding of the design and application of deep learning in image reconstruction.

Part II: Deep-Learning Architecture for Various Imaging Modalities

Much of the progress in the deployment of deep-learning methods is driven by experimentation. It is therefore crucial to practitioners and designers alike to be well aware of the implementation details of reconstruction pipelines and of the underlying imaging physics. As such, in Part II the leading pioneers in various imaging applications such as X-ray, CT, MRI, and ultrasound identify the features specific to each modality and emphasize the underlying modeling principles. The primary focus in Part II is how this field has evolved. Physics is adhered to whenever possible, thus merging model-based methods with deep learning.

Chapter 5 comprises a high-level overview of deep-learning methods for CT reconstruction. Chapter 6 then delves into the specifics of deep learning, both for direct reconstructions in the image domain and for reconstructions in differentiated back-projection domains. Deep learning has also had a tremendous impact on MRI, as explained in Chapter 7, which gives a broad overview of the field. This material is complemented with that of Chapters 8 and 9, which report on model-based strategies and on specific k-space methodologies for deep learning. Ultrasound imaging is another domain where deep learning has led to new opportunities and modality-specific developments. An extensive review of ultrasound beamforming is to be found in Chapter 10, where it is examined how deep learning can enhance classical beamformer pipelines. Subsequently, approaches for artifact removal based on deep learning are described in Chapter 11.

Part III: Generative Models for Biomedical Imaging

Deep generative models are attracting a considerable amount of attention because of their ability to encode the statistical distribution of images. The best-known instances are the generative adversarial networks (GANs), which learn the prior distribution of large datasets in a purely unsupervised manner. A particularly impressive application of GANs in computer vision is style transfer. In Chapter 12, it is shown how this methodology can be transposed to MRI for contrast transfer and synthesis from missing data.

Deep image prior (DIP) is a mechanism whose purpose is to regularize the solution space. It falls within the generative framework, with the important difference that it does not require any prior training. The key idea of DIP is to exploit the structure of a convolutional neural network as a strong inductive prior on the solution space. Under the assumption that the generator successfully captures the prior knowledge, one can then perform image reconstruction by constraining the reconstructed image to lie in the range of the generator. This principle is applied in Chapters 13 and 14 to address the difficult image-reconstruction problems originating from dynamic MRI and phase unwrapping, respectively.

While GANs were developed initially for the synthesis of images, they have also been found to be effective in the context of unsupervised image reconstruction. It is detailed in Chapter 15 how a GAN-based strategy can be applied to cryo-electron microscopy to perform three-dimensional reconstructions in the challenging scenario where the poses of the individual projections are unknown, as is the case in single-particle analysis.

Acknowledgements

The editors would like to thank the chapter authors – all experts in their field – for their enthusiastic support of this project and for their valuable and timely contributions to the book.

Part I

Theory of Deep Learning for Image Reconstruction

1 Formalizing Deep Neural Networks

Michael Unser

1.1 Introduction

Since the various contributions presented in this book rely heavily on deep neural networks, we felt that it would be useful to include some background material on such computational structures. Our intention with this chapter is to provide a short, self-contained introduction to deep neural networks that is aimed at mathematically inclined readers. Our primary inspiration for writing it was to demonstrate the usage of a vector–matrix formalism that is well suited to the compositional structure of these networks and that facilitates the derivation and description of the backpropagation algorithm. In what follows we first develop the formalism and then present a detailed analysis of supervised learning for the two most common scenarios: (i) multivariate regression, and (ii) classification; these rely on the minimization of least squares and cross-entropy criteria, respectively. The regression setting is the most relevant one for biomedical image reconstruction; see for instance [1].

1.2 Primary Components of Neural Networks

A deep neural network (DNN) is a parameterized computational structure that implements a multidimensional map generically denoted by $f_\theta : \mathbb{R}^{N_{\text{in}}} \to \mathbb{R}^{N_{\text{out}}}$, where N_{in} and N_{out} are the dimensions of the input and output spaces, respectively. The vector θ represents the parameters of the neural network, which are adjusted during training. A DNN results from the composition of simple computational modules: multidimensional linear (or affine) transformations and pointwise nonlinearities referred to as neuronal activations. This is often represented by a graph (see Fig. 1.1) where each node represents a neuron and where each arrow pointing to a neuron is associated with a linear (adjustable) weight.

To be more precise, a DNN is composed of L layers of neurons indexed (from left to right) by ℓ. The ℓth layer of the network has N_ℓ neurons indexed by n. In the case of a fully connected DNN, any given neuron (ℓ, n) of layer ℓ has arrows originating from all neurons of level $\ell-1$. The architecture of a fully connected DNN is therefore uniquely specified by its node descriptor $(N_0, N_1, \ldots, N_L)$, where $N_0 = N_{\text{in}}$ and $N_L = N_{\text{out}}$.

To describe the computations performed by the DNN, we denote the intermediate values in the network at the output of layer ℓ by $\mathbf{z}_\ell = (z_{\ell,1}, \ldots, z_{\ell,N_\ell})$. If the

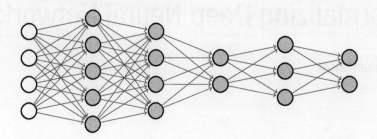

Figure 1.1 Diagram of a fully connected neural network $\mathbb{R}^4 \to \mathbb{R}^2$ with $L = 5$ layers of active neurons (gray circles) and node descriptor $(4, 5, 4, 2, 3, 2)$.

real-valued weights associated with the arrow $(\ell - 1, m) \to (\ell, n)$ are denoted by $\theta_{\ell,n,m}$ then the computation performed at neuron (n, ℓ) (a node of the graph) is simply

$$z_{\ell,n} = \sigma \left(\theta_{\ell,n} + \sum_{m=1}^{N_{\ell-1}} \theta_{\ell,n,m} z_{\ell-1,m} \right), \tag{1.1}$$

where $\sigma : \mathbb{R} \to \mathbb{R}$ is the activation function of the neuron, while $\theta_{\ell,n} \in \mathbb{R}$ is the bias parameter.

1.2.1 Vectorial Representation of a Deep Neural Network

To make the compositional structure of the DNN explicit, we shall now adopt an equivalent vectorial description. To that end, we collect the linear weights associated with layer ℓ in the matrix $\mathbf{W}_\ell = \left[(\theta_{\ell,n,m})_{n=1}^{N_\ell}, m = 1, \ldots, N_{\ell-1} \right] \in \mathbb{R}^{N_\ell \times N_{\ell-1}}$ and the biases in the vector $\mathbf{b}_\ell = (\theta_{\ell,n})_{n=1}^{N_\ell}$. Likewise, we denote the response of the nth neuron in layer ℓ by $\sigma_{\ell,n} : \mathbb{R} \to \mathbb{R}$. This then allows us to implement the DNN with Algorithm 1.1, which is sequential.

Algorithm 1.1 Feedforward DNN

---

1. Input: $\mathbf{z}_0 = \mathbf{x} \in \mathbb{R}^{N_0}$.
2. Layer-to-layer propagation with intermediate input variable $\mathbf{z}_{\ell-1} \in \mathbb{R}^{N_{\ell-1}}$, output variables $\mathbf{u}_\ell, \mathbf{z}_\ell \in \mathbb{R}^{N_\ell}$, linear network parameters $\mathbf{W}_\ell \in \mathbb{R}^{N_\ell \times N_{\ell-1}}$, and $\mathbf{b}_\ell \in \mathbb{R}^{N_\ell}$. For $\ell = 1, \ldots, L$, compute

$$\mathbf{u}_\ell = \mathbf{W}_\ell \mathbf{z}_{\ell-1} + \mathbf{b}_\ell, \tag{1.2}$$

$$\mathbf{z}_\ell = \boldsymbol{\sigma}_\ell(\mathbf{u}_\ell), \tag{1.3}$$

where the vector-valued function $\boldsymbol{\sigma}_\ell = (\sigma_{\ell,1}, \ldots, \sigma_{\ell,N_\ell}) : \mathbb{R}^{N_\ell} \to \mathbb{R}^{N_\ell}$ provides a concise representation of the pointwise nonlinearities.

3. Output: $\mathbf{f}_\theta(\mathbf{x}) = \mathbf{z}_L \in \mathbb{R}^{N_L}$.

The first processing step described by Eq. (1.2) is the part symbolized by the arrows in Fig. 1.1, which connect one layer of the network to the next. It takes the output $\mathbf{z}_{\ell-1} \in \mathbb{R}^{N_{\ell-1}}$ of the $(\ell-1)$th layer and applies a linear transformation followed by the addition of a constant vector $\mathbf{b}_\ell$ (the bias). The linear transformation is encoded in the matrix $\mathbf{W}_\ell$ of size $N_\ell \times N_{\ell-1}$. Note that the bias can also be encoded in terms of linear weights applied to a constant input common to all layers – structurally, this is equivalent to augmenting the matrix $\mathbf{W}_\ell$ by one line and expressing $\mathbf{z}_{\ell-1}$ in homogeneous coordinates by including the (dummy) constant 1 as an additional component. Consequently, this part of the processing is intrinsically linear and parameterized by the weights $(\mathbf{W}_\ell, \mathbf{b}_\ell)$ which are learned during the training of the network.

Equation (1.3) describes the combined effect of the neuronal activations at layer ℓ, which corresponds to the nodes (gray circles) in Fig. 1.1. This module is essential since it constitutes the nonlinear part of the processing. In practice, the responses of the individual neurons are often chosen to be the same, leading to $\sigma_{\ell,n} = \sigma : \mathbb{R} \to \mathbb{R}$, with one of the most popular choices being $\sigma(x) = \text{ReLU}(x) = \max(0, x)$ (a rectified linear unit).

1.3 Training

The parameters of the neural network are adjusted during the training process. This is done over the training data by minimizing some prescribed training loss with the help of a stochastic version of the steepest-descent algorithm, referred to as *stochastic gradient descent* (SGD). The latter requires the repeated calculation of the gradient of the training loss over data subsets – called batches – which are selected randomly.

As far as supervised learning is concerned, one needs to distinguish between two classes of problems. The first is *multivariate regression*, where the goal is to construct a multivariate mapping $\boldsymbol{f}_\theta : \mathbb{R}^{N_0} \to \mathbb{R}^{N_L}$ such that $\boldsymbol{f}_\theta(\mathbf{x}_m) \approx \mathbf{y}_m$, without overfitting, for a given set of training data $(\mathbf{x}_m, \mathbf{y}_m) \in \mathbb{R}^{N_0} \times \mathbb{R}^{N_L}$ with $m = 1\ldots, M$. The second is *classification*, where the training data are partitioned (or labeled) into K classes $(C_1, \ldots, C_K)$ and one wishes to construct a mapping $\boldsymbol{p}_\theta : \mathbb{R}^{N_0} \to [0, 1]^K$ that returns the posterior probabilities of class membership of an observed pattern $\mathbf{x} \in \mathbb{R}^{N_0}$, so that

$$\boldsymbol{p}_\theta(\mathbf{x}) \approx \big(\text{Prob}(C_1|\mathbf{x}), \ldots, \text{Prob}(C_K|\mathbf{x})\big).$$

Here, we shall first consider the problem of (multivariate) regression, which is a refinement of classical least squares data fitting and which lends itself naturally to the derivation of the celebrated backpropagation algorithm. In Section 1.4 we then show how the underlying computational structure and training algorithm should be modified to yield a classifier.

1.3.1 The Backpropagation Algorithm

The backpropagation algorithm is an efficient way to compute the partial derivatives (the gradient) of the loss function with respect to the parameters of the network. It is the workhorse of deep learning.

Appetizer: Deep Neural Network of Unit Width

To understand the procedure, it is helpful first to consider a simplified DNN with a unit width across all layers, as shown in Fig. 1.2. Such an elementary network is described recursively by the set of (scalar) equations

$$u_\ell = w_\ell z_{\ell-1} + b_\ell, \tag{1.4}$$

$$z_\ell = \sigma_\ell(u_\ell), \tag{1.5}$$

with $\ell = 1, \ldots, L$, where $z_0 = x$ is the input of the network and $\sigma_\ell : \mathbb{R} \to \mathbb{R}$ is the function that describes the neuronal response at layer ℓ. Moreover, σ_ℓ is assumed to be differentiable and its derivative is denoted by σ'_ℓ. This network implements a parametric function $f_\theta : \mathbb{R} \to \mathbb{R}$. The vector $\theta = (w_1, b_1, \ldots, w_L, b_L)$ collects the weights of the network, to be adjusted during training. Given a data batch $\{(x_m, y_m)\}_{m=1}^M$ and a quadratic cost

$$J(\theta) = \sum_{m=1}^M J_m(\theta) \quad \text{with} \quad J_m(\theta) = \frac{1}{2}\big(f_\theta(x_m) - y_m\big)^2, \tag{1.6}$$

the goal is now to efficiently evaluate the gradient $\nabla_\theta J$ of the criterion. The gradient components are the partial derivatives $\partial J/\partial w_\ell$ and $\partial J/\partial b_\ell$ for $\ell = 1, \ldots, L$. Since the cost J is additive, it is sufficient to consider the contribution to the gradient of a single data point (x_m, y_m) with associated elementary cost J_m. To that end, we first apply the neural network to x_m with the current set of parameter θ, which yields $f_\theta(x_m)$ and the corresponding intermediate variables $(u_\ell, z_\ell)_{\ell=1}^N$ defined by Eqs. (1.4) and (1.5). This step is called the "forward pass." To compute the required derivatives we then proceed backwards, starting from $\ell = L$, and propagate the differentiation within the network using the chain rule. Specifically, by substituting the equation for the last layer, we have that

$$J_m = \frac{1}{2}\big(\sigma_L(u_L) - y_m\big)^2 \quad \text{with} \quad u_L = w_L z_{L-1} + b_L \tag{1.7}$$

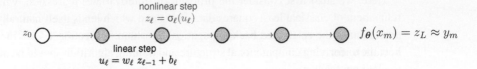

Figure 1.2 Minimal-width neural network $\mathbb{R} \to \mathbb{R}$ with $L = 5$ layers of active neurons (gray circles) and node descriptor $(1, 1, 1, 1, 1)$.

which yields

$$\delta_L = \frac{\partial J_m}{\partial b_L} = \big(\sigma_L(u_L) - y_m\big)\frac{\partial \sigma_L(u_L)}{\partial u_L}\frac{\partial u_L}{\partial b_L} = \big(\sigma_L(u_L) - y_m\big)\sigma'_L(u_L) \qquad (1.8)$$

$$\frac{\partial J_m}{\partial w_L} = \underbrace{\big(\sigma_L(u_L) - y_m\big)\sigma'_L(u_L)}_{\delta_L}\frac{\partial u_L}{\partial w_L} = \delta_L z_{L-1}, \qquad (1.9)$$

where, for computational efficiency, we have identified the common (re-weighted) error term δ_L. By inserting $z_{L-1} = \sigma_{L-1}\big(w_{L-1}z_{L-2} + b_{L-1}\big)$ into Eq. (1.7), we then proceed with the chain rule to the next layer and obtain

$$\delta_{L-1} = \frac{\partial J_m}{\partial b_{L-1}} = \underbrace{\big(\sigma_L(u_L) - y_m\big)\sigma'_L(u_L)}_{\delta_L}\frac{\partial u_L}{\partial b_{L-1}} = \delta_L w_L \sigma'_{L-1}(u_{L-1}), \qquad (1.10)$$

$$\frac{\partial J_m}{\partial w_{L-1}} = \big(\sigma_L(u_L) - y_m\big)\sigma'_L(u_L)\frac{\partial u_L}{\partial w_{L-1}} = \delta_{L-1} z_{L-2}. \qquad (1.11)$$

By repeating this process for ℓ down to 1, we uncover the simple recursion

$$\delta_\ell = \frac{\partial J_m}{\partial b_\ell} = w_{\ell+1}\delta_{\ell+1}\sigma'_\ell(u_\ell), \qquad (1.12)$$

$$\frac{\partial J_m}{\partial w_\ell} = \delta_\ell z_{\ell-1}, \qquad (1.13)$$

which is the celebrated backpropagation algorithm. Let us also note that Eq. (1.12) is consistent with Eq. (1.8) if we set $w_{L+1} = 1$ and $\delta_{L+1} = f_\theta(x_m) - y_m$. The backpropagation algorithm can therefore be interpreted as feeding the prediction error $f_\theta(x_m) - y_m$ backwards in a slightly modified version of the network in Fig. 1.2, where the biases have been suppressed and the pointwise nonlinearity replaced by a simple multiplication (re-weighting) by $\sigma'_\ell(u_\ell)$. The output of every layer then yields $\delta_\ell = \partial J_m/\partial b_\ell$, which is then used to compute $\partial J_m/\partial w_\ell$.

Backpropagation in Full Generality

In practice, of course, the layers are wider so that the scalar multiplications in Eqs. (1.12) and (1.13) need to be replaced by matrix–vector multiplication.

In the interest of clarity and to highlight the parallel with the scalar scenario that has just been considered, we shall make use of the vectorial–tensor calculus formalism. Given a data batch $\{(\mathbf{x}_m, \mathbf{y}_m)\}_{m=1}^M$ (in the multivariate regression setting), we are now aiming at training the neural network to minimize the quadratic cost

$$J(\boldsymbol{\theta}) = \sum_{m=1}^M J_m(\boldsymbol{\theta}) \quad \text{with} \quad J_m(\boldsymbol{\theta}) = \frac{1}{2}\|\boldsymbol{f}_\theta(\mathbf{x}_m) - \mathbf{y}_m\|_2^2, \qquad (1.14)$$

where the network parameters $\boldsymbol{\theta}$ are encoded in the weight matrices $\mathbf{W}_\ell$ and bias vectors $\mathbf{b}_\ell$ for $\ell = 1, \ldots, L$.

We now focus on some particular layer ℓ and write $\mathbf{W} = \mathbf{W}_\ell$ and $\mathbf{b} = \mathbf{b}_\ell$ for better readability, while keeping all the other network parameters fixed. Under those conditions, $J(\boldsymbol{\theta}) = J(\mathbf{b}, \mathbf{W})$ is a real-valued functional that depends on the vector and matrix parameters $\mathbf{b} = (b_n) \in \mathbb{R}^N$ and $\mathbf{W} \in \mathbb{R}^{M \times N}$, with $[\mathbf{W}]_{m,n} = w_{m,n}$. It is then helpful to represent the partial derivatives of J with respect to those parameters by the following vector- and matrix-valued functions:

$$\frac{\partial J(\mathbf{b})}{\partial \mathbf{b}} = \begin{bmatrix} \partial J(\mathbf{b})/\partial b_1 \\ \vdots \\ \partial J(\mathbf{b})/\partial b_N \end{bmatrix} \quad (1.15)$$

$$\frac{\partial J(\mathbf{W})}{\partial \mathbf{W}} = \begin{bmatrix} \partial J(\mathbf{W})/\partial w_{1,1} & \cdots & \partial J(\mathbf{W})/\partial w_{1,N} \\ \vdots & & \\ \partial J(\mathbf{W})/\partial w_{M,1} & \cdots & \partial J(\mathbf{W})/\partial w_{M,N} \end{bmatrix}. \quad (1.16)$$

We now have all the elements needed to present the backpropagation algorithm for a generic feedforward neural network that has L layers indexed by ℓ, each of which is composed of N_ℓ neurons. Since we are dealing with partial derivatives and making use of the chain rule, it should not come as a surprise that determining this backpropagation algorithm requires a knowledge of the "derivative" map $\boldsymbol{\sigma}'_\ell = (\sigma'_{\ell,1}, \ldots, \sigma'_{\ell,N_\ell})$: $\mathbb{R}^{N_\ell} \to \mathbb{R}^{N_\ell}$. Other than that, the sequence of computations is essentially the flowgraph transpose of Algorithm 1.1 with the recursion running backwards.

Algorithm 1.2 Gradient computation by backpropagation

1. Initialization:

$$\boldsymbol{\delta}_L = \frac{\partial J_m}{\partial \mathbf{b}_L} = \left(\boldsymbol{f}_{\boldsymbol{\theta}}(\mathbf{x}_m) - \mathbf{y}_m \right) \odot \boldsymbol{\sigma}'_L(\mathbf{u}_L) \quad \in \mathbb{R}^{N_L}, \quad (1.17)$$

where the symbol $\odot$ denotes the pointwise (or Hadamard) product of two vectors.

2. Backward propagation of the error through the network: For $\ell = L - 1$ down to 1, compute

$$\boldsymbol{\delta}_\ell = \frac{\partial J_m}{\partial \mathbf{b}_\ell} = \mathbf{W}_{\ell+1}^T \boldsymbol{\delta}_{\ell+1} \odot \boldsymbol{\sigma}'_\ell(\mathbf{u}_\ell) \quad \in \mathbb{R}^{N_\ell}, \quad (1.18)$$

$$\frac{\partial J_m}{\partial \mathbf{W}_\ell} = \boldsymbol{\delta}_\ell \cdot \mathbf{z}_{\ell-1}^T \quad \in \mathbb{R}^{N_\ell \times N_{\ell-1}}, \quad (1.19)$$

where (1.19) is valid for layer $\ell = L$ as well.

As in the scalar scenario, we can also extend the validity of (1.18) for $\ell = L$ by defining $\mathbf{W}_{L+1} = \mathbf{I}$ and $\boldsymbol{\delta}_{L+1} = \left(\boldsymbol{f}_{\boldsymbol{\theta}}(\mathbf{x}_m) - \mathbf{y}_m \right)$. The backpropagation algorithm therefore essentially amounts to feeding the prediction error $\left(\boldsymbol{f}_{\boldsymbol{\theta}}(\mathbf{x}_m) - \mathbf{y}_m \right)$ for each test datum $(\mathbf{x}_m, \mathbf{y}_m)$ backwards into the network. The only adjustment to the initial structure is that the nonlinear neuronal transformation at any given node (n, ℓ) is replaced by a pointwise multiplication of the backpropagated error with $\sigma'_{\ell,n}([\mathbf{u}_\ell]_n)$. The striking parallel between the forward and backward computations is best illustrated by the juxtaposition of Figs. 1.3 and 1.4. Practically, this translates into the determination of the gradient being essentially as fast as the application of the neural network to the

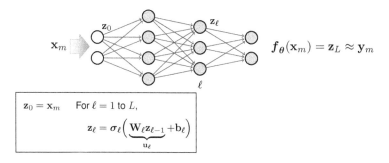

Figure 1.3 Feedforward computations in a DNN.

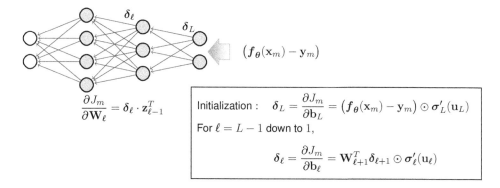

Figure 1.4 Efficient computation of the gradient of the elementary least squares term $\frac{1}{2}\|f_\theta(x_m) - y_m\|_2^2$ by backpropagation for the DNN of Fig. 1.3.

data. While these computations can be done very efficiently, the remaining limiting factor is the slow convergence of the gradient-descent algorithm and the necessity to process large amounts of data to ensure correct behavior. Fundamentally, it is the need for a massive number of iterations that is responsible for the large computational cost of the training of DNNs.

1.4 Categorical Loss for Classification

For cases where the DNN is being designed for a classification task, it is common to include an additional output layer that converts the real-valued output of the DNN into a set of pseudo-probabilities. This recoding is typically achieved with a softmax unit. Given the input vector $z = (z_1, \ldots, z_K)$, the softmax transformation $\mathbb{R}^K \to [0,1]^K$ is defined by

$$\text{Softmax}(z) = (p_1, \ldots, p_K), \quad p_k = \frac{\exp(z_k)}{\sum_{k=1}^{K} \exp(z_k)}, \quad (1.20)$$

the effect of which is to translate the z_k into a set of "probabilities" $\{p_k\}_{k=1}^{K}$, with $\sum_{k=1}^{K} p_k = 1$. These are intended to approximate the posterior probability distribution

$\{\text{Prob}(C_k|\mathbf{x})\}_{k=1}^{K}$ of the datum $\mathbf{x}$, where C_k denotes the occurrence of the kth class. The ultimate classification output of the neural network (the hard decision) is then given by

$$k(\mathbf{x}) = \arg\max p_k = \arg\max z_k.$$

Given a representative ensemble $(\mathbf{x}_m, C(m))_{m=1}^{M}$ of labeled instances of training data with class label $C(m) \in \{1, \ldots, K\}$, we now need to specify a suitable training criterion. The preferred choice for a categorical output is the cross-entropy

$$J = -\sum_{m=1}^{M}\sum_{k=1}^{K}[\mathbf{y}_m]_k \log\left(\text{Prob}(C_k|\mathbf{x}_m)\right) = -\sum_{m=1}^{M} \log \text{Prob}(C(m)|\mathbf{x}_m), \quad (1.21)$$

where $[\mathbf{y}_m]_k = \delta_{k,C(m)}$ is a binary variable (0 or 1) that indicates class membership and $\text{Prob}(C_k|\mathbf{x}_m) = p_k$, with p_k given by Eq. (1.20); here $\mathbf{z} = \mathbf{f}_\theta(\mathbf{x}_m)$ is the output of the neural network for the input $\mathbf{x}_m$. Let us note that the minimization of Eq. (1.21) is actually equivalent to the minimization of the empirical Kulback–Leibler divergence between the "true" distribution $[\mathbf{y}_m]_k = \delta_{k,C(m)}$ and the predicted distribution given by $\text{Prob}(C_k|\mathbf{x}_m)$.

In order to adapt the backpropagation scheme of Section 1.3.1 to the present scenario, we need to determine the Jacobian of the softmax transformation. Again, this is achieved by applying the chain rule. To that end we evaluate the partial derivatives of the probabilities (1.20) with respect to z_l, which yields

$$\frac{\partial p_k}{\partial z_l} = \begin{cases} p_k(1 - p_l), & k = l \\ -p_k p_l, & \text{otherwise} \end{cases}, \quad (1.22)$$

$$= p_k(\delta_{k,l} - p_l). \quad (1.23)$$

Let us now consider the contribution to the cross-entropy criterion (1.21) of sample m which belongs to the class $C(m)$. It is given by

$$J_m = -\sum_{k=1}^{K} y_k \log p_k = -\log p_{C(m)}$$

with $y_k = [\mathbf{y}_m]_k$ and $p_k = \text{Prob}(C_k|\mathbf{x}_m)$. By the chain rule and using the property that $\sum_{k=1}^{K} y_k = 1$, we obtain the formula

$$\frac{\partial J_m}{\partial z_l} = -\sum_{k=1}^{K} y_k \frac{p_k(\delta_{k,l} - p_l)}{p_k} = p_l - y_l$$

which is remarkable for its simplicity. The surprising aspect is that the gradient is essentially the same as if we had applied a least squares criterion to the probabilities – the difference, of course, is that we are backpropagating the gradient with respect to z_k and not p_k!

In effect, this means that one can adapt the backpropagation procedure (Algorithm 1.2) to the case of a categorical loss by a straightforward adjustment of the initialization step. More precisely, we substitute the term $(\mathbf{f}_\theta(\mathbf{x}_m) - \mathbf{y}_m)$ in Eq. (1.17) by $(p_\theta(x_m) - y_m)$, with $p_\theta(x_m) = \text{Softmax}(f_\theta(x_m))$, where the softmax operator is defined by Eq. (1.20).

There is also a probabilistic interpretation of criterion (1.21). It is used in what is called logistic regression. If we assume that the samples x_m are independent, with class probabilities $\text{Prob}(C_k|x_m)$, then the global probability associated with the dataset $(x_1, \ldots, x_M)$ is

$$\prod_{m=1}^{M} \prod_{k=1}^{K} \text{Prob}(C_k|x_m)^{[y_m]_k}, \tag{1.24}$$

which, upon taking the (negative) logarithm, yields Eq. (1.21). This shows that minimization of the cross-entropy criterion is actually equivalent to the search for a maximum-likelihood solution.

1.5 Good Practice for Training DNNs

Here is a short list of practical aspects to consider when designing and training neural networks.

1. **Start from an architecture that already works**. Neural networks require a lot of heuristics. It is therefore recommended to start from models that are known to perform well for a given task – e.g., convolutional neural networks for image processing or segmentation. Also, consider incorporating components such as batch normalization [2] and skip-connections [3], whose positive effect on performance is well documented.
2. **Split the data into training, validation, and test sets**. Knowledge of the validation loss of the trained network is helpful to optimize the hyperparameters (the number of layers, number of filters, regularization weight, etc.). The test set should only be used at the very end for evaluation purposes.

    When the data is scarce – as is often the case for medical applications – perform k-fold cross-validation, which makes use of the whole dataset.
3. **Regularize the network to improve generalization or promote sparsity**. For the first use-case, the most common practice is to add weight decay (an ℓ_2-penalty on the weights); other methods include early stopping and dropout [4]. For promoting sparsity, an ℓ_1-norm regularization can be added to the loss.
4. **Use prior knowledge to guide the construction of the network**. For example, choose filter sizes that are appropriate to capture the content of the images you are classifying.
5. **Progressively reduce the learning rate (the step size of SGD) as training progresses**. One should typically begin training with a larger learning rate – in order to obtain faster convergence and avoid local minima – and then decrease it when the loss starts to converge – to prevent oscillations around a minimum.
6. **Prefer the adaptive moment estimation (ADAM) optimizer to "vanilla" gradient descent methods**. Classical gradient descent methods have been mostly

replaced by optimization strategies that rely on higher-order moments of the gradient or include momentum. The most salient example is the ADAM optimizer [5], which has now become the *de facto* standard.
7. **Augment the training data to improve performance**. The training data can be extended by performing operations on the available data points which do not change their labels. Common operations of this kind which are often applied to images include flipping, cropping, and rigid-body or elastic deformations [6, 7].
8. **Consider alternatives to grid search for optimizing the hyperparameters**. Beware that there are better methods to optimize the network hyperparameters than simple intuition or grid search; these include Bayesian and evolutionary optimization [8, 9].
9. **Use tools such as Tensorboard**[1,2] to visualize the evolution of training and guide your decisions on hyperparameter optimization.

References

[1] K. H. Jin, M. T. McCann, E. Froustey, and M. Unser, "Deep convolutional neural network for inverse problems in imaging," *IEEE Transactions on Image Processing*, vol. 26, no. 9, pp. 4509–4522, 2017.

[2] S. Ioffe and C. Szegedy, "Batch normalization: Accelerating deep network training by reducing internal covariate shift," in *Proc. International Conference on Machine Learning*. PMLR, 2015, pp. 448–456.

[3] K. He, X. Zhang, S. Ren, and J. Sun, "Deep residual learning for image recognition," in *Proc. 2016 IEEE Conference on Computer Vision and Pattern Recognition*. IEEE, 2016, pp. 770–778.

[4] N. Srivastava, G. Hinton, A. Krizhevsky, I. Sutskever, and R. Salakhutdinov, "Dropout: A simple way to prevent neural networks from overfitting," *Journal of Machine Learning Research*, vol. 15, pp. 1929–1958, 2014.

[5] D. P. Kingma and J. Ba, "Adam: A method for stochastic optimization," in *Proc. International Conference on Learning Representations*, 2015.

[6] S. Zagoruyko and N. Komodakis, "Wide residual networks," *arXiv:1605.07146*, 2016.

[7] S. Bianco, C. Cusano, F. Piccoli, and R. Schettini, "Personalized image enhancement using neural spline color transforms," *IEEE Transactions on Image Processing*, vol. 29, pp. 6223–6236, 2020.

[8] J. Bergstra, D. Yamins, and D. Cox, "Making a science of model search: Hyperparameter optimization in hundreds of dimensions for vision architectures," in *Proc. International Conference on Machine Learning*. PMLR, 2013, pp. 115–123.

[9] J. Bergstra, R. Bardenet, Y. Bengio, and B. Kégl, "Algorithms for hyper-parameter optimization," in *Advances in Neural Information Processing Systems*, J. Shawe-Taylor, R. Zemel, P. Bartlett, F. Pereira, and K. Q. Weinberger, eds., vol. 24. Curran Associates, 2011.

[1] https://www.tensorflow.org/tensorboard/
[2] https:/pytorch.org/docs/stable/tensorboard.html

2 Geometry of Deep Learning

Jong Chul Ye and Sangmin Lee

2.1 Introduction

Since the groundbreaking performance improvement by AlexNet [1] at the ImageNet challenge, deep learning has provided significant gains over classical approaches in various fields of data science including imaging reconstruction. The availability of large-scale training datasets and advances in neural network research have resulted in the unprecedented success of deep learning in various applications.

Nonetheless, the success of deep learning appears very mysterious. The basic building blocks of deep neural networks consist of primitive tools of mathematics, which are convolution, pooling, and nonlinear activation functions. Interestingly, the cascaded connection of these primitive tools results in a superior performance over traditional approaches. To understand this mystery, one can go back to the basic idea of classical approaches to understand the similarities to and differences from modern deep-neural-network methods.

In this chapter, we explain the limitations of classical machine learning approaches and provide a review of mathematical foundations to explain why deep neural networks have successfully overcome these limitations.

2.2 Limitations of Classical Machine Learning

2.2.1 Kernel Machines

In classical learning, a feature space is *engineered* to have good mathematical properties. Here, "good" mathematical properties refer to well-defined structures such as the existence of the inner product, completeness, reproducing properties, etc.

In particular, the feature space with these properties is often called the *reproducing kernel Hilbert space* (RKHS) [2]. Although RKHS is only a small subset of Hilbert space, its mathematical properties are very versatile, which makes the algorithm development simpler. Specifically, a classifier design or regression problem for a given test dataset $\{(x_i, y_i)\}_{i=1}^n$ in the RKHS can be addressed by solving the following optimization problem:

$$\min_{f \in \mathcal{H}_k} \frac{1}{2}\|f\|_{\mathcal{H}}^2 + C \sum_{i=1}^n \ell(y_i, f(x_i)) \qquad (2.1)$$

where $\mathcal{H}_k$ denotes the RKHS with kernel $k(x, x')$, $\|\cdot\|_\mathcal{H}$ is the Hilbert-space norm, and $\ell(\cdot,\cdot)$ is the loss function. Here, the underlying kernel $k(x, x')$ is a nonlinear function that can be represented as

$$k(x, x') = \langle \varphi(x), \varphi(x') \rangle_\mathcal{H} \tag{2.2}$$

where $\varphi(\cdot)$ denotes the embedding function from the ambient space to the feature space. Then, one of the most celebrated results from classical machine learning is the *representer theorem* [3]. Precisely, this theorem says that the minimizer f for (2.1) has the following closed-form representation:

$$f(x) = \sum_{i=1}^{n} \alpha_i k(x_i, x) \tag{2.3}$$

where $\{\alpha_i\}_{i=1}^{n}$ are parameters learned from the training dataset. This is the basic idea of the kernel support vector machine (kernel SVM) and kernel regression [2].

Another important contribution of classical machine learning theory is the so-called "kernel trick" [2]. Specifically, the embedding function $\varphi(x)$ in Eq. (2.2), which is a mapping to a potentially higher-dimensional latent space, is not necessarily required. Instead, we need only a kernel that satisfies Eq. (2.2). Therefore, one main research thrust in classical machine learning approaches is to find appropriate kernels that are suitable for specific applications.

That said, the expression in Eq. (2.3) still has fundamental limitations. First, the RKHS $\mathcal{H}_k$ is specified by the engineered kernel in a top-down manner, and is a small subset of a Hilbert space. Thus, the representation power of kernel machines may be restricted. Second, once the kernel machine is trained, the parameters $\{\alpha_i\}_{i=1}^{n}$ are fixed, and it is not possible to adjust them during the test phase. These drawbacks lead to the limitations of the kernel machine.

2.2.2 Shallow Neural Networks

Another important machine learning approach is the shallow neural network, which has one hidden layer. This was developed from the perceptron, which is a classic binary classifier. Specifically, let $\varphi : \mathbb{R} \mapsto \mathbb{R}$ be a nonconstant, bounded, and continuous activation function. Then, for an input space $\mathcal{X} \subset \mathbb{R}^n$, a shallow neural network $f_\Theta : \mathcal{X} \mapsto \mathbb{R}$ can be represented by

$$f_\Theta(x) = \sum_{i=1}^{d} v_i \varphi\left(w_i^\top x + b_i\right), \quad x \in \mathcal{X}, \tag{2.4}$$

where $w_i \in \mathbb{R}^n$ is a weight vector, $v_i, b_i \in \mathbb{R}$ are real constants, and $\Theta = \{(w_i, v_i, b_i)\}_{i=1}^{d}$ represents the neural network parameters. Then, the parameters are estimated by solving the following optimization problem using the training data $\{(x_i, y_i)\}_{i=1}^{N}$:

$$\min_{\Theta} \sum_{i=1}^{N} \ell(y_i, f_\Theta(x_i)) + \lambda R(\Theta) \qquad (2.5)$$

where λ is a regularization parameter and $R(\Theta)$ is a regularization function with respect to the parameter set Θ.

When compared with the kernel machine (2.3), the pros and cons of the shallow neural network in Eq. (2.4) can be easily understood. Specifically, $\varphi\left(w_i^\top x + b_i\right)$ in Eq. (2.4) works similarly to a kernel function $k(x_i, x)$, and v_i in Eq. (2.4) is similar to the weight parameter α_i in Eq. (2.3). However, the nonlinear mapping in the shallow neural network, i.e., $\varphi\left(w_i^\top x + b_i\right)$, does not need to satisfy positive semi-definiteness of the kernel, thereby increasing the approximable functions beyond the RKHS to a larger function class in Hilbert space. Therefore, there exists potential for improving the expressivity.

In fact, one of the classical results on the representation power of shallow neural networks is the universal approximation theorem [4], which says that any continuous function on a compact set can be approximated by a shallow neural network containing a finite number of neurons. Although this theorem appears to show that a shallow neural network has a quite good expressive power, the theorem does not specify how many neurons are required for a given approximation error. Only recently people realized that the depth matters, i.e., there exists a function that a deep neural network can approximate but a shallow neural network with the same number of parameters cannot [5–9]. In fact, this is the main driving force in modern deep learning.

2.2.3 Frame Representation

Yet another important classical learning approach is closely related to *frame theory*. Specifically, a frame is a class of functions that can represent a vector in a vector space [10]. Formally, a set of functions

$$\Phi = [\phi_k]_{k \in \Gamma} = [\cdots \quad \phi_{k-1} \quad \phi_k \quad \cdots]$$

in a Hilbert space H is called a frame if there exist $0 < \alpha \leq \beta < \infty$ such that the following inequality holds [10]:

$$\alpha \|f\|^2 \leq \sum_{k \in \Gamma} |\langle f, \phi_k \rangle|^2 \leq \beta \|f\|^2, \quad \forall f \in H. \qquad (2.6)$$

Such constants α and β are called the frame bounds. If $\alpha = \beta$, the frame is said to be tight, and this orthogonal basis is a special case of tight frames. This implies that the energy of the expansion coefficients should be bounded by the original signal energy, and, for the case of a tight frame, the expansion coefficient energy is the same as the original signal energy up to a scaling factor.

Then, recovery of the original signal from the frame coefficient vector $c = \Phi^\top f$ can be achieved using the *dual frame operator* $\tilde{\Phi}$ given by

$$\tilde{\Phi} = [\cdots \ \tilde{\phi}_{k-1} \ \tilde{\phi}_k \ \cdots],$$

which satisfies the so-called *frame condition*

$$\tilde{\Phi}\Phi^\top = I, \tag{2.7}$$

because we have

$$\hat{f} := \tilde{\Phi}c = \tilde{\Phi}\Phi^\top f = f,$$

or equivalently,

$$f = \sum_{k \in \Gamma} c_k \tilde{\phi}_k = \sum_{k \in \Gamma} \langle f, \phi_k \rangle \tilde{\phi}_k. \tag{2.8}$$

Note that (2.8) is a linear signal expansion so that it is not useful for machine learning tasks. However, more interesting things occur when it combines with nonlinear regularization.

For example, consider a regression problem to estimate the noiseless signal from a noisy measurement y:

$$y = f + w$$

where w is the additive noise and f is the unknown signal to be estimated. If we formulate a loss function as follows,

$$\min_{f} \frac{1}{2} \|y - f\|^2 + \lambda \|\Phi^\top f\|_1 \tag{2.9}$$

where $\|\cdot\|_1$ is the l_1-norm, then the solution is given by the following form [11]:

$$\hat{f} = \sum_{k \in \Gamma} \rho_\lambda (\langle y, \phi_k \rangle) \tilde{\phi}_k \tag{2.10}$$

where $\rho_\lambda(\cdot)$ is a nonlinear thresholding function that depends on the regularization parameter λ. In particular, the main idea of wavelet shrinkage for signal denoising [12] is zeroing out the small-magnitude wavelet coefficients using a thresholding operation $\rho_\lambda(\cdot)$ and retaining the large wavelet coefficients beyond the threshold values that have important signal characteristics. This implies that the signal representation depends on the input y, since the set of coefficients $\langle y, \phi_k \rangle$ with small values varies depending on the input y, after which the signal is represented by only a small set of dual basis functions $\tilde{\phi}_k$ which correspond to the locations of the nonzero expansion coefficients.

Extending this idea beyond signal denoising, another successful tool in signal processing theory is the *compressed sensing* or *sparse recovery technique* [13]. By using a sparse frame representation, if there are very few measurements below the classical limits such as the Nyquist limit, one could obtain a stable solution of the inverse problem by searching for a sparse representation that generates a consistent output from the measured data. As a result, the goal of the image reconstruction problem is to find an optimal set of sparse basis functions suitable for the given measurement data. This is why the classical method is often called *basis pursuit* [13].

In contrast with the kernel machine in Eq. (2.3), basis pursuit using the frame representation has several unique advantages. First, the function space that the basis pursuit can generate is often larger than the RKHS from Eq. (2.3). In fact, this space is often called the *union of subspaces* [14], which is a large subset of a Hilbert space. Second, among the given frames, the choice of an *active* dual frame basis $\widetilde{\phi}_k$ is totally data-dependent. Therefore, the basis pursuit representation is an adaptive model. Moreover, the expansion coefficients $\rho_\lambda(\langle y, \phi_k \rangle)$ of basis pursuit are also totally dependent on the input y, thereby generating more diverse representations than the kernel machine with fixed expansion coefficients.

Having said this, one of the most fundamental limitations of the basis pursuit approach in Eq. (2.10) is that it is *transductive*, which means that it does not allow *inductive* learning from the training data. In general, the basis pursuit regression in Eq. (2.9) should be solved for each input, since the nonlinear thresholding function needs to be found by an optimization method. Therefore, it is difficult to transfer the learning from one set of data to another.

2.3 Understanding Deep Neural Networks

In this section, we will describe how deep neural networks differ from the classical machine learning approaches, and how they successfully overcome the limitations of the classical approaches.

2.3.1 Multi-Layer Perceptron

First, consider a multi-layer feedforward neural network. Specifically, let $o_j^{(l-1)}$ denote the jth output of the $(l-1)$th-layer neuron, which is given as the jth dendrite presynaptic potential input for the lth-layer neuron, and let $w_{ij}^{(l)}$ correspond to the synaptic weights at the lth layer. Then, without loss of generality, the neural network can be described by

$$\boldsymbol{o}^{(l)} = \sigma\left(\boldsymbol{g}^{(l)}\right), \quad \boldsymbol{g}^{(l)} := \boldsymbol{W}^{(l)} \boldsymbol{o}^{(l-1)} \tag{2.11}$$

where $\boldsymbol{W}^{(l)} \in \mathbb{R}^{d^{(l)} \times d^{(l-1)}}$ is the weight matrix whose (i, j) element is given by $w_{ij}^{(l)}$; $\sigma(\cdot)$ denotes the nonlinearity $\sigma(\cdot)$ applied to each element in the vector, and

$$\boldsymbol{o}^{(l)} = \begin{bmatrix} o_1^{(l)} & \cdots & o_{d^{(l)}}^{(l)} \end{bmatrix}^\top \in \mathbb{R}^{d^{(l)}}. \tag{2.12}$$

Here ⊤ indicates the transpose of the row vector. When there exists a pooling layer, the dimension $d^{(l)}$ is reduced by a half.

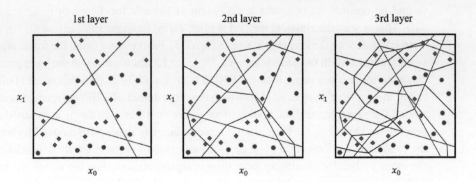

Figure 2.1 An example of the partitioning of input spaces by a deep neural network.

Among the various forms of activation functions, the most successful and popular in modern deep learning is the rectified linear unit (ReLU), which is defined by

$$\text{ReLU}(x) := \max\{0, x\} \tag{2.13}$$

[15]. For the case of an L-layer feedforward neural network, the neural network output for a given input x can be represented by

$$f_\Theta(x) := B_\Theta(x) x \tag{2.14}$$

where $\Theta = \begin{bmatrix} W^{(1)} & \cdots & W^{(L)} \end{bmatrix}$ and

$$B_\Theta(x) = W^{(L)} \Lambda^{(L-1)}(x) W^{(L-2)} \Lambda^{(L-2)}(x) \cdots \Lambda^{(1)}(x) W^{(1)}. \tag{2.15}$$

Here, $\Lambda^{(l)}(x)$ is a diagonal matrix with 0 and 1 elements indicating the ReLU activation patterns. Note that the matrix $B_\Theta(x)$ in Eq. (2.15) depends on the ReLU activation patterns $\Lambda^{(l)}, l = 1, \ldots, L - 1$, which are dependent upon the input x. In fact, this ReLU activation-dependent diagonal matrix provides a key role in enabling inductive learning. Specifically, the nonlinearity is applied after the linear operation, so the on-and-off activation pattern of each ReLU determines a binary partition of the feature space at each layer across the hyperplanes that is determined by the weight matrix.

This leads to an important insight. As shown in Fig. 2.1, deep neural networks partition the input space into multiple nonoverlapping regions for each value of the depth, so that input images for each region share the same linear representation, but not across the partition. Accordingly, once the network is trained, in contrast with classical optimization-based approaches, deep neural networks do not solve the optimization problem for a new input; rather, they switch to different linear representations only by changing the ReLU activation patterns. This is an important advance over the classical machine learning approaches.

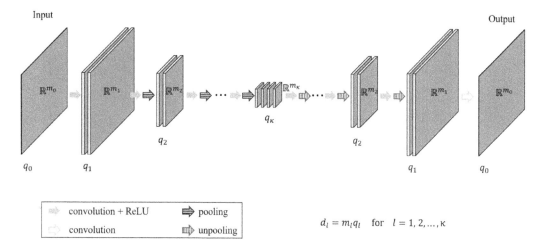

Figure 2.2 An example of encoder–decoder CNNs without skip connections.

2.3.2 Convolutional Neural Networks

Now we are ready to explain the multi-layer convolutional neural network (CNN). For simplicity, we consider encoder–decoder networks without skip connections as shown in Fig. 2.2, although the analysis can be applied equally well when skip connections are present. Furthermore, we assume a symmetric configuration so that both encoder and decoder have the same number of layers, say κ; the input and output dimensions for the encoder layer $\mathcal{E}^l$ and the decoder layer $\mathcal{D}^l$ are symmetric,

$$\mathcal{E}^l : \mathbb{R}^{d_{l-1}} \mapsto \mathbb{R}^{d_l}, \quad \mathcal{D}^l : \mathbb{R}^{d_l} \mapsto \mathbb{R}^{d_{l-1}}, \quad l \in [\kappa] \tag{2.16}$$

where $[n]$ denotes the set $\{1, \ldots, n\}$. At the lth layer, m_l and q_l denote the dimension of the signal and the number of channels, respectively. This encoder–decoder architecture is quite often used in image reconstruction [16].

We now define the lth-layer input signal for the encoder layer from q_{l-1} input channels:

$$z^{l-1} := \begin{bmatrix} z_1^{l-1\top} & \cdots & z_{q_{l-1}}^{l-1\top} \end{bmatrix}^\top \in \mathbb{R}^{d_{l-1}}, \tag{2.17}$$

where $z_j^{l-1} \in \mathbb{R}^{m_{l-1}}$ refers to the jth channel input with dimension m_{l-1}. The lth-layer output signal z^l is similarly defined.

The architecture of a simple encoder–decoder CNN *without* skip connection is shown in Fig. 2.2. First we start with the *linear* case, disregarding the ReLU nonlinearity. In [16] it was shown that this type of CNN has the following linear representation at the lth encoder layer:

$$z^l = E^{l\top} z^{l-1} \tag{2.18}$$

with

$$E^l = \begin{bmatrix} \Phi^l \circledast \psi^l_{1,1} & \cdots & \Phi^l \circledast \psi^l_{q_l,1} \\ \vdots & \ddots & \vdots \\ \Phi^l \circledast \psi^l_{1,q_{l-1}} & \cdots & \Phi^l \circledast \psi^l_{q_l,q_{l-1}} \end{bmatrix} \quad (2.19)$$

where Φ^l denotes the $m_l \times m_l$ matrix that represents the pooling operation at the lth layer, $\psi^l_{i,j} \in \mathbb{R}^r$ represents the lth-layer encoder filter that generates the ith channel output from the contribution of the jth channel input, and $\Phi^l \circledast \psi^l_{i,j}$ represents a single-input multi-output (SIMO) convolution [16]:

$$\Phi^l \circledast \psi^l_{i,j} = \begin{bmatrix} \phi^l_1 \circledast \psi^l_{i,j} & \cdots & \phi^l_n \circledast \psi^l_{i,j} \end{bmatrix} \quad (2.20)$$

Bias can be accounted for by including an additional row in E^l and augmenting z^{l-1} by a final element equal to 1.

Similarly, the lth decoder layer can be represented by

$$\tilde{z}^{l-1} = D^l \tilde{z}^l \quad (2.21)$$

with

$$D^l = \begin{bmatrix} \tilde{\Phi}^l \circledast \tilde{\psi}^l_{1,1} & \cdots & \tilde{\Phi}^l \circledast \tilde{\psi}^l_{1,q_l} \\ \vdots & \ddots & \vdots \\ \tilde{\Phi}^l \circledast \tilde{\psi}^l_{q_{l-1},1} & \cdots & \tilde{\Phi}^l \circledast \tilde{\psi}^l_{q_{l-1},q_l} \end{bmatrix} \quad (2.22)$$

where $\tilde{\Phi}^l$ denotes the $m_l \times m_l$ matrix that represents the unpooling operation at the lth layer and $\tilde{\psi}^l_{i,j} \in \mathbb{R}^r$ represents the lth-layer decoder filter that generates the ith channel output from the contribution of the jth channel input.

Then the output v of the encoder–decoder CNN with respect to an input z can be represented as follows [16]:

$$v = \mathcal{T}_\Theta(z) = \sum_i \langle b_i, z \rangle \tilde{b}_i \quad (2.23)$$

where Θ refers to all the encoder and decoder convolution filters, and b_i and $\tilde{b}_i$ denote the ith columns of the following matrices, respectively:

$$B = E^1 E^2 \cdots E^\kappa, \quad (2.24)$$
$$\tilde{B} = D^1 D^2 \cdots D^\kappa. \quad (2.25)$$

Note that this representation is completely linear.

In fact, the analysis of the CNN with ReLU nonlinearities turns out to be a simple modification, but it provides a fundamental insight into the geometry of the deep neural network. Specifically, in [16] the authors showed that the basis matrix has additional ReLU pattern blocks between encoder, decoder, and skip blocks. Moreover, Eqs. (2.24) and (2.25) are changed as follows:

$$B(z) = E^1 \Lambda^1(z) E^2 \Lambda^2(z) \cdots \Lambda^{\kappa-1}(z) E^\kappa, \tag{2.26}$$

$$\tilde{B}(z) = D^1 \tilde{\Lambda}^1(z) D^2 \tilde{\Lambda}^2(z) \cdots \tilde{\Lambda}^{\kappa-1}(z) D^\kappa, \tag{2.27}$$

where $\Lambda^l(z)$ and $\tilde{\Lambda}^l(z)$ are diagonal matrices with 0 and 1 elements indicating the ReLU activation patterns. Again, we can see that the ReLU-dependent diagonal matrices in Eqs. (2.26) and (2.27) provide a key role to enable inductive learning. Specifically, the nonlinearity is applied after the convolution operation, so the on-and-off activation pattern of each ReLU determines a binary partition of the feature space at each layer across the hyperplane that is determined by the convolution.

Accordingly, the linear representation in Eq. (2.23) should be modified as a nonlinear representation:

$$v = \mathcal{T}_\Theta(z) = \sum_i \langle b_i(z), z \rangle \, \tilde{b}_i(z), \tag{2.28}$$

which appears similar to the frame representation. However, we now have an explicit dependency on z for $b_i(z)$ and $\tilde{b}_i(z)$ owing to the input-dependent ReLU activation patterns, and this makes the representation nonlinear. Note that the basis and its dual basis are changed for each input, because of the input dependency. This is the main principle of deep neural networks that could express a variable output for each input.

Given the input-dependent geometry of the CNN, we can easily see that, with more input space partitions, a nonlinear function approximation by a piecewise linear frame representation becomes more accurate. Therefore, the number of piecewise linear regions is directly related to the expressivity or representation power of the deep neural network. If each ReLU activation pattern is independent of all the others, then the number of distinct ReLU activation pattern is $2^{\text{number of neurons}}$, where the number of neurons is determined by the total number of features. Therefore, the number of distinct linear representations increases exponentially with the depth, width, and skip connections. This again confirms the expressive power of deep neural networks, thanks to the ReLU activation function.

2.4 Generalization Capability of Deep Neural Networks

2.4.1 Deep Double Descent

Yet another reason for the enormous success of deep neural networks is their amazing ability to generalize, which seems mysterious from the perspective of classic machine learning. In particular, the number of trainable parameters in deep neural networks is often greater than the training dataset, this situation being notorious for overfitting from the point of view of classical statistical learning theory. However, empirical results have shown that a deep neural network generalizes well in the testing phase, resulting in high performance for unseen data.

This apparent contradiction has raised questions about the mathematical foundations of machine learning and their relevance to practitioners. A number of theoretical papers have been published aiming to understand this intriguing generalization phenomenon in deep-learning models [17–23]. The simplest approach to studying generalization in deep learning is to prove a generalization bound, which is typically an upper limit for the test error. A key component in these generalization bounds is the notion of *complexity measure*: a quantity that relates monotonically to some aspect of generalization. Unfortunately, it is difficult to find tight bounds for deep neural networks that can explain their fascinating generalization ability.

Recently, the authors in [24, 25] have delivered a groundbreaking work that can reconcile classical understanding and modern practice in a unified framework. The so-called "double descent" curve extends the classical U-shaped bias–variance trade-off curve by showing that increasing the model capacity beyond the interpolation point leads to improved performance in the test phase (see Fig. 2.3).

To understand this phenomenon, let $\mathcal{Q}$ be an arbitrary distribution over $z := (x, y)$, where $x \in \mathcal{X}$ and $y \in \mathcal{Y}$ denote the input and output of the learning algorithm, and $\mathcal{Z} := \mathcal{X} \times \mathcal{Y}$ refers to the sample space. Let $\mathcal{F}$ be a hypothesis class and let $\ell(f, z)$ be a loss function. For the case of regression with mean square error (MSE) loss, the loss can be defined as

$$\ell(f, z) = \frac{1}{2}\|y - f(z)\|^2.$$

Then, the estimated hypothesis from the popular empirical risk minimization (ERM) principle [26] is given by

$$f_{ERM} = \arg\min_{f \in \mathcal{F}} \hat{R}_N(f) \qquad (2.29)$$

where the empirical risk $\hat{R}_N(f)$ is defined by

$$\hat{R}_N(f) := \frac{1}{N} \sum_{n=1}^{N} \ell(f, z_n), \qquad (2.30)$$

which is assumed to uniformly converge to the population (or expected) risk, defined by

$$R(f) := \mathbb{E}_{z \sim \mathcal{Q}}[\ell(f, z)]. \qquad (2.31)$$

Empirical risk minimization can only be considered a solution to a machine learning problem if the difference between the training error and the generalization error, called the *generalization gap*, is small enough. This implies that the following probability should be sufficiently small,

$$\mathbb{P}\left\{\sup_{f \in \mathcal{F}} |R(f) - \hat{R}_N(f)| > \epsilon\right\}, \qquad (2.32)$$

which leads to the following upper bound for the ERM in Eq. (2.29):

$$R(f^*_{ERM}) \leq \underbrace{\hat{R}_N(f^*_{ERM})}_{\text{empirical risk (training error)}} + \underbrace{\mathcal{O}\left(\sqrt{\frac{c}{N}}\right)}_{\text{complexity penalty}} \qquad (2.33)$$

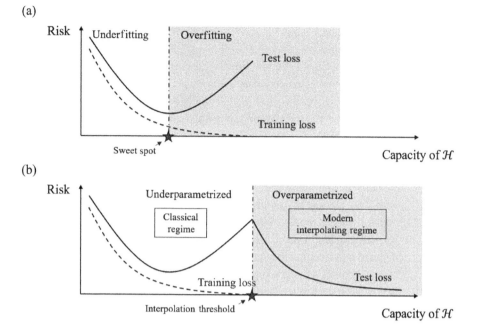

Figure 2.3 Curves for the training risk (dashed line) and test risk (solid line). (a) The classical U-shaped risk curve arising from the bias–variance trade-off. (b) The double descent risk curve, which incorporates the U-shaped risk curve (i.e., the "classical" regime) together with the observed behavior from using high-capacity function classes (i.e., the "modern" interpolating regime), separated by the interpolation threshold. The predictors to the right of the interpolation threshold have zero training risk.

where $\mathcal{O}(\cdot)$ denotes the "big O" notation and c refers to the model complexity such as the VC dimension, Radamacher complexity, etc.

In Eq. (2.33), with increasing complexity c the empirical risk or training error decreases as the complexity penalty increases. The functional class capacity can therefore be controlled explicitly by choosing the complexity c (e.g., the selection of the neural network architecture). This is summarized in the classic U-shaped risk curve, which is shown in Fig. 2.3(a) and has often been used as a guide for model selection. A widely accepted view deriving from this curve is that a model with zero training error is overfitted to the training data and will typically generalize poorly [26]. Classical thinking therefore deals with the search for the "sweet spot" between underfitting and overfitting (Fig. 2.3(a)).

Lately, this view has been challenged by empirical results that seem mysterious. For example, in [27] the authors trained several standard architectures on a copy of the data, with the true labels replaced by random labels. Their central finding can be summarized as follows: deep neural networks are easily fitted with random labels. More precisely, neural networks achieve zero training error if they are trained on a completely random labeling of the true data. While this observation is easy to formulate, it has profound implications from a statistical learning perspective: the effective capacity of neural networks is sufficient to store the entire dataset. According to the classical U-shape results in the generalization bound in Fig. 2.3(a), the neural network should not generalize well owing to its large complexity. Nonetheless, despite the high capacity of the functional classes and the almost perfect fit to training data, these predictors often give very accurate predictions for new data in the test phase.

The recent breakthrough in the "double descent" risk curve of Belkin *et al.* [24, 25] reconciles the classic bias–variance trade-off with behaviors that have been observed in overparameterized regimes for a large number of machine learning models. In particular, when the functional class capacity is below the interpolation threshold, learned predictors show the classic U-shaped curve from Fig. 2.3(a), where the function class capacity is identified with the number of parameters needed to specify a function within the class. The minimum of the U-shaped risk can be achieved at the sweet spot, which balances the fit to the training data and the susceptibility to overfitting. When we increase the function class capacity high enough by increasing the size of the neural network architecture, the learned predictors achieve (almost) perfect fits to the training data. Although the learned predictors obtained at the interpolation threshold typically have high risk, further increasing the function class capacity beyond this point leads to decreasing risk, which typically falls below the risk achieved at the sweet spot in the "classic" regime.

By considering larger functional classes that contain more candidate predictors that are compatible with the data, one could find interpolation functions that are "simpler." Therefore, increasing the capacity of the functional class in the overparameterized area improves the performance of the resulting classifiers.

Then, one of the remaining questions is: what is the underlying mechanism for a trained network to choose a simpler solution? This is closely related to the inductive bias (or implicit bias) of an optimization algorithm such as gradient descent, stochastic gradient descent (SGD), etc. [28–32]. Indeed, this is an active area of research. For example, the authors in [30] showed that gradient descent for the linear classifier for a specific loss function leads to the maximum-margin support vector machine (SVM) classifier. Other researchers have shown that gradient descent in deep-neural-network training leads to a simple solution [31, 32].

2.4.2 High-Dimensional Partition Geometry for Generalization

Recall that the deep neural network is trained to partition the input data manifold into piecewise linear regions that can effectively perform machine learning tasks such as classification, regression, etc. Furthermore, the network complexity or expressive power is dependent upon the number of piecewise-linear regions. Therefore, the clue to unveiling the mystery of the generalization capability of deep neural networks may come from understanding how the partitions can be controlled during the training of the neural network.

In fact, many machine learning theoreticians have been focusing on this, thereby generating intriguing theoretical and empirical observations [33–36]. Although the maximum number of linear regions can be very large, these researchers have observed that the actual number of piecewise linear representations is much smaller after the training. Namely, the number of piecewise linear regions is determined not only by the number of epochs but also by the choice of optimization algorithm.

This phenomenon can be understood as a data-driven adaptation to eliminate unnecessary partitions for machine learning tasks, because the partition boundary can collapse, resulting in a smaller number of partitions. It is therefore believed that there is a compromise between the approximation error and the robustness of the neural network in terms of the number of piecewise-linear areas. Many questions remain unanswered, and many researchers are still at work on them.

2.5 Summary

In this chapter, we have briefly reviewed the classical machine learning approaches and explained how modern deep-learning approaches have overcome their limitations. In addition, we have elucidated the role of nonlinearity in deep learning, which may be one of the most important aspects. In particular, for the case of ReLU, the nonlinearity is also related to the partitioning of the input domain and the assignment of a different linear representation for each partition, which is the origin of the emergence of inductive learning. We briefly reviewed the remarkable generalization ability of deep neural networks and presented a modern theory of double descent to fill in the theoretical gap. The important link of the high-dimensional partition geometry and the generalization bound were also discussed. These theoretical approaches are key ingredients to understanding and developing deep learning.

References

[1] A. Krizhevsky, I. Sutskever, and G. E. Hinton, "Imagenet classification with deep convolutional neural networks," in *Advances in Neural Information Processing Systems*, 2012, pp. 1097–1105.

[2] B. Scholkopf and A. J. Smola, *Learning with Kernels: Support Vector Machines, Regularization, Optimization, and Beyond*. MIT Press, 2001.

[3] B. Schölkopf, R. Herbrich, and A. J. Smola, "A generalized representer theorem," in *Proc. International Conference on Computational Learning Theory*. Springer, 2001, pp. 416–426.

[4] G. Cybenko, "Approximation by superpositions of a sigmoidal function," *Mathematics of Control, Signals and Systems*, vol. 2, no. 4, pp. 303–314, 1989.

[5] M. Telgarsky, "Representation benefits of deep feedforward networks," *arXiv:1509.08101*, 2015.

[6] R. Eldan and O. Shamir, "The power of depth for feedforward neural networks," in *Proc. Conference on Learning Theory*, 2016, pp. 907–940.

[7] M. Raghu, B. Poole, J. Kleinberg, S. Ganguli, and J. S. Dickstein, "On the expressive power of deep neural networks," in *Proc. 34th International Conference on Machine Learning*, vol. 70, JMLR, 2017, pp. 2847–2854.

[8] D. Yarotsky, "Error bounds for approximations with deep ReLU networks," *Neural Networks*, vol. 94, pp. 103–114, 2017.

[9] R. Arora, A. Basu, P. Mianjy, and A. Mukherjee, "Understanding deep neural networks with rectified linear units," *arXiv:1611.01491*, 2016.

[10] R. J. Duffin and A. C. Schaeffer, "A class of nonharmonic Fourier series," *Transactions of the American Mathematical Society*, vol. 72, no. 2, pp. 341–366, 1952.

[11] S. Mallat, *A Wavelet Tour of Signal Processing*. Academic Press, 1999.

[12] D. L. Donoho, "De-noising by soft-thresholding," *IEEE Transactions on Information Theory*, vol. 41, no. 3, pp. 613–627, 1995.

[13] ——, "Compressed sensing," *IEEE Transactions on Information Theory*, vol. 52, no. 4, pp. 1289–1306, 2006.

[14] Y. C. Eldar and M. Mishali, "Robust recovery of signals from a structured union of subspaces," *IEEE Transactions on Information Theory*, vol. 55, no. 11, pp. 5302–5316, 2009.

[15] V. Nair and G. E. Hinton, "Rectified linear units improve restricted Boltzmann machines," in *Proc. 27th International Conference on Machine Learning*, 2010, pp. 807–814.

[16] J. C. Ye and W. K. Sung, "Understanding geometry of encoder–decoder cnns," in *Proc. International Conference on Machine Learning*. PMLR, 2019, pp. 7064–7073.

[17] B. Neyshabur, R. Tomioka, and N. Srebro, "Norm-based capacity control in neural networks," in *Proc. Conference on Learning Theory*. PMLR, 2015, pp. 1376–1401.

[18] P. Bartlett, D. J. Foster, and M. Telgarsky, "Spectrally-normalized margin bounds for neural networks," *arXiv:1706.08498*, 2017.

[19] V. Nagarajan and J. Z. Kolter, "Deterministic PAC-Bayesian generalization bounds for deep networks via generalizing noise-resilience," *arXiv:1905.13344*, 2019.

[20] C. Wei and T. Ma, "Data-dependent sample complexity of deep neural networks via Lipschitz augmentation," *arXiv:1905.03684*, 2019.

[21] S. Arora, R. Ge, B. Neyshabur, and Y. Zhang, "Stronger generalization bounds for deep nets via a compression approach," in *Proc. International Conference on Machine Learning*. PMLR, 2018, pp. 254–263.

[22] N. Golowich, A. Rakhlin, and O. Shamir, "Size-independent sample complexity of neural networks," in *Proc. Conference On Learning Theory*. PMLR, 2018, pp. 297–299.

[23] B. Neyshabur, S. Bhojanapalli, and N. Srebro, "A PAC-Bayesian approach to spectrally-normalized margin bounds for neural networks," *arXiv:1707.09564*, 2017.

[24] M. Belkin, D. Hsu, S. Ma, and S. Mandal, "Reconciling modern machine-learning practice and the classical bias–variance trade-off," *Proceedings of the National Academy of Sciences*, vol. 116, no. 32, pp. 15 849–15 854, 2019.

[25] M. Belkin, D. Hsu, and J. Xu, "Two models of double descent for weak features," *SIAM Journal on Mathematics of Data Science*, vol. 2, no. 4, pp. 1167–1180, 2020.

[26] V. Vapnik, *The Nature of Statistical Learning Theory*. Springer, 2013.

[27] C. Zhang, S. Bengio, M. Hardt, B. Recht, and O. Vinyals, "Understanding deep learning requires rethinking generalization," *arXiv:1611.03530*, 2016.

[28] A. Bietti and J. Mairal, "On the inductive bias of neural tangent kernels," *arXiv:1905.12173*, 2019.

[29] B. Neyshabur, R. Tomioka, and N. Srebro, "In search of the real inductive bias: On the role of implicit regularization in deep learning," *arXiv:1412.6614*, 2014.

[30] D. Soudry, E. Hoffer, M. S. Nacson, S. Gunasekar, and N. Srebro, "The implicit bias of gradient descent on separable data," *Journal of Machine Learning Research*, vol. 19, no. 1, pp. 2822–2878, 2018.

[31] S. Gunasekar, J. Lee, D. Soudry, and N. Srebro, "Implicit bias of gradient descent on linear convolutional networks," *arXiv:1806.00468*, 2018.

[32] ——, "Characterizing implicit bias in terms of optimization geometry," in *Proc. International Conference on Machine Learning*. PMLR, 2018, pp. 1832–1841.

[33] B. Hanin and D. Rolnick, "Complexity of linear regions in deep networks," *arXiv:1901.09021*, 2019.

[34] ——, "Deep ReLU networks have surprisingly few activation patterns," in *Advances in Neural Information Processing Systems*, 2019, pp. 359–368.

[35] X. Zhang and D. Wu, "Empirical studies on the properties of linear regions in deep neural networks," *arXiv:2001.01072*, 2020.

[36] G. F. Montufar, R. Pascanu, K. Cho, and Y. Bengio, "On the number of linear regions of deep neural networks," in *Advances in Neural Information Processing Systems*, 2014, pp. 2924–2932.

3 Model-Based Reconstruction with Learning: From Unsupervised to Supervised and Beyond

Zhishen Huang, Siqi Ye, Michael T. McCann, and Saiprasad Ravishankar

3.1 Introduction

Imaging modalities such as magnetic resonance imaging (MRI), X-ray computed tomography (CT), positron-emission tomography (PET), and single-photon-emission computed tomography (SPECT) are regularly used in clinical practice for characterizing anatomical structures and physiological functions and to aid clinical diagnosis and treatment. Various sophisticated image reconstruction methods have been proposed for these modalities that help recover high-quality images, especially from limited or corrupted measurements. Classical analytical reconstruction methods include filtered back projection (FBP) for CT and the inverse Fourier transform for MRI. These methods rely on simple imaging models and have efficient implementations but can produce suboptimal reconstructions, especially when the number of measurements is limited.

Model-based image reconstruction (MBIR, alternatively called model-based iterative reconstruction) methods are popular in many medical imaging modalities. These methods exploit models of the imaging system's physics (forward models) along with statistical models of the measurements and noise and often simple object priors. They iteratively optimize model-based cost functions to estimate the underlying unknown image [1, 2]. Typically, such cost functions consist of a data-fidelity term (e.g., least squares or weighted least squares terms) capturing the imaging forward model and the measurement noise statistical model, and a regularizer term (e.g., smoothness or sparsity penalties) capturing presumed object properties. For example, sparsity- or low-rankness-based regularizers have been widely used for image reconstruction from limited data (e.g., in compressed sensing [3]).

3.2 Classical Model-Based Image Reconstruction

We start the discussion of MBIR with the mathematical formulation of the image reconstruction problem. An image is represented as an array of discrete pixels and denoted as a vector x of length mn (the number of pixels, where m and n are the numbers of rows and columns in the image), which can assume either real or complex values depending on the imaging modality. Classical imaging systems involve a measurement or forward operator $\mathcal{A}$ applied to an image for reconstruction, and one obtains corresponding observations $y = \mathcal{A}(x) + \varepsilon$, where ε is the noise in the measurements. When the measurement process is linear, the application of the forward operator to an image can be characterized by a matrix–vector product. Early publications on MBIR methods tended to exploit the mathematical Bayesian framework for image reconstruction. The dominant Bayesian approach [4] for reconstructing an image x from measurements y is maximum a posteriori (MAP) estimation, which finds the maximizer of the posterior distribution $p(x|y)$ by Bayes' law, i.e.

$$\hat{x} = \arg\min_{x}(-\log p(y|x) - \log p(x)), \tag{3.1}$$

where $-\log p(y|x)$ is the negative log-likelihood, which encompasses the physics of the imaging system and noise statistics, and $p(x)$ is the prior, which carries the assumptions one makes about the properties of the image under consideration. For any prior $p(x)$, one can define $\beta \mathcal{R}(x) = -\log p(x)$. Thus, the form of the Bayes reconstructor (3.1) can be cast as the solution to the following regularized optimization problem:

$$\min_{x} f(x, y) + \beta \mathcal{R}(x), \tag{3.2}$$

where $f(x, y)$ characterizes the data fidelity (e.g., $\frac{1}{2}\|\mathcal{A}(x) - y\|_2^2$) and $\mathcal{R}(x)$ is a regularizer that guides the reconstructed image x to have some assumed properties such as smoothness, sparsity, or low-rankness.

Regularization provides a mechanism of forcing the reconstructed images to have certain desired properties. When the measurement operator in the least squares datafidelity term has a large condition number, it makes the unregularized optimization problem ill-conditioned. Or when it is underdetermined, the corresponding unregularized optimization problem has infinitely many reconstruction results. In such cases, the Tikhonov regularizer $\mathcal{R}(x) = \|x\|_2^2$ with properly chosen magnitude parameter β will render the optimization problem well-posed. Recent methods use edge-preserving regularization involving non-quadratic functions of the differences between neighboring pixels [5], implicitly assuming that the image gradients are sparse. In two dimensions, an example of a difference-based regularizer would be $\mathcal{R}(x) = \sum_{l=2}^{m} \psi(x_{l,:} - x_{l-1,:}) + \sum_{l=2}^{n} \psi(x_{:,l} - x_{:,l-1})$, where $x_{l,:}$ and $x_{:,l}$ denote the lth row and column of the two-dimensional (2D) image respectively, and ψ is applied to vector elements (scalars) and summed, e.g., the hyperbola $\psi(z) = \sqrt{|z|^2 + \delta^2}$ or a generalized Gaussian function [6]. In practice, tweaks to the form of the regularizer have been proposed to address the issue of nonuniform spatial resolution in tomography (by introducing spatial weights in the penalty for neighboring pixels [7]) and the issue of sensitivity to the choice of hyperparameters (by penalizing the relative difference instead of the absolute difference between pixels [8]). These difference-based regularizers are relatives of the well-studied total variation (TV) regularizers, which use the nonsmooth absolute-value potential function $\psi(z) = |z|$. Such TV regularizers impose a strong assumption of gradient sparsity and may be less suitable for piecewise-smooth images. In particular, the TV regularizer may introduce CT images with undesirable patchy textures [9].

Another important aspect of classic image reconstruction techniques is the underlying *sampling* that constitutes the measurement; it is natural to ask what the minimum number of required measurements is to recover an image with prior assumptions in a given context. Conventional approaches to sampling signals follow Shannon's sampling theorem: the sampling rate must be at least twice the maximum frequency present in the signal. Many natural signals (and in particular images) have concise representations when expressed in a convenient basis, or, in other words, the information in a signal is concentrated on a few basis vectors in a domain, even though it may not necessarily be bandlimited. When an image can be *sparsely* represented in a certain domain and the measurements of the image are *incoherent* to the basis of its representation, such measurements may suffice to capture most of the information contained in that image and thus provide an accurate reconstruction of the image through convex optimization. This observation holds notwithstanding the fact that the sampling frequency may be lower than the Nyquist frequency [10]. To formulate this observation, consider a wide rectangular-shaped measurement matrix $A \in \mathbb{R}^{d \times mn}$ (e.g., in CT) or $A \in \mathbb{C}^{d \times mn}$ (e.g., in MRI), where typically $d < mn$. The image reconstruction can be mathematically formulated as

$$\hat{x} = \arg\min_{x} \frac{1}{2} \|Ax - y\|_2^2 + \beta \mathcal{R}(Wx), \tag{3.3}$$

where W is a sparsifying transform matrix that transforms the input image x to a domain where it is sparse.

The ℓ_1-norm is often used as the sparsity regularizer, and it can be viewed as a convex relaxation or convex envelope of the nonconvex "ℓ_0-norm" that counts the number of nonzero elements in a vector. Empirical observation reveals that images often have a sparse representation under Fourier or wavelet transforms, and the *compressed sensing* theory shows that random sensing matrices such as Gaussian matrices or Rademacher matrices are largely incoherent with fixed bases [11]. A signature result in compressed sensing theory states that with only $\mathcal{O}(S \log(N/S))$ measurements encoded in a matrix A with the incoherence property, the convex *basis pursuit* problem $\min_{\tilde{x} \in \mathbb{R}^N} \|\tilde{x}\|_1$ subject to $y = A\tilde{x}$ can recover any S-sparse signal vector $x \in \mathbb{R}^N$ exactly with high probability.

Low-rank models have also shown promise in imaging applications. The assumption of low-rank structure is particularly useful when processing dynamic or time-series data, and has been popular in dynamic MRI, where the underlying image sequences are correlated over time. Recent work has applied low-rank plus sparse (L+S) models to dynamic MRI reconstruction [12], and accurate reconstruction is reported when the underlying L and S components are distinguishable and the measurement is also sufficiently incoherent with these components. Low-rank components capture the background or slowly changing parts of the dynamic object, while the sparse components capture the dynamics in the foreground such as local motion or contrast changes. The L+S reconstruction problem is formulated as

$$\min_{x_L, x_S} \frac{1}{2} \|A(x_L + x_S) - y\|_2^2 + \lambda_L \|\text{Mat}(x_L)\|_* + \lambda_S \|Wx_S\|_1, \tag{3.4}$$

where $\|\cdot\|_*$ is the nuclear norm (the sum of the singular values), x_L and x_S are vectors of dimension mnt with t the number of temporal frames, $\text{Mat}(x_L)$ reshapes x_L into a space–time matrix of dimension $mn \times t$, and $\mathbf{W}$ is a sparsifying operator.

Another low-rankness-based image reconstruction approach proposed for the medical imaging setting uses a low-rank Hankel-structure matrix by exploiting the duality between spatial domain sparsity and the spectral domain Hankel matrix rank. The corresponding optimization problem is purely in the measurement domain. Given sparsely sampled spectral measurements on the index set $\Omega \subseteq \{0, 1, \ldots, n-2, n-1\}$, the missing spectrum estimation problem is cast as

$$\underset{\mathbf{m} \in \mathbb{C}^n}{\arg\min} \; \|H(n,d)(\mathbf{m})\|_*$$
$$\text{s.t.} \; P_\Omega(\mathbf{m}) = P_\Omega(\hat{x}), \tag{3.5}$$

where

$$H(n,d)(\mathbf{m}) = \begin{pmatrix} m(0) & m(1) & \cdots & m(d-1) \\ \vdots & \vdots & \vdots & \vdots \\ m(n-d) & m(n-d+1) & \cdots & m(n-1) \end{pmatrix} \in \mathbb{C}^{n \times d}$$

(d is a hyperparameter), $P_\Omega(\cdot)$ denotes the projection of the measured k-space samples onto the index set Ω, and $\hat{x}$ is the Fourier data of the signal. When $|\Omega| = \mathcal{O}(\text{rank}(H(n,d))d^{-1} \log n)$, the optimization problem (3.5) recovers the image exactly with high probability [13].

Iterative methods such as proximal gradient descent [14] and the alternating direction method of multipliers (ADMM) [15] are deployed for solving the aforementioned optimization problems. To minimize loss functions with nuclear norm, a surrogate relaxation and careful initialization are needed to deploy the conditional gradient method [16].

For classical MBIR models involving sparsity, transforms such as the Fourier transform and wavelet transform are fixed and the corresponding measurement operators are generic and nonadaptive. Also, the choice of hyperparameters can considerably affect the quality of the reconstructed images, and there is not always a systematic guideline for how to choose hyperparameters for imaging problems. In the following sections, object-adaptive MBIR methods will be highlighted.

3.3 MBIR with Unsupervised Learning

In this section we introduce several unsupervised learning techniques that can be used with MBIR to provide unsupervised prior knowledge of the target image. We first introduce two sparsity and compressed-sensing-inspired unsupervised learning techniques, dictionary learning and sparsifying transform learning, which can be incorporated into MBIR. Then we present some generic neural-network-based unsupervised learning techniques and their combinations with MBIR (see Fig. 3.1). The descriptions

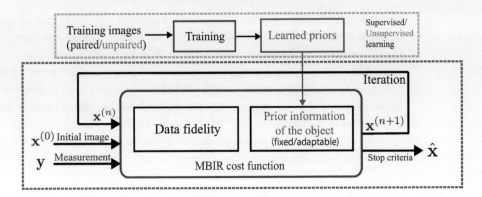

Figure 3.1 A general illustration of MBIR with learning.

are based on applications with real-valued images such as in CT, but they are readily extended to applications with complex-valued images such as MRI.

3.3.1 Synthesis Dictionary Learning

A synthesis dictionary provides a set of atoms, sparse linear combinations of which can represent a signal or an image. Let $D \in \mathbb{R}^{m \times n}$ denote a dictionary, whose columns are atoms. Usually, the dictionary is overcomplete, i.e., $n \gg m$. A signal $x \in \mathbb{R}^m$ can be represented as $x = Dz + \mathbf{e}$, where $z \in \mathbb{R}^n$ denotes sparse coefficients, and $\mathbf{e} \in \mathbb{R}^m$ denotes small differences between the synthesized signal and the original signal.

Given a matrix $X \in \mathbb{R}^{m \times S}$ composed of S vectorized training image patches, the dictionary learning problem can be formulated as

$$\arg\min_{D \in \mathcal{D}, Z} \sum_{i=1}^{S} \left(\|X_i - DZ_i\|_2^2 + \lambda_i \|Z_i\|_0 \right), \qquad (3.6)$$

where $\mathcal{D}$ denotes the set of feasible dictionaries (e.g., those with unit norm columns [17]), X_i and Z_i denote the ith columns of matrices X and Z respectively, and λ_i is a nonnegative regularizer parameter for the ith patch. The operator $\|\cdot\|_0$ is the ℓ_0 sparsity functional, which counts the number of nonzeros in a vector. Sparsity constraints (rather than penalties) are also often used [17, 18].

The dictionary learning problem (Eq. (3.6) or its variants) is often optimized via alternating between a sparse coding step, i.e., updating Z with a fixed dictionary D, and a dictionary learning step, i.e., updating D with fixed sparse coefficients Z. The algorithm K-SVD [17] is a well-known approximate algorithm for dictionary learning, but it involves computationally expensive sparse coding using the greedy orthogonal matching pursuit [19] algorithm. Recent work has efficiently tackled dictionary learning with sparsity penalties (e.g., the sum-of-outer-products dictionary learning method [20]) using block coordinate descent optimization with thresholding-based closed-form sparse code updates. However, [20] still optimizes rows of Z in a sequential manner, creating a bottleneck.

The image reconstruction problem incorporating a pre-learned dictionary-based regularizer can be formulated as

$$\min_{x,z} f(x, y) + \beta \sum_{j=1}^{N} \left(\|P_j x - Dz_j\|_2^2 + \gamma_j \|z_j\|_0 \right), \tag{3.7}$$

where $f(x, y)$ represents the data-fidelity term (e.g., $f(x, y) \triangleq \|Ax - y\|_2^2$) capturing the imaging measurement model and noise model, A and y are the system matrix and measurements, respectively, as described in the previous section, P_j is the patch extraction operator (the dictionary acts on image patches), $\beta > 0$ is the regularizer parameter, and the nonnegative γ_j's control sparsity levels. While using pre-learned (from a set of images) dictionaries has been shown to be useful for biomedical image reconstruction [21], the dictionary can also be directly learned or optimized at image reconstruction time [18, 21, 22], which can yield models highly adaptive to the underlying image. For example, Eq. (3.7) can be optimized by alternating between updating x (by solving a least squares problem) and updating the sparse codes. Dictionary learning methods provided one of the earliest breakthroughs for machine learning in biomedical image reconstruction [18, 21, 23].

3.3.2 Sparsifying Transform Learning

The sparsifying transform is a general-analysis operator that is applied to a signal to produce sparse or approximately sparse representations. For a sparsifying transform $\Omega \in \mathbb{R}^{m \times n}$, the model is

$$\Omega x = z + e, \quad \|z\|_0 \leq s, \tag{3.8}$$

where $x \in \mathbb{R}^n$ denotes the original signal, $z \in \mathbb{R}^m$ is a vector with several zeros (or up to s nonzeros), and $e \in \mathbb{R}^m$ represents the approximation error or residual in the transform domain. An advantage of the sparsifying transform model over the synthesis dictionary is that the signal's sparse coefficients can be cheaply obtained by thresholding [24]. This has led to very efficient sparsifying-transform learning methods for image recovery tasks, including image denoising and image reconstruction [22, 24–26].

In the following, we first introduce several types of sparsifying transforms and their learning, and then describe their application to biomedical image reconstruction.

Learning Square Transforms

Given a training matrix $X \in \mathbb{R}^{n \times S}$, whose columns are the S vectorized patches of length n drawn from training images, the learning problem is formulated as follows:

$$\arg\min_{\mathbf{\Omega}, \mathbf{Z}} \|\mathbf{\Omega X} - \mathbf{Z}\|_F^2 + \lambda \widetilde{\mathcal{R}}(\mathbf{\Omega}) + \sum_{i=1}^{S} \gamma^2 \|\mathbf{Z}_i\|_0, \qquad (3.9)$$

where $\mathbf{\Omega} \in \mathbb{R}^{n \times n}$ is the transform to be learned, $\mathbf{Z} \in \mathbb{R}^{n \times S}$ is the transform-domain sparse-coefficients matrix associated with $\mathbf{\Omega}$, $\lambda > 0$ is a weighting parameter for the regularizer $\widetilde{\mathcal{R}}(\mathbf{\Omega})$, and $\gamma > 0$ controls the sparsity of $\mathbf{Z}$. The regularizer $\widetilde{\mathcal{R}}(\mathbf{\Omega})$ is designed to avoid trivial solutions of $\mathbf{\Omega}$ such as a zero matrix or $\mathbf{\Omega}$ with repeated rows or arbitrarily large scaling [24]. A possible choice is $\widetilde{\mathcal{R}}(\mathbf{\Omega}) \triangleq \|\mathbf{\Omega}\|_F^2 - \log |\det \mathbf{\Omega}|$ [24]. One could also enforce $\mathbf{\Omega}$ to be unitary, i.e., $\mathbf{\Omega}^T \mathbf{\Omega} = \mathbf{I}$ [22]. Alternating-minimization strategies are commonly used to optimize (3.9). Specifically, when solving for $\mathbf{Z}$ with fixed $\mathbf{\Omega}$, we have the subproblem $\arg\min_{\mathbf{Z}} \|\mathbf{\Omega X} - \mathbf{Z}\|_F^2 + \sum_{i=1}^{S} \gamma^2 \|\mathbf{Z}_i\|_0$, which can be exactly solved via hard-thresholding, i.e., by setting $\hat{\mathbf{Z}} = H_\gamma(\mathbf{\Omega X})$, with $H_\gamma(\cdot)$ an elementwise hard-thresholding operator that zeros out elements smaller than γ. When solving for $\mathbf{\Omega}$ (fixed $\mathbf{Z}$), the corresponding subproblem can also have a closed-form solution [27]. These algorithms have proven convergence guarantees [27, 28]. The approach can be extended to learn overcomplete sparsifying transforms [29].

Learning Unions of Transforms

The union of learned transforms (ULTRA) model [30] is an extension of the single square-transform case (see Fig. 3.2). Here, we assume that image patches can be grouped into multiple classes, with a different transform for each class that sparsifies the patches in that class. The model enables flexible capturing of the diversity of

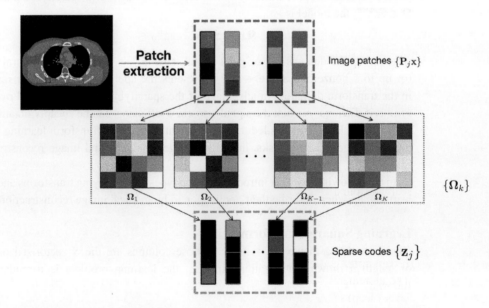

Figure 3.2 Illustration of ULTRA. Each patch extracted from an image is sparsified by a particular transform from the learned union of transforms that can sparsify it best.

features in image patches with a collection or union of transforms. The union of transforms can be learned as follows:

$$\arg\min_{\{\boldsymbol{\Omega}_k, \mathbf{Z}, C_k\}} \sum_{k=1}^{K} \sum_{i \in C_k} \left\{ \|\boldsymbol{\Omega}_k \mathbf{X}_i - \mathbf{Z}_i\|_2^2 + \gamma^2 \|\mathbf{Z}_i\|_0 \right\} + \sum_{k=1}^{K} \lambda_k \widetilde{\mathcal{R}}(\boldsymbol{\Omega}_k), \text{ s.t. } \{C_k\} \in \mathcal{G}, \quad (3.10)$$

where k is the cluster or group index, K is the total number of clusters, C_k is the set of indices of image patches grouped into the kth class with transform $\boldsymbol{\Omega}_k$, and $\mathcal{G}$ is the set containing all possible partitions of $\{1, \ldots, S\}$ into K disjoint subsets. The regularizers $\widetilde{\mathcal{R}}(\boldsymbol{\Omega}_k)$ $\forall k$ control the properties of the learned transforms. Problem (3.10) and its variants can be optimized by alternately optimizing the cluster memberships and sparse codes and the collection of transforms, with efficiently computed solutions at each step [26, 30].

Other variations of this model have also been explored, e.g., in [31], where the different transforms are related to each other by flipping and rotation operators, creating a more structured model.

Image Reconstruction with Learned Transforms

We formulate the image reconstruction problem regularized by a union of pre-learned sparsifying transforms as

$$\min_{\boldsymbol{x}, \{z_j, C_k\}} f(\boldsymbol{x}, \boldsymbol{y}) + \beta \sum_{k=1}^{K} \sum_{j \in C_k} \left\{ \|\boldsymbol{\Omega}_k \mathbf{P}_j \boldsymbol{x} - z_j\|_2^2 + \gamma \|z_j\|_0 \right\}, \text{ s.t. } \{C_k\} \in \mathcal{G}. \quad (3.11)$$

When $K = 1$, this is equivalent to using a single-square sparsifying transform model for reconstruction. The algorithm for Eq. (3.11) in [26] alternates between an image update step, i.e., updating $\boldsymbol{x}$ with the other variables fixed, and a sparse coding and clustering step, i.e., updating all the z_j and C_k jointly with $\boldsymbol{x}$ fixed.

While the sparsifying transforms can be pre-learned from images and used in Eq. (3.11) (see [26] for an application in CT), they can also be learned at the time of image reconstruction (see [22, 32] for examples in MRI) in order to capture more image-adaptive features. The latter involves optimizing more variables at reconstruction time and more computation. A hybrid between the two approaches such as pre-learning a sparsifying model and then adapting it further at reconstruction time could combine the benefits of both regimes. As a side note, while sparsifying transforms can be learned from and applied to any type of signal, including image patches from whole images, they can be equivalently applied directly to input images via convolutional operations [22, 33].

Extensions: Multi-Layer and Online Sparsifying Transform Learning

The aforementioned (single-layer) sparsifying transforms can be extended to multi-layer (deep) setups to further sparsify signals or images [34, 35]. Here, the transform-domain residuals (obtained by subtracting the transformed data and their sparse approximations) in each layer of the multi-layer transform model are fed as input to the subsequent layer to be further sparsified. The learning problem (using a single transform rather than a union of transforms per layer) can be formulated as

$$\min_{\{\mathbf{\Omega}_l, \mathbf{Z}_l\}} \sum_{l=1}^{L} \{\|\mathbf{\Omega}_l \mathbf{R}_l - \mathbf{Z}_l\|_F^2 + \gamma_l^2 \|\mathbf{Z}_l\|_0\}, \text{ s.t. } \mathbf{R}_{l+1} = \mathbf{\Omega}_l \mathbf{R}_l - \mathbf{Z}_l, \mathbf{\Omega}_l^T \mathbf{\Omega}_l = \mathbf{I}, \forall l,$$
(3.12)

where $\mathbf{R}_l$ denotes the transform-domain residuals in the lth layer (defined iteratively) for $l = 2, \ldots, L$, in the first layer $\mathbf{R}_1$ denotes a matrix with the original training signals or image patches, and $\{\gamma_l > 0\}$ are layer wise parameters controlling sparsity levels. For learning simplicity, a unitary constraint is used for the transforms. Once learned, the multi-layer model can be used to regularize the image reconstruction cost function; it can outperform single-layer models [34, 35]. Recently, the union of transforms model has been combined with the multi-layer framework with $L = 2$ layers [36] for CT image reconstruction.

Finally, recent works have also studied the learning of sparsifying transforms and dictionaries for dynamic image reconstruction in a fully online (or time-sequential) manner [37, 38]. In the online learning framework, the learned transform (or dictionary) is evolved over time while simultaneously a time-series of images from limited or corrupted measurements is reconstructed. This allows the transform to change with the data depending on the underlying dynamics, and is particularly suitable for applications such as dynamic MRI [37]. Often an optimization problem for learning and reconstruction needs to be solved every time that measurements for a new frame are received. However, using "warm starts" for the variables based on previous estimates speeds up the convergence of the algorithms (necessitating only few iterations per frame), making online algorithms both effective and efficient for dynamic imaging.

3.3.3 Autoencoder

An autoencoder usually consists of an encoding pipeline (or encoder), which extracts features from the input signal, and a decoding pipeline (or decoder) which reconstructs the signal from the extracted features. The basic structure of an autoencoder can be described as follows:

$$\text{Encoding:} \quad z = E_\theta(x),$$
$$\text{Decoding:} \quad \hat{x} = D_\phi(z),$$
(3.13)

where $x \in \mathbb{R}^n$ is an input signal or image, $z \in \mathbb{R}^m$ represents encoded features from the encoder $E_\theta(\cdot)$ with parameters θ, $D_\phi(\cdot)$ denotes the decoder with parameters ϕ, and $\hat{x} \in \mathbb{R}^n$ is the reconstructed signal.

Autoencoders are usually trained with loss functions that describe reconstruction errors such as the mean squared error (MSE) loss or the cross-entropy loss. A representative type of autoencoder using the reconstruction error loss is the *denoising autoencoder* (DAE) [39]. A DAE is trained in an unsupervised manner with a set of given reference images. Noisy samples are generated by applying a random perturbation to the reference images, e.g., adding Gaussian noise. The DAE is then trained to map between noisy images and their corresponding reference images. Additional constraints can be considered in training an autoencoder, which leads to various types of autoencoders. For example, *contractive autoencoders* (CAE) constrain the encoder by adding the Frobenius norm of the Jacobian of the encoder to the basic reconstruction-error training loss, in order to improve the robustness of feature extraction with the encoder [40]; *sparse autoencoders* take into account the sparsity constraint during training via setting the majority of the hidden units to be zeros in the encoding process [41].

Autoencoders have been incorporated into MBIR frameworks to provide learning-based priors, but in these cases the autoencoders have mostly been trained in a supervised manner based on paired noisy and reference images [42–45]. An MBIR framework with unsupervised learning-based DAE was proposed recently for MRI and sparse-view CT image reconstruction [46–48]. Other types of unsupervised autoencoders may also have the potential to be combined with MBIR in a similar way.

3.3.4 Generative Adversarial Network (GAN)-Based Methods

A generative adversarial network (GAN) is composed of two models that compete with each other during training: a generator $G(\cdot)$ and a discriminator $D(\cdot)$. The generator attempts to generate realistic data, while the discriminator attempts to discriminate between the generated data and the real data [49]. Let the inputs to a GAN be distributed as $z \sim p_z(z)$, where z denotes the noise variables and $p_z(z)$ is the probability distribution of z. The generator maps the noise z to the data space as $G(z)$. The generated data is expected to have a distribution close to that of the real data x, i.e., $p_{\text{data}}(x)$. The discriminator outputs a single scalar that represents the probability of its input coming from real data rather than from generated data.

This idea leads to a cross-entropy-based GAN optimization cost:

$$\min_G \max_D \mathbb{E}_{x \sim p_{\text{data}}(x)}[\log D(x)] + \mathbb{E}_{z \sim p_z(z)}[\log(1 - D(G(z)))], \quad (3.14)$$

where D is trained to maximize the probability of correct labeling, i.e., to maximize $D(x)$ and $1 - D(G(z))$ (in logarithmic form), and G is trained to minimize the term $\log(1 - D(G(z)))$, all in an expected sense over the respective distributions. Various improvements have been made to GANs in terms of loss functions and network architectures. Examples include: WGAN [50], which replaces the cross-entropy-based loss with the Wasserstein-distance-based loss; LSGAN [51], which uses the least squares measure for the loss function; the VGG19 network-induced GAN [52]; and cyclic-consistency-based cycleGAN [53].

In medical imaging, GANs have shown promising results for applications such as low-dose CT image denoising [54–57], multiphase coronary CT image denoising [58], and compressive sensing MRI [59]. Recently, GANs have been combined with model-based methods for image reconstruction. For example, a method for PET image reconstruction uses a pre-trained GAN to constrain an MBIR problem by encouraging the iterative reconstructions to lie in a feasible set created by outputs of the trained generator with noisy inputs. To force the noisy inputs to have a noise level similar to that of the training images, an additional regularizer involving the noisy inputs is introduced into MBIR and the inputs to the generator are optimized along with the iterative reconstructions [60]. It is also possible to train a GAN using model-based costs. In [61], an MBIR cost that embeds a generative network was used to formulate a training loss for the generator on the basis of optimal transport [62]. In particular, the generator takes measurements as inputs and outputs images, which models the inverse path of the imaging problem. In the testing phase, images are simply reconstructed by feeding the measurements to the trained generator.

3.3.5 Deep Image Prior

The deep image prior (DIP) method was proposed to implicitly capture prior information from deep networks for inverse image reconstruction problems such as denoising and super-resolution [63]. Rather than utilizing explicit regularizers, DIP implicitly regularizes the problem via *parameterization*, i.e., $x = f_\theta(z)$ [63], where $f_\theta(\cdot)$ represents the network with parameters θ, z is some fixed input such as random noise, and x is the unknown image to be restored. The network in DIP is task-specific and network training does not rely on external prior data. Therefore, it has great potential in medical imaging tasks, which lack large amounts of training pairs.

The DIP technique has been applied to reconstruct medical images, e.g., in PET [64], dynamic MRI [65], and CT [66]. A general formulation for image reconstruction tasks with DIP is

$$\theta^* = \arg\min_\theta \| y - A f_\theta(z) \|, \qquad \widehat{x} = f_{\theta^*}(z), \tag{3.15}$$

where A and y denote the system matrix and measurements respectively, as described in the previous section, z is the input to the network, $f_\theta(\cdot)$ is the network that is updated from scratch during "training," and $\widehat{x}$ is the final restored image. Additional regularizers can be combined with this data-fidelity-based cost function, i.e.,

$$\theta^* = \arg\min_\theta \| y - A f_\theta(z) \| + \beta \mathcal{R}(f_\theta(z)), \qquad \widehat{x} = f_{\theta^*}(z), \tag{3.16}$$

where β is the regularizer parameter and $\mathcal{R}(f_\theta(z))$ is an additional regularizer such as the total variation [66].

Image reconstruction performance using DIP largely depends on the choice of network architectures [63]. In [64], the authors used a modified three-dimensional U-Net [67] for PET image reconstruction; in [66], a U-Net without skip connections showed satisfying reconstruction results for CT; in [65], the authors proposed a compound network consisting of a manifold-based network and a generative convolutional neural network (CNN) for dynamic MRI reconstruction. The manifold-based network was designed to learn a mapping from a one-dimensional manifold to a more expressive latent space that could better represent temporal dependencies of the dynamic measurements.

Deep image prior methods are quite similar to synthesis dictionary and sparsifying-transform learning-based reconstruction methods, when the sparsifying models are adapted directly on the basis of the measurements [22] (also called blind compressed sensing). However, unlike the latter case, wherein iterative algorithms are designed to adequately optimize MBIR costs with convergence guarantees, DIP methods (e.g., for Eq. (3.15)) may involve the ad hoc early stopping of iterations to prevent overfitting.

3.4 MBIR with Supervised Learning

Deep-learning methods are being increasingly deployed as alternatives to regularized image reconstruction owing to their adaptiveness in various contexts. The direct-inversion end-to-end approaches, which use a neural network as an autoencoder, rely on a deep CNN to recover images from initial reconstructions (e.g., undersampled gridding reconstruction [68] or undersampled filtered back-projection reconstruction [69]).

Neural networks can also be trained to replace the role of priors in classical MBIR methods. Instead of using manually fixed priors in the methods discussed in the previous section, it is tempting to tailor the prior to a given problem without compromising the generalizability of such a learned prior. One representative work of constructing a learned prior by neural networks is the *Neumann network* [70], an end-to-end data-driven method to solve the optimization problem (3.2). Motivated by expanding the matrix inverse term in the analytical solution to the first-order critical condition of optimizing a penalized least squares objective function as a Neumann series, the Neumann network has iterative blocks entailing steps of the gradient descent method. These gradient descent blocks are cascaded in the form of a residual network, and the neural network which represents the gradient of the prior is assumed to be piecewise linear.

As well as the use of neural networks as a direct end-to-end pipeline and a latent prior, we describe two major categories where neural networks are combined with MBIR in a supervised manner.

3.4.1 Plug-and-Play

Plug-and-play (PnP) regularization, originally proposed in [71], involves the use of a standard, off-the-shelf denoising algorithm as a replacement for the proximal operator inside an algorithm for MBIR, e.g., proximal gradient descent or ADMM. Several algorithms to solve an optimization problem with the general form (3.2) involve alternating between an operation that promotes data fidelity (i.e., reduces $f(x, y)$) and an operation that accounts for regularization (i.e., reduces $\mathcal{R}(x)$). For example, a proximal gradient algorithm iterates between a gradient descent step that drives the current estimate $x^{(k)}$ to be a better fit to the observation y and a following proximal operator step that outputs a well-regulated estimate in the proximity of the resulting estimate from the gradient descent step. The proximal operator can be interpreted as a Gaussian denoising step in the maximum a posteriori sense. Thus there have been attempts to use neural networks to learn the proximal operator, essentially rendering the neural network function a prior.

The PnP idea has been extended in various ways, notably with the consensus equilibrium framework [72], which allows the free combination of data terms and regularizers without any reference to any variational problem. The main advantage of the PnP model is that it provides a simple way to combine physical-system and noise models with existing state-of-the-art *denoisers* in a principled manner with some theoretical guarantees [73] (although weaker than those frequently offered by conventional variational approaches based on convex optimization). This flexibility has been used by several authors to combine deep-learning-based methods with model-based reconstruction; we discuss several of these works here.

The work in [74] demonstrates a straightforward approach to using a CNN as a PnP regularizer. The authors use half-quadratic splitting to arrive at an iterative reconstruction algorithm of the form

$$\mathbf{x}_{k+1} = \arg\min_{\mathbf{x}} \|\mathbf{y} - \mathbf{A}\mathbf{x}\|_2^2 + \alpha \|\mathbf{x} - \mathbf{z}_k\|_2^2, \qquad (3.17)$$

$$\mathbf{z}_{k+1} = \arg\min_{\mathbf{z}} \alpha \|\mathbf{z} - \mathbf{x}_{k+1}\|_2^2 + \beta \mathcal{R}(\mathbf{z}), \qquad (3.18)$$

where k denotes the iteration number, $\mathcal{R}$ is an arbitrary regularization functional, and α increases at each iteration. The upper equation may be viewed as enforcing data consistency (because it involves the forward model, $\mathbf{A}$). Following the PnP methodology, rather than solving the minimization problem to update $\mathbf{z}$, the authors of [75] viewed the $\mathbf{z}$ update as a denoising step with noise level proportional to $\sqrt{\alpha/\beta}$. They accomplished this denoising via a set of CNNs, each trained to remove a different level of Gaussian noise from images to account for the increasing values of α during reconstruction. They validated their approach on image denoising, deblurring, and super-resolution, but the same method would immediately extend to any modality with a linear forward operator. For example, one approach [75] tackles low-dose X-ray CT by plugging a CNN-based denoiser into an ADMM algorithm and [76] addresses super-resolution with a similar approach.

Once one begins to use PnP with CNNs, it is natural to explore training the CNN in conjunction with the data consistency in a supervised manner. There is theoretical justification for this: even though (3.18) is a denoising problem, the effective noise may not be Gaussian and the formulation suggests that a MAP rather than minimum mean square error denoiser is required [77].

One approach to training a CNN inside a PnP structure end-to-end is to unroll a fixed number of iterations of the use PnP optimization (see the next subsection for a detailed discussion of the unrolling technique). If the data-fidelity-updated version of Eq. (3.17) can be computed in closed form [78], or by conjugate gradient (CG) [79], it is simple to compute gradients through the unrolled PnP. Otherwise, the data-fidelity minimization (3.17) may have to be limited to a small number of steps in order to permit gradient computations. An important design decision here is whether to use the same or different CNN weights in each PnP iteration; the first possibility is explored in [78, 79] and the second in [80]. Other works following the PnP theme include [81], where the proximal step is replaced by a general denoising method such as nonlocal means or BM3D [82], or [83], where the proximal step is designed to be a projection network which aims to project the input onto the target image space and is trained in the framework of generative adversarial networks.

3.4.2 Unrolling

The deep network approach is closely related to the idea of *unrolling*, where the network is not necessarily a neural network. "Unrolling" an optimization method refers to treating a series of iterates of an iterative optimization method as a single operation block to be applied to an input. In other words, a network considered in the unrolling context is a cascade of such blocks that embodies the iterative steps of an optimization method (e.g., ADMM or block coordinate descent) with tunable parameters, and this cascade of unrolled blocks is then an end-to-end reconstruction pipeline to be applied to the observed images. The earliest proposed unrolled inverse problem solver was LISTA [84], where the authors unrolled the iterative shrinkage and thresholding algorithm ISTA [85], which aimed to learn the dictionary for sparse encoding, which is the same across all unrolling layers. Another work [86] considers the image reconstruction problem (3.2) and chooses as prior a power function with filtered patches of the image as input. Blocks of the half-quadratic method (HQ method) are cascaded as a network, and the shrinkage step in the HQ method is replaced by a linear combinations of radial basis function (RBF) kernels. To train the network, one aims to learn the filters used to convolve patches, the weight parameter for each RBF kernel, and other parameters occurring in the HQ methods.

Other typical examples of unrolling inverse problem solvers include: *ADMM-Net* [80], which generalizes operations in classical ADMM algorithms for the case of learnable parameters, encloses each operation in a node, and connects the nodes into a data flow graph as an end-to-end network; *BCD-Net* [42, 43], which involves learned parameters in unrolled block coordinate descent algorithms for reconstruction; *primal–dual methods* [87], where proximal operators are realized by a parameterized residual neural network with parameters learned from training; *iterative reweighted least squares* [88], where the weight update is executed through a CNN denoiser; and *approximate message passing* [89], where the denoising step in the algorithm is carried out by a trained CNN.

3.5 Case Study: Combined Supervised–Unsupervised (SUPER) Learning

As a case study that illustrates many of the principles in this chapter, we take the SUPER algorithm proposed in [90, 91]. SUPER combines regularization learned without supervision with supervised learned regularization in the form of a CNN. Experimental results (Fig. 3.3) show that such a model can outperform both of its constituent parts.

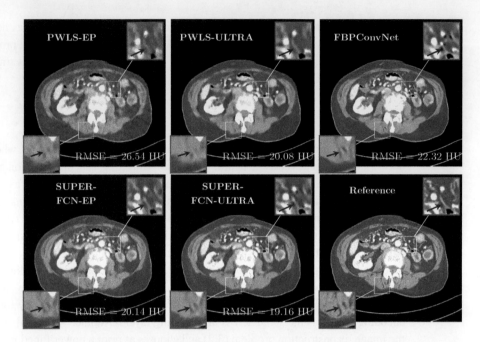

Figure 3.3 Low-dose X-ray CT reconstruction results with a classical MBIR method (PWLS-EP), an MBIR method with unsupervised learned prior (PWLS-ULTRA), a supervised method (FBPConvNet), and combinations of them with the SUPER framework (SUPER-FCN-EP and SUPER-FCN-ULTRA, where FCN refers to FBPConvNet). The upper right inset in each panel of the figure shows an area where unsupervised reconstruction oversmooths some details and supervised reconstruction adds spurious features. SUPER reconstructions sharpen some details and reduce spurious features. The experiments are based on the NIH AAPM Mayo Clinic Low Dose CT Grand Challenge dataset. © 2021 IEEE. Reprinted, with permission, from [91].

In the following, we will discuss how SUPER reconstruction works and how SUPER is trained and present a few results comparing SUPER with its unsupervised and supervised components in the context of X-ray CT image reconstruction.

3.5.1 SUPER Reconstruction

SUPER reconstruction involves two main ingredients: a variational reconstruction formulation and algorithm, and a supervised, image-domain, learned reconstruction model. The choice of both parts is flexible. Here, we will focus on a combination of PWLS-ULTRA [26] (involving unsupervised learning) and FBPConvNet [69] (learning with supervision), which we discuss in the following.

The PWLS-ULTRA method reconstructs images from measurements by optimizing a weighted least squares data term combined with an ULTRA regularizer (discussed in Section 3.3.2):

$$\underbrace{\underbrace{\arg\min_{x} \|y - Ax\|^2_{W(y)}}_{f(x,y)} + \underbrace{\beta \min_{\{z_j\},\{C_k\}} \sum_{k=1}^{K} \sum_{j \in C_k} \tau_j(\|\mathbf{\Omega}_k P_j x - z_j\|^2_2 + \gamma^2 \|z_j\|_0)}_{\beta \mathcal{R}(x)}}_{J(x,y)}.$$
(3.19)

The first term in Eq. (3.19) is the data-fidelity term, where $\| \cdot \|_{W(y)}$ denotes the weighted Euclidean norm with diagonal weight matrix $W(y) \in \mathbb{R}^{m \times m}$ containing the estimated inverse variance of y (see [26] for further details). The second term in Eq. (3.19) forms a regularization term that measures the error when the patches of x are sparse-coded by a union of transforms (see Section 3.3.2 for more details). The PWLS-ULTRA algorithm reconstructs an image x from its measurements y and simultaneously clusters (in an unsupervised way) the patches of x by minimizing a functional of the form $f(x, y) + \beta \mathcal{R}(x)$. Note also that the sparsifying transforms $\{\mathbf{\Omega}_k\}$ are learned as discussed in Section 3.3.2.

FBPConvNet reconstruction [69] proceeds by applying a trained CNN to an initial reconstruction as follows:

$$x^* = G_\theta(x_0(y)),$$
(3.20)

where G_θ denotes a CNN with parameters θ (we discuss learning these parameters in the next section). As originally proposed in [69], the initial reconstruction, $x_0(y)$, is a pseudoinverse of A, e.g., filtered back-projection in the case of X-ray CT. SUPER reconstruction makes use of the fact that this initial reconstruction is flexible; the input to the CNN can be anything, as long as it is the proper shape, i.e., an element of the image domain rather than the measurement domain.

Given a variational reconstruction formulation, e.g., the PWLS-ULTRA problem in Eq. (3.19), and an image-domain, learned reconstruction, model, e.g., the FBPConvNet in Eq. (3.20), SUPER combines the two in a sequential, alternating, manner,

$$\widehat{x}_\theta^{(l)}(y) = \arg\min_{x} J(x, y) + \mu \left\| x - G_\theta\left(\widehat{x}_\theta^{(l-1)}(y)\right) \right\|_2^2.$$
(3.21)

Equation (3.21) may be understood as starting from an initialization $\widehat{x}_\theta^{(0)}(y) = \widehat{x}^{(0)}(y)$, e.g., an FBP reconstruction. Reconstruction proceeds by applying a CNN to $\widehat{x}^{(0)}(y)$ and then solving a variational problem to produce $\widehat{x}_\theta^{(1)}(y)$. The procedure repeats for several iterations, resulting in a final reconstruction. Each iteration is called a SUPER layer. Because the character of the $\widehat{x}_\theta^{(l)}(y)$ functions changes with each SUPER layer, different network weights can be used in each layer, leading at testing time to the following SUPER reconstruction strategy:

$$\widehat{x}_{\theta^{(l)}}^{(l)}(y) = \arg\min_{x} J(x, y) + \mu \left\| x - G_{\theta^{(l)}}\left(\widehat{x}_{\theta^{(l-1)}}^{(l-1)}(y)\right) \right\|_2^2.$$
(3.22)

This algorithm strategy arises from an underlying mathematical framework, to be presented in Sections 3.5.2 and 3.5.3.

3.5.2 SUPER Training

SUPER training involves two datasets: a dataset of high-quality images without measurements, $\{x_n\}_{n=1}^{N_{\text{unsup}}}$, and a dataset of high-quality reference images (coming from regular-dose measurements) with corresponding noisy (e.g., low-dose, sparse-view, etc.) measurements, $\{x_n, y_n\}_{n=1}^{N_{\text{sup}}}$. These datasets may have different sizes, depending on the application. Complementing the models may allow the exploitation of limited training data. The first dataset is used to train the unsupervised regularization term (e.g., ULTRA, as described in the previous subsection), following standard procedures from the literature (for ULTRA, see [26]). Subsequently, the latter dataset is used to train the supervised part of SUPER. We focus here on the supervised training.

SUPER uses a greedy layerwise approach to learning weights, solving

$$\theta^{(l)} = \arg\min_{\theta} \sum_{n=1}^{N_{\text{sup}}} \left\| G_{\theta^{(l)}}\left(\widehat{x}_{\theta^{(l-1)}}^{(l-1)}(y_n)\right) - x_n \right\|_2^2, \quad (3.23)$$

for $l = 1, 2, \ldots, L$. In short: the training alternates between solving the variational problem (3.22), with a fixed CNN-based reconstruction appearing in a quadratic penalty term, and performing the CNN training (3.23), with a fixed variational reconstruction as input to the CNN. Note that the variational problem may be nonconvex or it may involve discrete variables (e.g., clusters), as with PWLS-ULTRA.

3.5.3 Mathematical Underpinnings of SUPER

We describe three interpretations of the above SUPER framework: as a fixed point iteration, as an unrolled PnP reconstruction, and as a bilevel optimization.

To view SUPER as a fixed point iteration, note that the SUPER reconstruction (3.21) is a fixed point iteration that solves the equation

$$\widehat{x}(y) = \arg\min_{x} J(x, y) + \mu \|x - G_\theta(\widehat{x}(y))\|_2^2. \quad (3.24)$$

Thus, SUPER can be viewed as seeking a reconstruction $\widehat{x}$ that is the solution to a variational problem with the solution itself involved as a regularizer. Roughly speaking, an $\widehat{x}$ that solves this equation falls in between the model-based reconstruction ((3.24) with $\mu = 0$) and the result of applying a CNN to the same, $G_\theta(\widehat{x}(y))$. While the fixed point perspective gives a compact (if not complicated) view of SUPER, it cannot easily account for changing the CNN parameters in each layer, nor for the specifics of training.

To view SUPER as a generalized unrolled PnP reconstruction, recall the half-quadratic splitting in Eqs. (3.17) and (3.18), which, just like Eq. (3.21), alternates between a variational problem with a quadratic penalty and (assuming a CNN has been plugged in) a CNN. However, Eq. (3.21) combines unsupervised regularizers with the data-fidelity term and, because of the choice of the regularizer (e.g., nonconvex, nonsmooth, with discrete variables) in SUPER, gradients may not easily propagate through the variational step. Rather than unrolling the variational step, SUPER follows the greedy training approach discussed in the previous subsection. The generalized PnP perspective accounts nicely for the actual SUPER algorithm, but (as with all PnP regularizers) it does not map cleanly back to a variational problem unless the learned component happens to solve a specific denoising problem, which is probably not the case.

Finally, we can view the SUPER training scheme as a heuristic for solving the bilevel optimization

$$\arg\min_{\theta} \sum_{n=1}^{N_{\text{sup}}} \|G_\theta(\widehat{x}_\theta(y_n)) - x_n\|_2^2 \qquad (3.25)$$
$$\text{s.t.} \quad \widehat{x}_\theta(y_n) = \arg\min_{x} J(x, y_n) + \mu \|x - G_\theta(x)\|_2^2.$$

Specifically, training involves alternating between solving the upper-level problem with the lower-level result fixed and then solving the lower-level problem approximately. Roughly, we seek CNN parameters θ, such that, when the CNN is used as regularization in a variational reconstruction, the CNN applied to the minimizer is close to the ground truth. Such an approach makes sense when the CNN acts as a projector onto the space of high-quality reconstruction images. This interpretation is similar to the fixed point one except that, here, the focus is on the CNN parameters and the training process, as opposed to on the actual reconstruction.

Bilevel optimization is related to unrolling (Section 3.4.2). Whereas in unrolling the arg min is approximated with a finite number of steps of a *specific* minimization algorithm with parametric weights, some authors (in other settings) have begun to develop the tools to attack bilevel problems directly. For example, [92] and [93] provide differentiable solvers for convex optimization problems and [94–97] attack specific lower-level problems. To our knowledge, this formulation has not been used for performing supervised biomedical image reconstruction, thus opening a new frontier in the area. A potential advantage is that explicitly learning components of image reconstruction formulations in a bilevel manner could then enable the direct optimization of these formulations at testing time with any suitable algorithm (until convergence), thereby decoupling the training process from the finite unrolling of a specific algorithm.

3.6 Discussion and Future Directions

This chapter has focused on biomedical image reconstruction methods at the intersection of MBIR and machine learning. After briefly reviewing classical MBIR methods for image reconstruction, we discussed combinations of MBIR with unsupervised learning, supervised learning, or both. Such combinations offer potential advantages for learning even with limited data.

Multiple categories of unsupervised learning-based methods were reviewed, including sparsity-based synthesis dictionary learning and sparsifying transform learning-based methods and generic deep-neural-network methods based on autoencoders, GANs, or the deep image prior approach.

Dictionary and transform-based methods can use a variety of structures such as unions of models, rotation invariance, filterbank models, multi-layer models, and so on. Methods such as deep image prior adapt a network to reconstruct specific images.

Among methods at the intersection of MBIR and supervised learning, we reviewed unrolling-based methods (which learn the parameters in conventional iterative MBIR algorithms with limited iterations, in a supervised way) and CNN-based plug-and-play methods. Finally, we discussed a recent combined supervised–unsupervised learning approach called SUPER that brings together both unsupervised priors (e.g., the recently introduced transform-based or dictionary-learning-based priors) and supervised learned neural networks in a common MBIR-based framework. We interpreted this framework from the perspective of fixed point iterations, plug-and-play models, and a challenging bilevel learning formulation. SUPER learning with limited training data obtains improved image reconstructions (see [91]) compared with the individual supervised or unsupervised learning-based methods by themselves, i.e., a unified approach performs better than the individual parts.

Given the growing interest in research at the intersection of model-based and learning-based image reconstruction, we expect continued development of methods and the corresponding theory in this area for years to come. First, the systematic unification of physics-based models (e.g., forward models, partial-differential-equation-based models, etc.) and mathematical and statistical models with models learned from both simulation and experimental data is of critical importance for the better leveraging of datasets and domain knowledge in computational imaging applications. Second, developing provably correct algorithms to directly and effectively learn MBIR regularizers in a bilevel optimization framework may provide an interesting counterpoint or alternative to unrolling-based (supervised) methods, where the choice of the algorithm to unroll and the number of iterations to unroll is currently made in a heuristic manner. Finally, while we focused on machine learning methods for image reconstruction, there is budding interest in machine learning for designing data acquisition as well [98–100], where the integration of model-based and learnable components could play a key role.

References

[1] K. Sauer and C. Bouman, "A local update strategy for iterative reconstruction from projections," *IEEE Transactions on Signal Processing*, vol. 41, no. 2, pp. 534–48, 1993.

[2] J.-B. Thibault, C. A. Bouman, K. D. Sauer, and J. Hsieh, "A recursive filter for noise reduction in statistical iterative tomographic imaging," in *Proc. SPIE Conference on Computational Imaging IV*, 2006, p. 60 650X.

[3] D. L. Donoho, "Compressed sensing," *IEEE Transactions on Information Theory*, vol. 52, no. 4, pp. 1289–1306, 2006.

[4] K. M. Hanson, "Introduction to Bayesian image analysis," in *Proc. Medical Imaging 1993: Image Processing*, M. H. Loew, ed., vol. 1898, International Society for Optics and Photonics. SPIE, pp. 716–731.

[5] S. Ahn, S. Ross, E. Asma, J. Miao, X. Jin, L. Cheng, S. Wollenweber, and R. Manjeshwar, "Quantitative comparison of osem and penalized likelihood image reconstruction using relative difference penalties for clinical PET," *Physics in Medicine and Biology*, vol. 60, pp. 5733–5751, 07 2015.

[6] C. Bouman and K. Sauer, "A generalized gaussian image model for edge-preserving map estimation," *IEEE Transactions on Image Processing*, vol. 2, no. 3, pp. 296–310, 1993.

[7] J. A. Fessler and W. L. Rogers, "Spatial resolution properties of penalized-likelihood image reconstruction: Space-invariant tomographs," *IEEE Transactions on Image Processing*, vol. 5, no. 9, pp. 1346–1358, 1996.

[8] J. Nuyts, D. Beque, P. Dupont, and L. Mortelmans, "A concave prior penalizing relative differences for maximum-a-posteriori reconstruction in emission tomography," *IEEE Transactions on Nuclear Science*, vol. 49, no. 1, pp. 56–60, 2002.

[9] D. Liang, H. Wang, Y. Chang, and L. Ying, "Sensitivity ecoding reconstruction with nonlocal total variation regularization," *Magnetic Resonance Med.*, vol. 65. no. 5, pp. 1384–1392.

[10] E. J. Candes, J. Romberg, and T. Tao, "Robust uncertainty principles: Exact signal reconstruction from highly incomplete frequency information," *IEEE Transactions on Information Theory*, vol. 52, no. 2, pp. 489–509, 2006.

[11] S. Foucart and H. Rauhut, *A Mathematical Introduction to Compressive Sensing*. Birkhäuser, 2013.

[12] R. Otazo, E. Candès, and D. K. Sodickson, "Low-rank plus sparse matrix decomposition for accelerated dynamic MRI with separation of background and dynamic components," *Magnetic Resonance in Medicine*, vol. 73, no. 3, pp. 1125–1136, 2015.

[13] J. C. Ye, J. M. Kim, K. H. Jin, and K. Lee, "Compressive sampling using annihilating filter-based low-rank interpolation," *IEEE Transactions on Information Theory*, vol. 63, no. 2, pp. 777–801, 2017.

[14] Z. Huang and S. Becker, "Perturbed proximal descent to escape saddle points for non-convex and non-smooth objective functions," in *Recent Advances in Big Data and Deep Learning*, L. Oneto, N. Navarin, A. Sperduti, and D. Anguita, eds. Springer, 2020, pp. 58–77.

[15] A. Beck, *First-Order Methods in Optimization*. Society for Industrial and Applied Mathematics, 2017.

[16] A. W. Yu, W. Ma, Y. Yu, J. Carbonell, and S. Sra, "Efficient structured matrix rank minimization," in *Advances in Neural Information Processing Systems*, Z. Ghahramani, M. Welling, C. Cortes, N. Lawrence, and K. Q. Weinberger, eds., vol. 27. Curran Associates, 2014, pp. 1350–1358.

[17] M. Aharon, M. Elad, and A. Bruckstein, "K-SVD: An algorithm for designing overcomplete dictionaries for sparse representation," *IEEE Transactions on Signal Processing*, vol. 54, no. 11, pp. 4311–4322, 2006.

[18] S. Ravishankar and Y. Bresler, "MR image reconstruction from highly undersampled k-space data by dictionary learning," *IEEE Transactions on Medical Imaging*, vol. 30, no. 5, pp. 1028–1041, 2011.

[19] J. A. Tropp, "Greed is good: Algorithmic results for sparse approximation," *IEEE Transactions on Information Theory*, vol. 50, no. 10, pp. 2231–2242, 2004.

[20] S. Ravishankar, R. R. Nadakuditi, and J. A. Fessler, "Efficient sum of outer products dictionary learning (soup-dil) and its application to inverse problems," *IEEE Transactions on Computational Imaging*, vol. 3, no. 4, pp. 694–709, 2017.

[21] Q. Xu, H. Yu, X. Mou, L. Zhang, J. Hsieh, and G. Wang, "Low-dose X-ray CT reconstruction via dictionary learning," *IEEE Transactions on Medical Imaging*, vol. 31, no. 9, pp. 1682–97, 2012.

[22] B. Wen, S. Ravishankar, L. Pfister, and Y. Bresler, "Transform learning for magnetic resonance image reconstruction: From model-based learning to building neural networks," *IEEE Signal Processing Magazine*, vol. 37, no. 1, pp. 41–53, 2020.

[23] S. G. Lingala and M. Jacob, "Blind compressive sensing dynamic MRI," *IEEE Transactions on Medical Imaging*, vol. 32, no. 6, pp. 1132–1145, 2013.

[24] S. Ravishankar and Y. Bresler, "Learning sparsifying transforms," *IEEE Transactions on Signal Processing*, vol. 61, no. 5, pp. 1072–1086, 2013.

[25] L. Pfister and Y. Bresler, "Model-based iterative tomographic reconstruction with adaptive sparsifying transforms," in *Proc. Conference on Computational Imaging XII*, vol. 9020. International Society for Optics and Photonics, 2014, p. 90 200H.

[26] X. Zheng, S. Ravishankar, Y. Long, and J. A. Fessler, "PWLS-ULTRA: An efficient clustering and learning-based approach for low-dose 3D CT image reconstruction," *IEEE Transactions on Medical Imaging*, vol. 37, no. 6, pp. 1498–1510, 2018.

[27] S. Ravishankar and Y. Bresler, "ℓ_0 sparsifying transform learning with efficient optimal updates and convergence guarantees," *IEEE Transactions on Signal Processing*, vol. 63, no. 9, pp. 2389–2404, 2015.

[28] S. Ravishankar, A. Ma, and D. Needell, "Analysis of fast structured dictionary learning," *Information and Inference: A Journal of the IMA*, vol. 9, no. 4, pp. 785–811, 2019.

[29] S. Ravishankar and Y. Bresler, "Learning overcomplete sparsifying transforms for signal processing," in *Proc. 2013 IEEE International Conference on Acoustics, Speech and Signal Processing*, 2013, pp. 3088–3092.

[30] B. Wen, S. Ravishankar, and Y. Bresler, "Structured overcomplete sparsifying transform learning with convergence guarantees and applications," *International Journal of Computer Vision*, vol. 114, nos. 2–3, pp. 137–167, 2015.

[31] ——, "FRIST- flipping and rotation invariant sparsifying transform learning and applications," *Inverse Problems*, vol. 33, no. 7, p. 074 007, 2017.

[32] S. Ravishankar and Y. Bresler, "Data-driven learning of a union of sparsifying transforms model for blind compressed sensing," *IEEE Transactions on Computational Imaging*, vol. 2, no. 3, pp. 294–309, 2016.

[33] L. Pfister and Y. Bresler, "Learning filter bank sparsifying transforms," *IEEE Transactions on Signal Processing*, vol. 67, no. 2, pp. 504–519, 2019.

[34] S. Ravishankar and B. Wohlberg, "Learning multi-layer transform models," in *Proc. 2018 56th Annual Allerton Conference on Communication, Control, and Computing*, 2018, pp. 160–165.

[35] X. Yang, Y. Long, and S. Ravishankar, "Multi-layer residual sparsifying transform (mars) model for low-dose CT image reconstruction," *arXiv:2010.06144*, 2020.

[36] ——, "Two-layer clustering-based sparsifying transform learning for low-dose CT reconstruction," *arXiv:2011.00428*, 2020.

[37] B. E. Moore, S. Ravishankar, R. R. Nadakuditi, and J. A. Fessler, "Online adaptive image reconstruction (OnAIR) using dictionary models," *IEEE Transactions on Computational Imaging*, vol. 6, pp. 153–166, 2020.

[38] B. Wen, S. Ravishankar, and Y. Bresler, "VIDOSAT: High-dimensional sparsifying transform learning for online video denoising," *IEEE Transactions on Image Processing*, vol. 28, no. 4, pp. 1691–1704, 2019.

[39] P. Vincent, H. Larochelle, I. Lajoie, Y. Bengio, P.-A. Manzagol, and L. Bottou, "Stacked denoising autoencoders: Learning useful representations in a deep network with a local denoising criterion." *Journal of Machine Learning Research*, vol. 11, no. 12, 2010.

[40] S. Rifai, P. Vincent, X. Muller, X. Glorot, and Y. Bengio, "Contractive auto-encoders: Explicit invariance during feature extraction," in *Proc. International Conference on Machine Learning*, 2011.

[41] A. Makhzani and B. Frey, "K-sparse autoencoders," *arXiv:1312.5663*, 2013.

[42] S. Ravishankar, I. Y. Chun, and J. A. Fessler, "Physics-driven deep training of dictionary-based algorithms for MR image reconstruction," in *Proc. 2017 51st Asilomar Conference on Signals, Systems, and Computers*, 2017, pp. 1859–1863.

[43] I. Y. Chun and J. A. Fessler, "Deep BCD-Net using identical encoding–decoding CNN structures for iterative image recovery," in *Proc. 2018 IEEE 13th Image, Video, and Multidimensional Signal Processing Workshop*, 2018, pp. 1–5.

[44] I. Y. Chun, X. Zheng, Y. Long, and J. A. Fessler, "BCD-Net for low- dose CT reconstruction: Acceleration, convergence, and generalization," in *Proc. Conference on Medical Image Computing and Computer Assisted Interventions*, Shenzhen, China, 2019.

[45] I. Y. Chun, Z. Huang, H. Lim, and J. A. Fessler, "Momentum-Net: Fast and convergent iterative neural network for inverse problems," *IEEE Transactions on Pattern Analysis and Machine Intelligence*, published online, 2020.

[46] S. Wang, J. Lv, Y. Hu, D. Liang, M. Zhang, and Q. Liu, "Denoising auto-encoding priors in undecimated wavelet domain for MR image reconstruction," *arXiv:1909.01108*, 2019.

[47] Q. Liu, Q. Yang, H. Cheng, S. Wang, M. Zhang, and D. Liang, "Highly undersampled magnetic resonance imaging reconstruction using autoencoding priors," *Magnetic Resonance in Medicine*, vol. 83, no. 1, pp. 322–336, 2020.

[48] Z. He, J. Zhou, D. Liang, Y. Wang, and Q. Liu, "Learning priors in high-frequency domain for inverse imaging reconstruction," *arXiv:1910.11148*, 2019.

[49] I. J. Goodfellow, J. Pouget-Abadie, M. Mirza, B. Xu, D. Warde-Farley, S. Ozair, A. Courville, and Y. Bengio, "Generative adversarial networks," *arXiv:1406.2661*, 2014.

[50] M. Arjovsky, S. Chintala, and L. Bottou, "Wasserstein generative adversarial networks," in *Proc. International Conference on Machine Learning*. PMLR, 2017, pp. 214–223.

[51] X. Mao, Q. Li, H. Xie, R. Y. Lau, Z. Wang, and S. Paul Smolley, "Least squares generative adversarial networks," in *Proc. IEEE International Conference on Computer Vision*, 2017, pp. 2794–2802.

[52] C. Ledig, L. Theis, F. Huszár, J. Caballero, A. Cunningham, A. Acosta, A. Aitken, A. Tejani, J. Totz, Z. Wang et al., "Photo-realistic single image super-resolution using a generative adversarial network," in *Proc. IEEE Conference on Computer Vision and Pattern Recognition*, 2017, pp. 4681–4690.

[53] J.-Y. Zhu, T. Park, P. Isola, and A. A. Efros, "Unpaired image-to-image translation using cycle-consistent adversarial networks," in *Proc. IEEE International Conference on Computer Vision*, 2017, pp. 2223–2232.

[54] J. M. Wolterink, T. Leiner, M. A. Viergever, and I. Išgum, "Generative adversarial networks for noise reduction in low-dose CT," *IEEE Transactions on Medical Imaging*, vol. 36, no. 12, pp. 2536–2545, 2017.

[55] H. Shan, Y. Zhang, Q. Yang, U. Kruger, M. K. Kalra, L. Sun, W. Cong, and G. Wang, "3-D convolutional encoder–decoder network for low-dose CT via transfer learning from a

2-D trained network," *IEEE Transactions on Medical Imaging*, vol. 37, no. 6, pp. 1522–1534, 2018.

[56] Q. Yang, P. Yan, Y. Zhang, H. Yu, Y. Shi, X. Mou, M. K. Kalra, Y. Zhang, L. Sun, and G. Wang, "Low-dose CT image denoising using a generative adversarial network with wasserstein distance and perceptual loss," *IEEE Transactions on Medical Imaging*, vol. 37, no. 6, pp. 1348–1357, 2018.

[57] Y. Ma, B. Wei, P. Feng, P. He, X. Guo, and G. Wang, "Low-dose CT image denoising using a generative adversarial network with a hybrid loss function for noise learning," *IEEE Access*, vol. 8, pp. 67519–67529, 2020.

[58] E. Kang, H. J. Koo, D. H. Yang, J. B. Seo, and J. C. Ye, "Cycle-consistent adversarial denoising network for multiphase coronary CT angiography," *Medical Physics*, vol. 46, no. 2, pp. 550–562, 2019.

[59] T. M. Quan, T. Nguyen-Duc, and W.-K. Jeong, "Compressed sensing MRI reconstruction using a generative adversarial network with a cyclic loss," *IEEE Transactions on Medical Imaging*, vol. 37, no. 6, pp. 1488–1497, 2018.

[60] Z. Xie, R. Baikejiang, T. Li, X. Zhang, K. Gong, M. Zhang, W. Qi, E. Asma, and J. Qi, "Generative adversarial network based regularized image reconstruction for PET," *Physics in Medicine & Biology*, vol. 65, no. 12, p. 125016, 2020.

[61] B. Sim, G. Oh, S. Lim, and J. C. Ye, "Optimal transport, cycleGAN, and penalized ls for unsupervised learning in inverse problems," *arXiv:1909.12116*, 2019.

[62] C. Villani, *Optimal Transport: Old and New*. Springer Science & Business Media, 2008, vol. 338.

[63] D. Ulyanov, A. Vedaldi, and V. Lempitsky, "Deep image prior," in *Proc. IEEE Conference on Computer Vision and Pattern Recognition*, 2018, pp. 9446–9454.

[64] K. Gong, C. Catana, J. Qi, and Q. Li, "PET image reconstruction using deep image prior," *IEEE Transactions on Medical Imaging*, vol. 38, no. 7, pp. 1655–1665, 2018.

[65] K. H. Jin, H. Gupta, J. Yerly, M. Stuber, and M. Unser, "Time-dependent deep image prior for dynamic MRI," *arXiv:1910.01684*, 2019.

[66] D. O. Baguer, J. Leuschner, and M. Schmidt, "Computed tomography reconstruction using deep image prior and learned reconstruction methods," *Inverse Problems*, vol. 36, no. 9, p. 094004, 2020.

[67] Ö. Çiçek, A. Abdulkadir, S. S. Lienkamp, T. Brox, and O. Ronneberger, "3D U-Net: Learning dense volumetric segmentation from sparse annotation," in *Proc. International Conference on Medical Image Computing and Computer-Assisted Intervention*. Springer, 2016, pp. 424–432.

[68] Y. Han, L. Sunwoo, and J. C. Ye, "k-space deep learning for accelerated MRI," *IEEE Transactions on Medical Imaging*, vol. 39, no. 2, pp. 377–386, 2020.

[69] K. H. Jin, M. T. McCann, E. Froustey, and M. Unser, "Deep convolutional neural network for inverse problems in imaging," *IEEE Transactions on Image Processing*, vol. 26, no. 9, pp. 4509–22, 2017.

[70] D. Gilton, G. Ongie, and R. Willett, "Neumann networks for linear inverse problems in imaging," *IEEE Transactions on Computational Imaging*, vol. 6, pp. 328–343, 2020.

[71] S. V. Venkatakrishnan, C. A. Bouman, and B. Wohlberg, "Plug-and-play priors for model based reconstruction," in *Proc. 2013 IEEE Global Conference on Signal and Information Processing*. IEEE, 2013.

[72] G. T. Buzzard, S. H. Chan, S. Sreehari, and C. A. Bouman, "Plug-and-play unplugged: Optimization-free reconstruction using consensus equilibrium," *SIAM Journal on Imaging Sciences*, vol. 11, no. 3, pp. 2001–2020, 2018.

[73] E. Ryu, J. Liu, S. Wang, X. Chen, Z. Wang, and W. Yin, "Plug-and-play methods provably converge with properly trained denoisers," in *Proc. 36th International Conference on Machine Learning*, vol. 97, PMLR, 2019, pp. 5546–5557.

[74] K. Zhang, W. Zuo, S. Gu, and L. Zhang, "Learning deep CNN denoiser prior for image restoration," *arXiv:1704.03264*.

[75] D. H. Ye, S. Srivastava, J.-B. Thibault, K. Sauer, and C. Bouman, "Deep residual learning for model-based iterative CT reconstruction using plug-and-play framework," in *Proc. 2018 IEEE International Conference on Acoustics, Speech and Signal Processing*. IEEE, 2018.

[76] S. Zhao and H. Liang, "Multi-frame super resolution via deep plug-and-play CNN regularization," *Journal of Inverse and Ill-posed Problems*, vol. 28, no. 4, pp. 533–555, 2020.

[77] S. Bigdeli, D. Honzátko, S. Süsstrunk, and L. A. Dunbar, "Image restoration using plug-and-play cnn map denoisers," in *Proc. International Joint Conference on Computer Vision, Imaging and Computer Graphics Theory and Applications*, 2019.

[78] C. Qin, J. Schlemper, J. Caballero, A. N. Price, J. V. Hajnal, and D. Rueckert, "Convolutional recurrent neural networks for dynamic MR image reconstruction," *IEEE Transactions on Medical Imaging*, vol. 38, no. 1, pp. 280–290, 2019.

[79] H. K. Aggarwal, M. P. Mani, and M. Jacob, "MoDL: Model-based deep learning architecture for inverse problems," *IEEE Transactions on Medical Imaging*, vol. 38, no. 2, pp. 394–405, 2019.

[80] Y. Yang, J. Sun, H. Li, and Z. Xu, "Deep ADMM-Net for compressive sensing MRI," in *Proc. 30th International Conference on Neural Information Processing Systems*. Curran Associates, 2016, pp. 10–18.

[81] S. V. Venkatakrishnan, C. A. Bouman, and B. Wohlberg, "Plug-and-play priors for model based reconstruction," in *Proc. 2013 IEEE Global Conference on Signal and Information Processing*, 2013, pp. 945–948.

[82] U. S. Kamilov, H. Mansour, and B. Wohlberg, "A plug-and-play priors approach for solving nonlinear imaging inverse problems," *IEEE Signal Processing Letters*, vol. 24, no. 12, pp. 1872–1876, 2017.

[83] J. H. R. Chang, C. Li, B. Póczos, B. V. K. Vijaya Kumar, and A. C. Sankaranarayanan, "One network to solve them all – solving linear inverse problems using deep projection models," in *Proc. 2017 IEEE International Conference on Computer Vision*, 2017, pp. 5889–5898.

[84] K. Gregor and Y. LeCun, "Learning fast approximations of sparse coding," in *Proc. 27th International Conference on Machine Learning*. Omnipress, 2010, pp. 399–406.

[85] A. Beck and M. Teboulle, "A fast iterative shrinkage-thresholding algorithm for linear inverse problems," *SIAM Journal of Imaging Science*, vol. 2, no. 1, pp. 183–202, 2009.

[86] U. Schmidt and S. Roth, "Shrinkage fields for effective image restoration," in *Proc. 2014 IEEE Conference on Computer Vision and Pattern Recognition*, 2014, pp. 2774–2781.

[87] J. Adler and O. Öktem, "Learned primal–dual reconstruction," *IEEE Transactions on Medical Imaging*, vol. 37, no. 6, pp. 1322–1332, 2018.

[88] H. K. Aggarwal, M. P. Mani, and M. Jacob, "MoDL-MUSSELS: Model-based deep learning for multishot sensitivity-encoded diffusion MRI," *IEEE Transactions on Medical Imaging*, vol. 39, no. 4, pp. 1268–1277, 2020.

[89] C. A. Metzler, A. Mousavi, and R. G. Baraniuk, "Learned D-AMP: Principled neural network based compressive image recovery," in *Proc. 31st International Conference on Neural Information Processing Systems*. Curran Associates, 2017, pp. 1770–1781.

[90] Z. Li, S. Ye, Y. Long, and S. Ravishankar, "SUPER learning: A supervised–unsupervised framework for low-dose CT image reconstruction," in *Proc. 2019 IEEE/CVF International Conference on Computer Vision Workshop*. IEEE, 2019.

[91] S. Ye, Z. Li, M. T. McCann, Y. Long, and S. Ravishankar, "Unified supervised–unsupervised (SUPER) learning for X-ray CT image reconstruction," *IEEE Transactions on Medical Imaging*, vol. 40, no. 11, pp. 2986–3001, 2020.

[92] B. Amos and J. Z. Kolter, "OptNet: Differentiable optimization as a layer in neural networks," in *Proc. 34th International Conference on Machine Learning*. ser. PMLR, 2017, pp. 136–145.

[93] A. Agrawal, B. Amos, S. Barratt, S. Boyd, S. Diamond, and J. Z. Kolter, "Differentiable convex optimization layers," in *Advances in Neural Information Processing Systems*, vol. 32, H. Wallach, H. Larochelle, A. Beygelzimer, F. d'Alché-Buc, E. Fox, and R. Garnett, eds. Curran Associates, 2019, pp. 9562–9574.

[94] G. Peyré and J. M. Fadili, "Learning analysis sparsity priors," in *Proc. Conference on Sampling Theory and Applications*, 2011, p. 4.

[95] J. Mairal, F. Bach, and J. Ponce, "Task-driven dictionary learning," *IEEE Transactions on Pattern Analysis and Machine Intelligence*, vol. 34, no. 4, pp. 791–804, 2012.

[96] P. Sprechmann, R. Litman, T. Ben Yakar, A. M. Bronstein, and G. Sapiro, "Supervised sparse analysis and synthesis operators," in *Advances in Neural Information Processing Systems*, vol. 26, 2013, pp. 908–916.

[97] Y. Chen, T. Pock, and H. Bischof, "Learning ℓ_1-based analysis and synthesis sparsity priors using bi-level optimization," *arXiv:1401.4105 [cs.CV]*, 2014.

[98] S. Ravishankar and Y. Bresler, "Adaptive sampling design for compressed sensing MRI," in *Proc. 2011 Annual International Conference of the IEEE Engineering in Medicine and Biology Society*, 2011, pp. 3751–3755.

[99] B. Gözcü, R. K. Mahabadi, Y. Li, E. Ilıcak, T. Çukur, J. Scarlett, and V. Cevher, "Learning-based compressive MRI," *IEEE Transactions on Medical Imaging*, vol. 37, no. 6, pp. 1394–1406, 2018.

[100] H. K. Aggarwal and M. Jacob, "J-MoDL: Joint model-based deep learning for optimized sampling and reconstruction," *IEEE Journal of Selected Topics in Signal Processing*, vol. 14, no. 6, pp. 1151–1162, 2020.

4 Deep Algorithm Unrolling for Biomedical Imaging

Yuelong Li, Or Bar-Shira, Vishal Monga, and Yonina C. Eldar

4.1 Introduction

Model-based inversion played a dominant role in biomedical imaging prior to deep learning gaining widespread popularity and broad recognition. Model-based techniques rely on a *forward model* derived by modeling the imaging process analytically on the basis of physical laws. Typically, the forward model is formulated as

$$\mathbf{y} = \mathcal{F}(\mathbf{x}) + \mathbf{n},$$

where $\mathbf{y} \in \mathbb{R}^m$ is the observation, $\mathbf{x} \in \mathbb{R}^n$ is the underlying image data to be recovered, $\mathcal{F}$ is the mapping from image domain to observations, and $\mathbf{n} \in \mathbb{R}^m$ is the corruptive noise process.

Under many circumstances, the problem of estimating $\mathbf{x}$ from $\mathbf{y}$ is ill-posed, because the imaging process is often compressive owing to practical constraints [1]. For instance, in magnetic resonance imaging (MRI) one may undersample the image in order to accelerate the acquisition process; in computed tomography (CT) a reduction of dose through sparse views in X-ray radiation is generally preferred for patients' safety, which translates into an underdetermined forward model. Therefore, to reliably estimate $\mathbf{x}$, a prior structure is often incorporated either by capturing physical principles or through manual handcrafting. The underlying image $\mathbf{x}$ can then be estimated by solving a regularized optimization problem:

$$\min_{\mathbf{x}} \rho(\mathbf{y}, \mathcal{F}(\mathbf{x})) + \lambda \mathcal{R}(\mathbf{x}), \qquad (4.1)$$

where ρ is a metric measuring the deviation in the measurements, $\mathcal{R}$ is a regularization functional capturing the prior structure, and λ is a regularization parameter controlling the regularization strength. Problem (4.1) is generally solved via an iterative optimization algorithm.

In contrast with the model-based framework, $\mathbf{x}$ can alternatively be estimated by learning a regression mapping from $\mathbf{y}$ (or its transformed version such as compressive samples or Fourier coefficients) to $\mathbf{x}$ through a data-driven approach. The regression mapping can be chosen from various machine learning models, among which deep neural networks (DNNs) are increasingly popular nowadays. In particular, a purely data-driven approach adopts a generic DNN without incorporating any physical laws or prior structure. Instead, it relies on abundant training data to learn a huge number of network parameters and, in turn, the form of the regression mapping. When the training data is adequate, this approach can be highly successful because DNNs are extremely expressive and can learn to adapt to complicated mappings which are difficult to characterize and design manually. In addition, with the aid of highly optimized deep-learning platforms and hardware accelerators such as graphics processing units (GPUs), inference via DNNs can be performed rapidly, which gives rise to fast execution speed in practice. This is in contrast with traditional iterative algorithms, which need to repeat certain operations a large number of times sequentially, and can be comparatively quite slow.

Nevertheless, modern DNNs typically carry a deep hierarchical architecture composed of many layers and network parameters (there can be millions) and are thus often difficult to interpret. It is hard to discover what is the exact form that is learned by the DNNs and what are the roles of the individual parameters. In other words, as DNNs are intact learning machines, it is hard for human agents to analyze their operations. The capability of a DNN to be understood sufficiently for its inner mechanism to be manipulated in a principled, predictable, way is referred to as its *interpretability*. Lack of interpretability is an important concern as interpretability is usually the key to conceptual understanding and advancing knowledge frontiers. In addition, in practical biomedical applications, interpretability is key to ensuring trust in the reconstruction algorithm, as experts need to have a proper understanding of the origin of the artifacts introduced by the algorithm and to identify potential failure cases [2]. In contrast, traditional iterative algorithms are usually highly interpretable because they have been developed via modeling underlying physical processes and/or capturing prior domain knowledge.

In addition to interpretability, *generalizability* is another fundamental requirement in biomedical applications. It is well known that generic DNNs rely heavily on the quantity and quality of the training data in order to achieve empirically superior performance. In other words, when the training data are deficient, the issue of overfitting may manifest itself in the form of significantly degraded network performance. Indeed, this issue can be so serious that the DNNs may underperform traditional iterative algorithms. This phenomenon is especially apparent for DNNs with high model capacity, such as very deep or ultrawide neural networks. Unlike in the case of natural images, in biomedical imaging the data collection is an expensive and time-consuming procedure, and thus generalizability is especially important.

In the seminal work of Gregor and LeCun [3], a promising technique called algorithm unrolling was developed that has helped connect iterative algorithms such as sparse coding techniques to DNNs. This work has inspired a growing list of follow-ups in different fields of biomedical research: compressive sensing (CS) [4], computed tomography (CT) [5], ultrasound [6, 7], to name a few. Figure 4.1 provides a high-level illustration of this framework. Starting with an iterative procedure, each iteration of the algorithm is represented as one layer of a network. Concatenating these layers forms a deep neural network whose architecture borrows from the optimization method. Passing through the network is equivalent to executing the iterative algorithm a finite number of times. In addition, the algorithm parameters (such as the model parameters

4 Deep Algorithm Unrolling for Biomedical Imaging

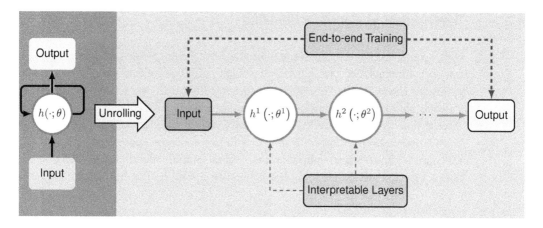

Figure 4.1 A high-level overview of algorithm unrolling: given an iterative algorithm (left), a corresponding deep network (right) can be generated by cascading its iterations h. The iteration step h (left) is executed a number of times, resulting in the network layers $h^1, h^2, \ldots$ (right). Each iteration h depends on the algorithm parameters θ, which become network parameters $\theta^1, \theta^2, \ldots$ and are learned from training datasets through end-to-end training. In this way, the resulting network can achieve better performance than the original iterative algorithm. In addition, the network layers naturally inherit interpretability from the iteration procedure.

and regularization coefficients) are transferred to become the network parameters. The network may be trained using backpropagation, resulting in model parameters that are learned from real-world training sets. In this way, the trained network can be naturally interpreted as a parameter-optimized algorithm, effectively overcoming the lack of interpretability in most conventional neural networks [8].

Traditional iterative algorithms generally entail significantly fewer parameters compared with popular neural networks. Therefore, the unrolled networks are highly parameter efficient and require less training data. In addition, unrolled networks naturally inherit prior structure and domain knowledge rather than learning them from intensive training data. Consequently, they tend to generalize better than generic networks and can be computationally faster as long as each algorithmic iteration (or the corresponding layer) is not overly expensive.

In this chapter we review biomedical applications and breakthroughs made via the leveraging of algorithm unrolling, an important technique that bridges between traditional iterative algorithms and modern deep-learning techniques. To provide context, we start by tracing the origin of algorithm unrolling and giving a comprehensive tutorial on how to unroll iterative algorithms into deep networks. We then extensively cover algorithm unrolling in a wide variety of biomedical imaging modalities and delve into several representative recent works in detail. Indeed, there is a rich history of iterative algorithms for biomedical image synthesis, which makes the field ripe for unrolling techniques. In addition we put algorithm unrolling into a broad perspective, in order to understand why it is particularly effective, and discuss recent trends. Finally, we conclude the chapter by discussing open challenges and suggesting future research directions.

4.2 Development of Algorithm Unrolling

This section introduces algorithm unrolling. We begin by reviewing the learned iterative shrinkage and thresholding algorithm (LISTA), the first piece of work that employed the algorithm unrolling strategy. To this end, we review the classical iterative shrinkage and thresholding algorithm (ISTA) (Section 4.2.1), and then discuss how it can be unrolled into a deep network (Section 4.2.2). After that we consider several related theoretical studies to gain a deeper understanding of LISTA (Section 4.2.3). We then introduce the general idea of algorithm unrolling in an abstract fashion and extend it to generic iterative algorithms (Section 4.2.4).

4.2.1 Iterative Shrinkage and Thresholding Algorithm

In many practical problems the signal of interest cannot be observed directly but must be inferred from observable quantities. In the simplest approximation, which often suffices, there is a linear relationship between the signal of interest and the observed quantities [9]. If we model the signal of interest by a vector $\mathbf{x}$, and the derived quantities by another vector $\mathbf{y}$, we can cast the problem of inferring $\mathbf{x}$ from $\mathbf{y}$ as a linear inverse problem. A basic linear inverse problem admits the following form:

$$\mathbf{y} = \mathbf{A}\mathbf{x} + \mathbf{n}, \tag{4.2}$$

where $\mathbf{A} \in \mathbb{R}^{n \times m}$ and $\mathbf{y} \in \mathbb{R}^n$ are known, $\mathbf{n}$ is an unknown noise vector, and $\mathbf{x}$ is the unknown signal of interest to be estimated.

In practice, it is often the case that $\mathbf{A}$ has fewer rows than columns ($n < m$) so that recovery of $\mathbf{x}$ from $\mathbf{y}$ becomes an ill-posed problem. It is therefore impossible to recover $\mathbf{x}$ without introducing additional assumptions on its structure. A popular strategy is to assume that $\mathbf{x}$ is sparse or admits a compressible (sparse) representation, meaning that it is sparse in a transformed domain [1].

In sparse coding we seek a vector $\mathbf{x} \in \mathbb{R}^m$ such that $\mathbf{y} \approx \mathbf{W}\mathbf{x}$ and such that as many coefficients in $\mathbf{x}$ as possible are zero (or small in magnitude) [10]. Here $\mathbf{W}$ is commonly called a *dictionary*, whose columns, typically named *atoms*, can either be taken from a standard basis (or frame) such as a Fourier or wavelet basis or be learned from real data. A popular technique to recover $\mathbf{x}$ is by solving an unconstrained convex minimization problem:

$$\min_{\mathbf{x} \in \mathbb{R}^m} \frac{1}{2} \|\mathbf{y} - \mathbf{W}\mathbf{x}\|_2^2 + \lambda \|\mathbf{x}\|_1, \tag{4.3}$$

where $\lambda > 0$ is a regularization parameter that controls the sparsity of the solution. A well-known class of methods for solving (4.3) is the class of proximal methods such as ISTA [9], which perform the following iterations:

$$\mathbf{x}^{l+1} = \mathcal{S}_\lambda \left\{ \left(\mathbf{I} - \frac{1}{\mu} \mathbf{W}^T \mathbf{W} \right) \mathbf{x}^l + \frac{1}{\mu} \mathbf{W}^T \mathbf{y} \right\}, \quad l = 0, 1, \ldots \tag{4.4}$$

Here, $\mathbf{I} \in \mathbb{R}^{m \times m}$ is the identity matrix, μ is a positive parameter that controls the iteration step size, $\mathcal{S}_\lambda(\cdot)$ is the soft-thresholding operator, defined as

$$\mathcal{S}_\lambda(x) = \text{sign}(x) \max\{|x| - \lambda, 0\}, \tag{4.5}$$

for a scalar x; $\mathcal{S}_\lambda(\cdot)$ operates element wise on vectors and matrices.

4.2.2 LISTA: Learned Iterative Shrinkage and Thresholding Algorithm

The slow convergence rate of ISTA can be problematic in real-time applications. Furthermore, the matrix $\mathbf{W}$ may not be known exactly. In their seminal work, Gregor et al. proposed a highly efficient learning-based method that computes good approximations of optimal sparse codes in a fixed amount of time, with the help of a matrix $\mathbf{W}$ learned from real data [3].

Specifically, the authors developed deep algorithm unrolling as a strategy for designing neural networks which integrate domain knowledge. In this approach, the network architecture is tailored to a specific problem, based on a well-founded iterative mathematical formulation for solving the problem at hand. This leads to increased convergence speed and accuracy relative to the standard iterative solution, and interpretability and robustness relative to a black-box large-scale neural network.

To unroll ISTA, the iteration (4.4) can be recast into a single network layer as depicted in Fig. 4.2. This layer comprises a series of analytical operations (matrix–vector multiplication, summation, soft-thresholding), and so is of the same nature as a neural network. A diagrammatic representation of one iteration step reveals its resemblance to a single network layer. Executing ISTA L times can be interpreted as cascading L such layers, which essentially forms an L-layer-deep network. Note that, in the unrolled network, an implicit substitution of parameters has been made: $\mathbf{W}_t = \mathbf{I} - (1/\mu)\mathbf{W}^T\mathbf{W}$ and $\mathbf{W}_e = (1/\mu)\mathbf{W}^T$.

After unrolling ISTA into a network, named learned ISTA (LISTA), the network is trained through backpropagation using real datasets to optimize the parameters $\mathbf{W}_t$, $\mathbf{W}_e$, and λ. Training is performed in a supervised manner, meaning that for every input vector $\mathbf{y}^t \in \mathbb{R}^n, t = 1, \ldots, T$, its corresponding sparse output $\mathbf{x}^{*t} \in \mathbb{R}^m, t = 1, \ldots, T$, is known (by choosing an appropriate λ which is fixed and might be different from the learned value). The sparse codes $\mathbf{x}^{*t}$ can be determined by, for example, executing ISTA when $\mathbf{W}$ is known. Inputting the vector $\mathbf{y}^t$ into the network results in a predicted output $\hat{\mathbf{x}}^t(\mathbf{y}^t; \mathbf{W}_t, \mathbf{W}_e, \lambda)$. The network training loss function is formed by comparing the prediction with the known sparse output $\mathbf{x}^{*t}$:

$$\ell(\mathbf{W}_t, \mathbf{W}_e, \lambda) = \frac{1}{T} \sum_{t=1}^{T} \left\| \hat{\mathbf{x}}^t \left(\mathbf{y}^t; \mathbf{W}_t, \mathbf{W}_e, \lambda\right) - \mathbf{x}^{*t} \right\|_2^2. \tag{4.6}$$

The network is trained through loss minimization, using popular gradient-based learning techniques such as stochastic gradient descent to learn the unknown parameters $\mathbf{W}_t, \mathbf{W}_e$, and λ [11].

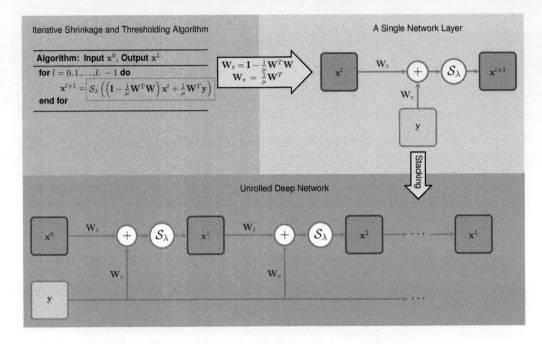

Figure 4.2 Illustration of LISTA: one iteration of ISTA executes a linear operation and then a nonlinear operation and thus can be recast into a network layer; by stacking the layers a deep network is formed. The network is subsequently trained using paired inputs and outputs by backpropagation to optimize the parameters $\mathbf{W}_e, \mathbf{W}_t$, and λ.

The learned version, LISTA, may achieve higher efficiency compared with ISTA. It is also useful when it is analytically hard to determine $\mathbf{W}$. Furthermore, it has been shown empirically that, the number of layers L in (trained) LISTA can be an order of magnitude smaller than the number of iterations required for ISTA to achieve convergence corresponding to a new observed input, thus dramatically boosting the sparse coding efficiency [3]. In practice, $\mathbf{W}_t, \mathbf{W}_e$, and λ can be untied and so can vary in each layer.

4.2.3 Towards a Theoretical Understanding of Algorithm Unrolling

While LISTA empirically achieves much higher efficiency compared with ISTA, a theoretical understanding of its behavior is still lacking. In particular, the conditions offering convergence guarantees, rates of convergence, structure of the optimal solutions, and the factors contributing to higher empirical performance are some of the important problems that need to be answered. In recent years, researchers have actively analyzed LISTA and its variants and made significant progress.

Xin et al. [12] studied the unrolled iterative hard thresholding (IHT) algorithm, which has been widely applied in various sparsity-constrained estimation problems. The IHT algorithm largely resembles ISTA except that an ℓ^0 norm is employed instead of the ℓ^1 norm. Formally, IHT solves the following optimization problem instead of (4.3):

$$\min_{\mathbf{x}\in\mathbb{R}^m} \frac{1}{2}\|\mathbf{y}-\mathbf{W}\mathbf{x}\|_2^2 \quad \text{s.t.} \quad \|\mathbf{x}\|_0 \leq k, \tag{4.7}$$

for some positive integer k. The algorithm essentially performs the following iterations:

$$\mathbf{x}^{l+1} = \mathcal{H}_k\left\{\left(\mathbf{I}-\frac{1}{\mu}\mathbf{W}^T\mathbf{W}\right)\mathbf{x}^l + \frac{1}{\mu}\mathbf{W}^T\mathbf{y}\right\}, \quad l=0,1,\ldots, \tag{4.8}$$

where μ is a positive parameter controlling the step size and $\mathcal{H}_k$ is a hard-thresholding operator which zeros out all but the largest (in magnitude) k coefficients of the vector. Similarly to LISTA, in [12] $\mathbf{W}_t = \mathbf{I} - (1/\mu)\mathbf{W}^T\mathbf{W}$ and $\mathbf{W}_e = (1/\mu)\mathbf{W}^T$ were used instead of $\mathbf{W}$ itself, which reduces the iteration (4.8) to the following form:

$$\mathbf{x}^{l+1} = \mathcal{H}_k\left\{\mathbf{W}_t\mathbf{x}^l + \mathbf{W}_e\mathbf{y}\right\}, \quad l=0,1,\ldots \tag{4.9}$$

The authors then proved that a necessary condition for the iteration (4.9) to recover a feasible solution to Eq. (4.7) is that the weight-coupling scheme $\mathbf{W}_t = \mathbf{I} - \mathbf{W}_e\mathbf{W}$ is satisfied, for some $\mathbf{W}$ introduced as a free variable. Note that this formula is reminiscent of the implicit re-parameterization scheme $\mathbf{W}_t = (1/\mu)\mathbf{W}^T\mathbf{W}$. Furthermore, under the weight-coupling constraint, when $\mathbf{W}$ and $\mathbf{W}_e$ additionally obey technical conditions, the iteration (4.9) recovers the underlying solution at a linear rate. In particular, $\mathbf{W}$ needs to satisfy certain restricted-isometry-property (RIP) conditions. The authors also articulated the exact forms of $\mathbf{W}_e$ needed in order to ensure linear convergence. Compared with classical IHT, the iteration (4.9) gives a much more relaxed requirement on the RIP condition, which implies that the unrolled network is capable of recovering sparse signals from dictionaries with relatively more coherent columns.

Chen et al. [13] studied the LISTA network with layer-specific parameters $\{\mathbf{W}_t^l, \mathbf{W}_e^l, \lambda^l\}_{l\in\mathbb{N}}$ and proved that, if LISTA recovers the underlying sparse solution and if the sequence $\mathbf{W}_t^1, \mathbf{W}_t^2, \ldots$ is bounded, a similar weight-coupling scheme must be satisfied asymptotically:

$$\mathbf{W}_t^l - (\mathbf{I} - \mathbf{W}_e^l\mathbf{W}) \to 0, \quad \text{as} \quad l \to \infty,$$
$$\lambda^l \to 0, \quad \text{as} \quad l \to \infty.$$

When LISTA adopts the weight-coupling parametrization $\mathbf{W} = \mathbf{I} - \mathbf{W}_e^l\mathbf{W}$ and employs a dedicated support-selection technique, the resulting network is called LISTA-CPSS, where "CP" stands for weight coupling and "SS" stands for support selection. Chen et al. proved that, if the underlying solution $\mathbf{x}^*$ is sufficiently sparse and bounded, the sequence $\{\mathbf{W}_e^l, \lambda^l\}_{l\in\mathbb{N}}$ can be chosen according to certain values such that LISTA-CPSS recovers $\mathbf{x}^*$ at a linear rate. As a follow-up, Liu et al. [14] introduced certain mutual-coherence conditions and analytically characterized optimal network parameters on the basis of those conditions. Similarly to networks with trained weights, networks adopting analytic weights converge at a linear rate, which implies that analytic weights can be as efficient as learned weights. In addition, analytic weights are of much lower dimensionality compared with trained weights. However, determining the analytic weights can be a nontrivial task, as typically another optimization problem has to be solved.

4.2.4 Unrolling Generic Iterative Algorithms

Although the initial motivation of the work of Gregor and LeCun [3] was to increase the efficiency of sparse-coding techniques via training, it was found that the underlying principles could be easily generalized. More specifically, provided with a certain iterative algorithm, we can unroll it into a corresponding deep network, following the procedures depicted in Fig. 4.3. The first step is to identify the analytical operations per iteration, which we represent abstractly as an h function, and the associated parameters, which we denote collectively as θ^l. The next task is to generalize the functional form of h into a more generic version $\widehat{h}$ and, correspondingly, expand the parameters θ^l into an enlarged version $\widehat{\theta}^l$ if necessary. For instance, in LISTA the parameter $\mathbf{W}$ is substituted by $\mathbf{W}_t$ and $\mathbf{W}_e$ through the formulas $\mathbf{W}_t = \mathbf{I} - (1/\mu)\mathbf{W}^T\mathbf{W}$ and $\mathbf{W}_e = (1/\mu)\mathbf{W}^T$. After this procedure each iteration can be recast into a network layer in the same spirit as in LISTA. By stacking the mapped layers together we obtain a deep network with undetermined parameters, and then obtain optimal parameters through end-to-end training using real-world datasets.

The exact approach to generalizing h and θ^l towards $\widehat{h}$ and $\widehat{\theta}^l$ is largely case specific. An extreme scenario is to follow strictly the original functional forms and parameters, i.e., to take $\widehat{h} = h$ and $\widehat{\theta}^l = \theta^l, \forall l$. In this way, the trained network corresponds exactly to the original algorithm with finite truncation and optimal parameters. In addition to efficiency enhancement thanks to training [3], the unrolled networks

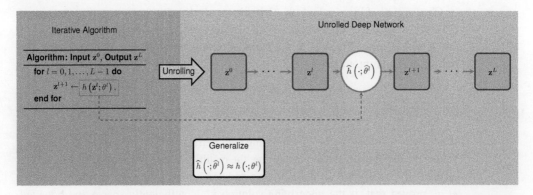

Figure 4.3 Illustration of the general idea of algorithm unrolling: given an iterative algorithm, we map one iteration (described as the function h parameterized by $\theta^l, l = 0, \ldots, L-1$) into a single network layer, and stack a finite number of layers to form a deep network. Feeding the data forward through an L-layer network is equivalent to executing the iteration L times (in the case of finite truncation). The parameters $\theta^l, l = 0, \ldots, L-1$, are learned from real datasets by training the network end-to-end to optimize the performance. They can either be shared across different layers or vary from layer to layer.

can aid with estimating structured parameters such as filters [15] or dictionaries [16], which are hard to design either analytically or by handcrafting. Alternatively, some operations may be replaced with a stand-alone deep neural network such as a convolutional neural network (CNN) or recurrent neural network (RNN). For instance, in [5] the authors replaced a proximal-gradient update step with a CNN. In addition, the parameters can be layer specific instead of being shared across different layers. In [17] the authors plugged in a CNN in each iteration step (layer) and allowed the network parameters to differ. As it is, networks with shared parameters generally resemble RNNs, while those with layer-specific parameters mimic CNNs, especially when there are convolutional structures embedded in layer wise operations. It is important to note that such custom modifications may potentially sacrifice certain conceptual merits: they may invalidate convergence guarantees, introducing departures from the original iterative algorithms, or may undermine the interpretability to some extent. Nevertheless, they are practically beneficial and critical for performance improvement because the representation capacity of the network can be significantly extended.

In addition to performance and efficiency benefits, unrolled networks can potentially reduce the number of parameters and hence storage footprints. Conventional generic neural networks typically reuse essentially the same architectures across different domains and thus require a large number of parameters to ensure their representation power. In contrast, unrolled networks generally deliver significantly fewer parameters, as they implicitly transfer problem structures (domain knowledge) from iterative algorithms to unrolled networks and their structures are more specifically tailored towards target applications. These benefits not only ensure higher efficiency but also provide better generalizability, especially under limited training schemes [4], as will be demonstrated through concrete examples in Section 4.3.

An unrolled network can share inter layer parameters, thus resembling an RNN, or carry over layer-specific parameters. Although parameter sharing could further reduce parameters, the RNN-like architecture significantly increases the difficulty in training. In particular, it may suffer from gradient explosion and vanishing problems. On the other hand, using layer-specific parameters may lead to deviations from the original iterative algorithm and may no longer inherit the convergence guarantees. However, these networks can have larger capacity and become easier to train.

Another important concern when designing unrolled networks is to determine the optimal number of layers, which corresponds to the number of iterations in the iterative algorithm. Most existing approaches generally treat this number as a hyperparameter and determine it through cross-validation. However, recent theoretical breakthroughs (some of which are reviewed in Section 4.2.3) regarding the convergence rate of the unrolled network can shed some light and guide future efforts.

4.3 Deep Algorithm Unrolling for Biomedical Imaging

In recent years, algorithm unrolling has found wide application in many areas of biomedical imaging. Traditionally, iterative algorithms have played a dominant role in solving various biomedical imaging problems. Recently, learning-based approaches, especially deep neural networks, have become increasingly popular. Algorithm unrolling makes a bridge between well-grounded iterative algorithms and contemporary deep networks and has attracted growing research attention. In this section, we discuss how algorithm unrolling can be successfully applied to biomedical image synthesis for several concrete cases.

4.3.1 Applications of Unrolling in Computed Tomography

In general, CT refers to a class of imaging techniques that reconstruct the original signal from its directional projections (slices). For instance, in X-ray CT, a mobilized X-ray source rotates around the patient to emit narrow beams of X-rays, which are then received by X-ray detectors located opposite the X-ray source. The X-rays pass through the patient to create image slices. In it simplest form, this procedure can be described by the well-known Radon transform. In two dimensions, the Radon transform is given by the following formula:

$$p(\xi,\phi) = \int f(x,y)\delta(x\cos(\phi) + y\sin(\phi) - \xi)dxdy,$$

where f is the imaged signal and δ is the Dirac delta function. The function p is often referred to as a *sinogram*. To invert the Radon transform, the classical central slice theorem plays a critical role, as it relates the Fourier transform of the original signal and its sinogram. This relationship gives rise to the filtered back-projection (FBP) algorithm: the sinogram is first filtered by a so-called ramp filter and then back-projected to obtain the original signal. The practical challenges of CT include the need for short scanning times and low doses and also the need to avoid motion and noise.

In an early work that employs deep CNN for CT [18], the authors observed the architectural similarity between the popular U-net [19] and the unrolled ISTA network. They then combined FBP with U-net to construct a deep network called FBPConvNet. Through extensive experimental studies on sparse-view X-ray CT reconstruction, FBPConvNet offers clear benefits over traditional iterative reconstruction techniques such as the total-variation (TV) regularization approach. In particular, FBPConvNet achieved over 3 dB signal-to-noise ratio (SNR) improvement over TV on a biomedical dataset that comprised 500 real *in vivo* CT images, and reduced the running time from several minutes to less than a second. Compared with conventional deep networks such as residual networks, FBPConvNet also achieves more than 1 dB higher SNR on the same dataset.

An emerging CT technique is photo-acoustic tomography (PAT), which provides high-resolution three-dimensional (3D) images by sensing laser-generated ultrasound. In [5] the authors developed a deep-learning technique to reconstruct high-resolution 3D images from restricted photoacoustic measurements. Modeling the acoustic signal using the wave equation, the initial pressure **x** and the measured photoacoustic signal **y** satisfy a linear mapping,

$$\mathbf{y} = \mathbf{W}\mathbf{x},$$

4 Deep Algorithm Unrolling for Biomedical Imaging

and thus PAT can be carried out by solving a linear inverse problem. A typical approach is to solve the following regularized problem:

$$\min_{\mathbf{x}} \rho(\mathbf{y}, \mathbf{W}\mathbf{x}) + \lambda \mathcal{R}(\mathbf{x}), \tag{4.10}$$

where $\rho(\cdot, \cdot)$ is a metric measuring the data consistency, $\mathcal{R}$ is a regularizer, and λ is the regularization coefficient. A simple approach to solving Eq. (4.10) is the proximal gradient descent (PGD) algorithm, which performs the following iterations:

$$\mathbf{x}_{k+1} \leftarrow \text{prox}_{\mathcal{R}} \left(\mathbf{x}_k - \gamma_{k+1} \nabla \rho(\mathbf{y}, \mathbf{W}\mathbf{x}_k); \lambda \gamma_{k+1} \right), \tag{4.11}$$

where $\gamma_{k+1} > 0$ controls the step size. Hauptmann et al. [5] replaced the proximal operator with a CNN and learned the arithmetic operations (subtraction, multiplication) rather than fixing them according to the gradient descent rule (4.11). In other words, they proposed unrolling the following update procedures:

$$\mathbf{x}_{k+1} = G_{\theta_k} \left(\nabla \rho(\mathbf{y}, \mathbf{W}\mathbf{x}_k), \mathbf{x}_k \right),$$

where G denotes a CNN and θ_k are its parameters. The corresponding unrolled network is depicted visually in Fig. 4.4. The network is dubbed deep gradient descent (DGD). To train the network, a stagewise scheme is employed: at the kth stage, θ_k is optimized by minimizing the mean square error (MSE) loss between $\mathbf{x}_k$ and the ground truth $\mathbf{x}^*$ while holding other $\theta_j (j \neq k)$ fixed. Experimentally, DGD achieves nearly 1 dB improvement in peak signal-to-noise ratio (PSNR) over U-net and more than 3 dB over TV reconstruction on an *in vivo* dataset from a human palm. Furthermore, compared with conventional deep networks such as U-net, DGD proves its superior robustness against perturbations of measurement procedures or targets by showing only slight deterioration in reconstruction quality under small perturbations such as varying subsampling patterns, noise levels, or deviations in sound speed.

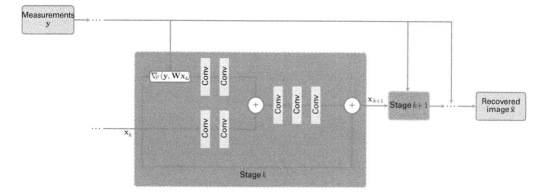

Figure 4.4 Representation of DGD [5]: each stage comprises a gradient descent iteration with the proximal operator replaced by a CNN. The CNN takes as input a concatenation of $\mathbf{x}_k$ and the gradient $\nabla \rho (\mathbf{y}, \mathbf{W}\mathbf{x}_k)$. It also adopts a skip connection from $\mathbf{x}_k$ to $\mathbf{x}_{k+1}$, which implies that it learns a residual mapping. "Conv" stands for convolutional layer.

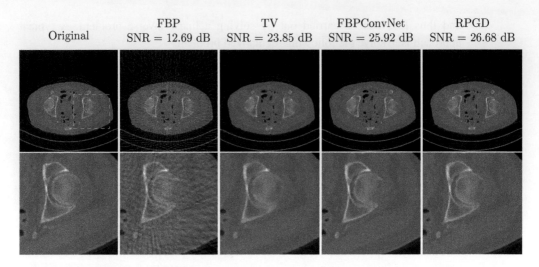

Figure 4.5 Experimental comparisons between RPGD [20] and state-of-the art reconstruction algorithms (including FBPConvNet from [18]) over a test sample from noiseless measurements with 45 views (×16 dosage reduction). The SNR values are also included for quantitative comparisons. The first row shows the full images and the second row shows magnified portions, respectively. The images courtesy of Michael Unser at EPFL [20].

In a closely related technique [20], Gupta *et al.* unrolled the PGD algorithm, and similarly replaced the proximal operator with a CNN. However, they did not learn the arithmetic operations and instead followed the update rule in Eq. (4.11). In addition, they introduced a relaxation to the PGD update, in order to maintain its convergence guarantee even when the proximal operator is substituted with a CNN. The unrolled network, dubbed relaxed projected gradient descent (RPGD), yields substantially improved CT recovery, as depicted in Fig. 4.5.

Low-dose X-ray CT is an important technique as X-ray CT causes potential cancer risks due to radiation exposure. A drawback of this approach, however, is the low SNR of the projections due to noise, and thus power-imaging techniques are required to retrieve high-quality reconstructed images. To overcome the challenge of noisy measurements, Kang *et al.* [21] built a wavelet residual network (WavResNet) based on a theory of deep convolutional framelets for deep-learning-based denoising. Suppose we seek to estimate a ground-truth signal $\mathbf{x}^* \in \mathbb{R}^n$ from its noisy measurement $\mathbf{y} \in \mathbb{R}^n$:

$$\mathbf{y} = \mathbf{x}^* + \mathbf{n},$$

where $\mathbf{n} \in \mathbb{R}^n$. A traditional approach is frame-based denoising, which shrinks the frame coefficient $\boldsymbol{\alpha}$ of the signal by solving

$$\min_{\mathbf{x},\boldsymbol{\alpha}} \frac{\mu}{2} \|\mathbf{y} - \mathbf{x}\|_2^2 + \frac{1-\mu}{2} \left\{ \|\mathbf{W}\mathbf{x} - \boldsymbol{\alpha}\|_2^2 + \lambda \|\boldsymbol{\alpha}\|_1 \right\},$$

where $\lambda, \mu > 0$ are the regularization coefficients and $\mathbf{W}$ is the analysis operator of the frame. The corresponding proximal update equation is given by

$$\mathbf{x}_{k+1} = \mu\mathbf{y} + (1-\mu)\mathbf{W}^T \mathcal{S}_\lambda(\mathbf{W}\mathbf{x}_k), \tag{4.12}$$

where $\mathcal{S}_\lambda$ is the soft-thresholding operator with thresholding value λ [22, 23].

Kang et al. [21] claimed that the term $\mathbf{W}^T \mathcal{S}_\lambda(\mathbf{W}\mathbf{x}_k)$ can be regarded as a CNN based on deep convolutional framelet theory [24]. In short, the analysis and synthesis operators of a frame can be concatenated to form an encoder–decoder layer structure which corresponds to a CNN, and the shrinkage operation can be implicitly obtained by reducing the channels in the CNN. The update rule (4.12) is then substituted with

$$\mathbf{x}_{k+1} = \mu\mathbf{y} + (1-\mu)\mathcal{Q}\left(\mathbf{x}_k; \phi, \widetilde{\phi}\right) \tag{4.13}$$

where $\mathcal{Q}$ is the CNN corresponding to the framelet, with undetermined filter coefficients $\phi, \widetilde{\phi}$ corresponding to the primal and dual frames, respectively. The unrolled networks are formed out of the update rules (4.13) and the unknown parameters are learned; the parameters are shared across different iterations, giving rise to an RNN-like architecture. In addition, Kang et al. [21] slightly modified the update rules (4.13) to ensure convergence. Through extensive experimental results, WavResNet shows its advantage in noise reduction and in preserving texture details of the organ under observation while maintaining the lesion information over state-of-the-art techniques.

A related technique that also focuses on low-dose CT reconstruction is found in the work by Adler and Öktem [17], where the proximal-dual hybrid gradient (PDHG) algorithm was generalized and unrolled into a learnable network. Technically, the PDHG algorithm solves problems of the following form:

$$\min_{\mathbf{x}} \mathcal{P}(\mathcal{K}(\mathbf{x})) + \mathcal{D}(f), \tag{4.14}$$

where $\mathcal{K}$ is an operator over signal $\mathbf{x}$ and $\mathcal{P}, \mathcal{D}$ are functionals on the primal and dual spaces, respectively. Problem (4.14) has many interesting instances; in particular, TV-regularized CT can be instantiated by setting $\mathcal{K}(\mathbf{x}) := [\mathcal{R}(\mathbf{x}), \nabla\mathbf{x}]$, $\mathcal{P}([\mathbf{h}_1, \mathbf{h}_2]) := \|\mathbf{h}_1 - \mathbf{y}\|_2^2 + \|\mathbf{h}_2\|_1$, and $\mathcal{D} := 0$, which gives rise to the following optimization problem:

$$\min_{\mathbf{x}} \|\mathcal{R}(\mathbf{x}) - \mathbf{y}\|_2^2 + \lambda\|\nabla\mathbf{x}\|_1,$$

where $\mathcal{R}$ is the Radon transform operator, ∇ is the gradient operator, and $\lambda > 0$ is a regularization coefficient. The PDHG algorithm essentially performs the following iterations for $k = 1, 2, \ldots$:

$$\mathbf{z}_{k+1} \leftarrow \mathrm{prox}_{\mathcal{P}^*}(\mathbf{z}_k + \sigma\mathcal{K}(\bar{\mathbf{x}}_k); \sigma),$$
$$\mathbf{x}_{k+1} \leftarrow \mathrm{prox}_{\mathcal{D}}\left(\mathbf{x}_k - \tau[\partial\mathcal{K}(\mathbf{x}_k)]^*(\mathbf{z}_{k+1}); \tau\right),$$
$$\bar{\mathbf{x}}_{k+1} \leftarrow \mathbf{x}_{k+1} + \gamma(\mathbf{x}_{k+1} - \mathbf{x}_k),$$

where $*$ is the Fenchel conjugate and ∂ is the Fréchet derivative. For the definitions of these terms, see [25]. The operator $\mathrm{prox}_{\mathcal{F}}(\cdot; \lambda)$ is the proximal operator over functional $\mathcal{F}$ and parameter λ, given by

$$\mathrm{prox}_{\mathcal{F}}(\mathbf{x}; \lambda) = \arg\min_{\mathbf{x}'} \mathcal{F}(\mathbf{x}') + \frac{1}{2\lambda}\|\mathbf{x}' - \mathbf{x}\|_2^2.$$

In a similar spirit to [5], Adler and Öktem generalized the update rules by replacing the proximal operators with CNNs and substituting the arithmetic operations (summation, multiplication, etc.) as learnable operations, giving rise to the following iterations:

$$z_k \leftarrow \Gamma_{\theta_k^d}\left(z_{k-1}, \mathcal{K}\left(x_{k-1}^{(2)}\right), y\right),$$

$$x_k \leftarrow \Lambda_{\theta_k^p}\left(x_{k-1}, \left[\partial \mathcal{K}\left(x_{k-1}^{(1)}\right)\right]^* \left(z_k^{(1)}\right)\right),$$

where $\Lambda_{\theta_k^p}$ and $\Gamma_{\theta_k^d}$ are CNNs replacing the primal and dual operators, with parameters θ_k^p and θ_k^d, respectively. The notation $x^{(1)}$ and $x^{(2)}$ selects the first and second blocks of coefficients in x to permit "memory" across iterations. A deep network, dubbed learned primal–dual, can then be formed by stacking iterations (layers) together. The parameters θ_k^p and θ_k^d are learned by minimizing the MSE loss.

Compared with FBP, TV reconstruction, and conventional deep networks such as U-Net, learned primal–dual obtains more than 6 dB improved PSNR for the Shepp–Logan phantom, and over 2 dB improvement on human phantoms on average. In terms of the structural similarity index (SSIM), it also outperforms its competitors by a large margin. The runtime is on the same order as that of FBP, rendering it amenable to time-critical applications.

Another technique for low-dose CT reconstruction is BCD-net [26, 27], where the authors generalized the block coordinate descent algorithm for model-based inversion by integrating a CNN as a learnable denoiser. As a follow-up of the learned primal–dual network, Wu et al. [28] concatenated it with a detection network, and applied joint fine-tuning after training both networks individually. Their jointly fine-tuned network achieves comparable performance a detector trained on the fully sampled data and outperforms detectors trained on the reconstructed images.

4.3.2 Unrolling in Super-Resolution Microscopy

Optical microscopy generates magnified images of small objects by using visible light and a system of lenses, thus enabling research into fields such as cell biology and microbiology. However, the spatial resolution is limited by the physics of light, which poses a hard limit on the resolution. The diffraction of light makes structures too blurry to be resolved once they are smaller than approximately half the wavelength of the light [15, 29]. This limitation is circumvented using photo-activated fluorescent molecules. In two-dimensional (2D) single-molecule localization microscopy (SMLM), a sequence of diffraction-limited images, produced by a sparse set of emitting fluorophores with minimally overlapping point spread functions (PSFs) is acquired, allowing the emitters to be localized with high precision [30]. However, the need for a low emitter density results in poor temporal resolution.

In [30], the authors unrolled the sparsity-based super-resolution correlation microscopy (SPARCOM) [15] method to incorporate domain knowledge, resulting in learned SPARCOM (LSPARCOM). Specifically, LSPARCOM aims to recover a single $M \times M$ high-resolution image X, corresponding to the locations of the emitters on a fine grid, from a set of T low-resolution frames of size $N \times N$ with $N < M$. A single frame taken at time t is denoted as $\mathbf{Y}(t)$. The method exploits the sparse nature of the fluorophores' distribution, alongside a statistical prior of uncorrelated emissions. The sparsity assumption implies that the overall number of emitter-containing pixels is significantly smaller than the total number of pixels in the high-resolution grid, yet, as opposed to classical methods for SMLM, it can account for scenarios with overlapping PSFs formed by adjacent emitters.

The authors of [30] formulated a sparse recovery problem on the temporal variance image given by

$$\mathbf{g}_Y = \mathbf{W}\mathbf{x}, \tag{4.15}$$

where $\mathbf{g}_Y \in \mathbb{R}^{N^2}$ comprises the temporal variances of the set of T low-resolution frames, $\mathbf{W}$ is a dictionary matrix based on the point spread function, and $\mathbf{x}$ comprises the variances of the emitter fluctuation on a high-resolution grid. The unknown $\mathbf{x}$ can be determined by solving a sparsity-constrained optimization problem similar to Eq. (4.3) with an additional nonnegativity constraint over $\mathbf{x}$, which in turn can be solved via the ISTA algorithm presented in Section 4.2.1. The algorithm can be further unrolled into a network whose architecture resembles LISTA. In contrast with LISTA, the authors propose the following proximal operator instead of the soft-thresholding operator defined in Eq. (4.5):

$$\mathcal{S}^+_{\alpha,\beta}(x) \triangleq \frac{\max\{0, x\}}{1 + \exp\left[-\beta(|x| - \alpha)\right]}, \tag{4.16}$$

where α and β are hyperparameters. When applied to vectors and matrices, $\mathcal{S}^+_{\alpha,\beta}(\cdot)$ operates elementwise. This is a sigmoid-based approximation of the positive hard-thresholding operator, which is the proximal operator for the ℓ^0-norm. Consequently, the use of such an approximation induces increased sparsity. Furthermore, it performs one-sided thresholding owing to the nonnegativity constraint.

For training, the authors of [30] introduce a compatible loss function given by

$$\text{Loss} = \frac{1}{M^2} \sum_{i,j=1}^{M} \mathbf{B}(i,j) |\hat{\mathbf{X}}(i,j) - \mathbf{X}^*(i,j)|^2 + \lambda \left[1 - \mathbf{B}(i,j)\right] |\hat{\mathbf{X}}(i,j)|, \tag{4.17}$$

where $\hat{\mathbf{X}}$ is the output of the network, $\mathbf{X}^*$ is the ground-truth image, and $\mathbf{B}$ is a binary mask, created by binarizing $\mathbf{X}^*$, which eliminates entries that do not contain emitters. This encourages the network to output signals coming only from emitters while suppressing the background.

Experimental results across different test sets show that the unrolled network can obtain super-resolution images from a small number of high-emitter-density frames without knowledge of the optical system. All the methods that were compared in [30] were evaluated on the quality of the reconstruction for exactly the same input frames, while the input for LSPARCOM is a single image, constructed by calculating the temporal variance of all the high-density frames. Figure 4.6 provides an example of the performance evaluation of SPARCOM, LSPARCOM, and Deep-STORM [31].

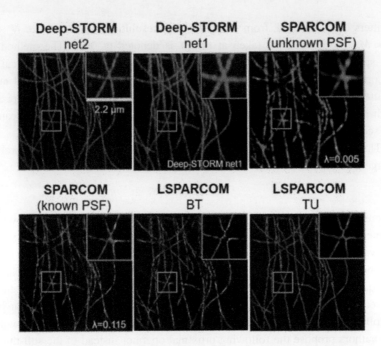

Figure 4.6 Performance evaluation for experimental tubulin sequence, composed of 500 high-density frames [35]. A difficult area for reconstruction is shown magnified in the larger boxes. From upper left to lower right: reconstructions using Deep-STORM architecture trained on different datasets (net1 and net2), SPARCOM reconstruction with unknown PSF, SPARCOM reconstruction using explicit knowledge of the PSF, and reconstructions using the LSPARCOM architecture trained on different datasets (BT, TU). Figure reproduced from [30] with the authors' permission.

Deep-STORM is another learning-based method, devised for the problem at hand. Two examples are brought forward for each method: the first incorporates prior knowledge (Deep-STORM net2 and LSPARCOM TU, which were trained for similar imaging parameters as in the test set and SPARCOM with known PSF), and the second does not incorporate prior knowledge. While all three methods achieve good results when incorporating prior knowledge, LSPARCOM is superior when generalizing to new data. Quantitatively speaking, the SPARCOM reconstruction took 39.32 seconds, the LSPARCOM reconstruction took 10.8 seconds, and Deep-STORM took 280.38 seconds for 500 128×128 input frames. Moreover, LSPARCOM is highly parameter-efficient: it had 9058 parameters in 10 folds of LSPARCOM as against 1.3 million trainable parameters in Deep-STORM. Consequently, the network achieves runtime savings compared with other learning-based methods.

4.3.3 Applications of Unrolling in Ultrasound

An important ultrasound-based modality is found in contrast enhanced ultrasound (CEUS) [32], which allows the detection and visualization of small blood vessels.

The main idea behind CEUS is the use of encapsulated gas microbubbles, serving as ultrasound contrast agents (UCAs). These are injected intravenously and can flow throughout the vascular system owing to their small size. To visualize them, strong clutter signals originating from stationary or slowly moving tissues must be removed, as they introduce significant artifacts in the resulting images. The latter pose a major challenge in ultrasonic vascular imaging [33].

In [6], the authors suggested a method for clutter suppression by applying the well-known low-rank and sparse-matrix decomposition technique, robust principal component analysis (RPCA), and unrolling the corresponding algorithm into a deep network named convolutional robust principal component analysis (CORONA). Specifically, they modeled the acquired contrast-enhanced ultrasound signal as a combination of a low-rank matrix (accounting for the tissue) and a sparse outlier signal (the UCAs).

In ultrasound imaging, a series of pulses is transmitted to the imaged medium. The resulting echoes from the medium are received in each transducer element and then combined in a process called beamforming to produce a focused image. If we denote the vertical and axial coordinates as x and z and the frame number as t, the observed signal from a specific point in space–time, denoted $Y(x,z,t)$, can be described as the sum of echoes returned from the blood and CEUS signals $S(x,z,t)$, as well as from the tissue $L(x,z,t)$, contaminated by additive noise $N(x,z,t)$. When acquiring a series of movie frames $t = 1, \ldots, T$, and stacking them as columns in a matrix $\mathbf{Y}$, the relation between the measurements and the signals can be described via

$$\mathbf{Y} = \mathbf{H}_1 \mathbf{L} + \mathbf{H}_2 \mathbf{S} + \mathbf{N}, \qquad (4.18)$$

where $\mathbf{H}_1$ and $\mathbf{H}_2$ are measurement matrices of appropriate dimensions. Similarly to $\mathbf{Y}$, $\mathbf{L}$ and $\mathbf{S}$ are matrices whose columns contain per-frame data. The tissue matrix $\mathbf{L}$ can be described as a low-rank matrix owing to its high spatio-temporal coherence. The CEUS-echoes matrix $\mathbf{S}$ is assumed to be sparse, as typically blood vessels only sparsely populate the imaged medium. Under the assumptions on the matrices $\mathbf{L}$ and $\mathbf{S}$, their recovery can be obtained by solving the following convex optimization problem:

$$\min_{\mathbf{L},\mathbf{S}} \frac{1}{2} \|\mathbf{Y} - (\mathbf{H}_1 \mathbf{L} + \mathbf{H}_2 \mathbf{S})\|_F^2 + \lambda_1 \|\mathbf{L}\|_* + \lambda_2 \|\mathbf{S}\|_{1,2}, \qquad (4.19)$$

where $\|\cdot\|_*$ stands for the nuclear norm (i.e., the sum of the singular values of $\mathbf{L}$), $\|\cdot\|_{1,2}$ stands for the mixed $\ell^{1,2}$ norm (i.e., the sum of the ℓ^2 norms of each row of $\mathbf{S}$), and λ_1 and λ_2 are regularization coefficients promoting the low-rank structure of $\mathbf{L}$ and the sparsity of $\mathbf{S}$, respectively.

Problem (4.19) can be solved using a generalized version of ISTA over matrices, by utilizing the proximal mapping corresponding to the nuclear norm and mixed $\ell_{1,2}$ norm [8]. In each iteration l, the estimations for both $\mathbf{S}$ and $\mathbf{L}$ are updated as follows:

$$\begin{aligned} \mathbf{L}^{l+1} &= \mathcal{T}_{\lambda_1/\mu} \left\{ \left(\mathbf{I} - \frac{1}{\mu}\mathbf{H}_1^H \mathbf{H}_1\right) \mathbf{L}^l - \mathbf{H}_1^H \mathbf{H}_2 \mathbf{S}^l + \mathbf{H}_1^H \mathbf{Y} \right\}, \\ \mathbf{S}^{l+1} &= \mathcal{S}^{1,2}_{\lambda_2/\mu} \left\{ \left(\mathbf{I} - \frac{1}{\mu}\mathbf{H}_2^H \mathbf{H}_2\right) \mathbf{S}^l - \mathbf{H}_2^H \mathbf{H}_1 \mathbf{L}^l + \mathbf{H}_2^H \mathbf{Y} \right\}. \end{aligned} \qquad (4.20)$$

Here $\mathcal{T}_\lambda\{\mathbf{X}\}$ is the singular-value thresholding operator that performs soft thresholding over the singular values of $\mathbf{X}$ with threshold λ, $\mathcal{S}_\lambda^{1,2}$ performs row-wise soft thresholding with parameter λ, and μ is the step-size parameter for ISTA.

To construct the unrolled network, Solomon et al. [6] replaced matrix multiplication with convolutional layers that are subsequently fed into the nonlinear proximal mappings. The lth iteration then becomes

$$\mathbf{L}^{l+1} = \mathcal{T}_{\lambda_1^l}\left\{\mathbf{W}_5^l * \mathbf{L}^l + \mathbf{W}_3^l * \mathbf{S}^l + \mathbf{W}_1^l * \mathbf{Y}\right\},$$
$$\mathbf{S}^{l+1} = \mathcal{S}_{\lambda_2^l}^{1,2}\left\{\mathbf{W}_6^l * \mathbf{S}^l + \mathbf{W}_4^l * \mathbf{L}^l + \mathbf{W}_2^l * \mathbf{Y}\right\}, \quad (4.21)$$

where $\mathbf{W}_i^l, i = 1, \ldots, 6$, are a series of convolution filters that are learned from the data in the lth layer, $*$ acts on both spatial and temporal dimensions and performs multi-channel convolution, and λ_1^l, λ_2^l are thresholding parameters for the lth layer. The architecture is illustrated in Fig. 4.7, in which $\mathbf{L}^0$ and $\mathbf{S}^0$ are set to some initial guess, and the output after L iterations, denoted $\mathbf{S}^L$, is used as an estimate of the desired sparse matrix $\mathbf{S}$.

The network is trained using backpropagation in a supervised manner. The training examples consists of observed matrices along with their corresponding ground-truth $\mathbf{L}^*$ and $\mathbf{S}^*$ matrices (refer to [6] for more details on how the dataset is generated). The loss function is chosen as the sum of the MSEs between $\mathbf{L}^*$ and $\mathbf{S}^*$ in comparison to the values predicted by the network.

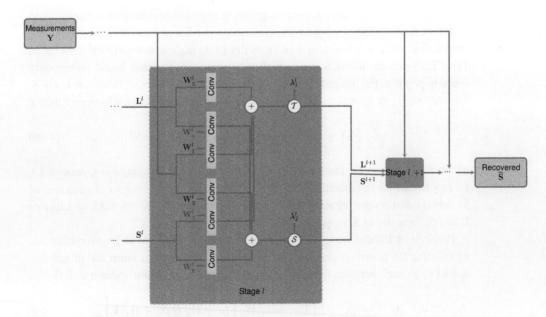

Figure 4.7 Diagrammatic representation of CORONA. The learned network draws its architecture from the iterative algorithm. Here $\mathbf{Y}$ is the input measurement matrix, and $\mathbf{S}^l$ and '$\mathbf{L}^l$ are the estimated sparse and low-rank matrices in each layer, respectively.

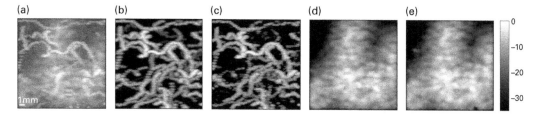

Figure 4.8 Sample experimental results demonstrating recovery of UCAs from cluttered MIP images. (a) MIP image of the input movie, composed from 50 frames of simulated UCAs cluttered by tissue. (b) Ground-truth UCA MIP image. (c) Recovered UCA MIP image via CORONA. (d) Ground-truth tissue MIP image. (e) Recovered tissue MIP image via CORONA. The gray-scale bar is in dB. Figure reproduced from [6] with the authors' permission.

Compared with state-of-the-art approaches, CORONA demonstrates vastly improved reconstruction quality and has many fewer parameters. Figure 4.8 provides visual results, demonstrating the power of CORONA to properly separate the low-rank matrix and the sparse matrix. Figure 4.8(a) shows the temporal maximum intensity projection (MIP) image of the input movie (50 frames). It can be seen that the ultrasound contrast agent (UCA) signal, depicted as randomly twisting lines, is considerably masked by the simulated tissue signal. Figures 4.8(b) and 4.8(d) illustrate the ground-truth MIP images of the UCA signal and tissue signal, respectively, whereas Figs. 4.8(c) and 4.8(e) present the MIP images of the UCA and tissue signals recovered via CORONA.

Another approach that leverages deep algorithm unrolling is used for super-resolution US [33, 34], where UCAs are pinpointed and tracked through a sequence of US frames to yield super-resolution images of the microvasculature, thus circumventing the diffraction-limited resolution of conventional ultrasound. Unrolling is used to derive a robust method for super-resolution US that is suited to various imaging conditions and does not depend on prior knowledge such as the PSF of the system.

By leveraging the signal structure, i.e., assuming that the UCA's distribution is sparse on a sufficiently high-resolution grid and assuming that the US frames contain only UCA signals (i.e., the absence of tissue clutter and noise), the model has the same form as (4.2), where $\mathbf{x}$ is a vector that describes the sparse microbubble distribution on a high-resolution image grid, $\mathbf{y}$ is the vectorized image frame of the ultrasound sequence, $\mathbf{W}$ is a dictionary matrix derived from the PSF of the system, and $\mathbf{n}$ is a noise vector. Again ISTA can be employed to solve this problem.

The unrolled network, named deep unrolled ultrasound localization microscopy (ULM), is similar to LISTA except that the soft-thresholding operator is substituted with a sigmoid-based soft-thresholding operation to avoid the vanishing of gradients during training [35]. The network is trained using the backpropagation under supervision of synthetically generated 2D images of point sources.

The authors trained a ten-layer deep network comprising 5×5 convolutional kernels. Figure 4.9 provides visual results of deep unrolled ULM for the super-resolution vascular imaging of a rat's spinal cord, exhibiting a smooth reconstruction of the

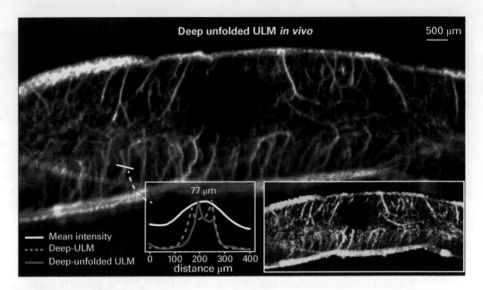

Figure 4.9 Deep unrolled ULM for super-resolution vascular imaging of a rat's spinal cord. Right-hand rectangle: Standard maximum intensity projection across a sequence of frames. Left-hand rectangle: Spatial resolution comparison between deep unrolled ULM and deep-ULM. Figure reproduced from [33] the with authors' permission.

vasculature. Compared with an encoder–decoder approach devised for the same task and trained on the same data, named deep-ULM as proposed in [33], deep unrolled ULM exhibited a drastically lower memory footprint and reduced power consumption in addition to achieving higher inference rates and improved generalization when encountering new data. This can be seen in the spatial resolution comparison in Fig. 4.9, showing the superior capability of deep unrolled ULM to separate close structures. Recently, the method was harnessed to improve breast lesion characterization [34]. Recoveries of three *in vivo* human scans of lesions in the breasts of three patients were shown. Figure 4.10 presents the recoveries of the three lesions along with the B-mode US images that are used in clinical practice. The results reveal vascular structures that are not seen in the B-mode images. Furthermore, the recoveries show a unique vascular structure for each lesion, thus assisting differentiation between the lesions.

4.3.4 Applications of Unrolling in Magnetic Resonance Imaging

A typical MRI system comprises the following components: a primary magnet which exerts the primary magnetic field, gradient magnets which create a spatial variation of the magnetic field and allow spatial encoding of the signal, radio-frequency (RF) coils which send out and receive a series of RF pulses and magnetic field gradients, and a computer system which decodes the received signal and reconstructs an image of the original item of interest. Formally speaking, the signal $\mathbf{y}(t)$ received in the receiver at time t is given by

4 Deep Algorithm Unrolling for Biomedical Imaging

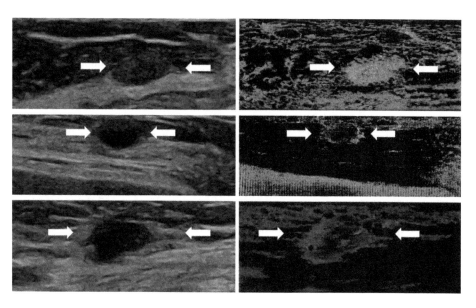

Figure 4.10 Super-resolution demonstrations in human scans of three lesions in the breasts of three patients. Left: B-mode images. Right: Super-resolution recoveries. The white arrows point to the lesions. Top: Fibroadenoma (benign). The super-resolution recovery shows an oval, well-circumscribed mass with homogeneous high vascularization. Middle: Cyst (benign). The super-resolution recovery shows a round structure with high concentration of blood vessels at the periphery of the lesion. Bottom: Invasive ductal carcinoma (malignant). The super-resolution recovery shows an irregular mass with ill-defined margins, high concentration of blood vessels at the periphery of the mass, and low concentration of blood vessels at the center of the mass. Figure reproduced from [34] with the authors' permission.

$$\mathbf{y}(\mathbf{k},t) = \int \mathbf{x}(x,y,z) \exp\{-j2\pi(k_x x + k_y y + k_z z)\}\,dx\,dy\,dz,$$

where $\mathbf{x}$ is the image, x, y, z gives the spatial location, and $\mathbf{k} = (k_x, k_y, k_z)$ is the k-space location at time t. In practice, a discretized model is commonly used:

$$\mathbf{y}(\mathbf{k}_i) = \sum_l \mathbf{x}(\mathbf{r}_l)\exp\{-j2\pi \mathbf{k}_i \cdot \mathbf{r}_l\},$$

where $\mathbf{k}_i$ and $\mathbf{r}_l$ are k-space and spatial locations and the multiplication dot denotes the scalar product between 3D vectors. That is, the measured signal $\mathbf{y}$ comprises a Fourier series encoding of the signal $\mathbf{x}$. In compressed MRI, only a subset of Fourier samples is observed, which gives rise to the following linear compressive imaging model:

$$\mathbf{y} = \mathbf{SFx},$$

where $\mathbf{F} \in \mathbb{C}^{m \times m}$ is the Fourier matrix and $\mathbf{S} \in \{0,1\}^{n \times m}$ is the subsampling matrix. In the MRI literature, more advanced imaging models have been developed to account for other factors such as inhomogeneity of the magnetic fields, the coupling between gradient coils, timing inaccuracies, and motion. For a thorough review of these models, see [36].

To accelerate data acquisition in dynamic MRI, Qin et al. [37] proposed a convolutional recurrent neural network (CRNN) which captures temporal dependencies and reconstructs high-quality cardiac MR images from highly undersampled k-space data. They first analyzed the following classical regularized inverse problem:

$$\min_{\mathbf{x}} \mathcal{R}(\mathbf{x}) + \lambda \|\mathbf{y} - \mathbf{SFx}\|_2^2, \tag{4.22}$$

where $\mathbf{x}$ comprises a sequence of complex-valued MR images, $\mathcal{R}$ is the regularization functional, and λ is the (reciprocal) regularization coefficient. By introducing an extra variable, $\mathbf{z}$, problem (4.22) can be relaxed into

$$\min_{\mathbf{x},\mathbf{z}} \lambda \|\mathbf{y} - \mathbf{SFx}\|_2^2 + \mu \|\mathbf{x} - \mathbf{z}\|_2^2 + \mathcal{R}(\mathbf{z}), \tag{4.23}$$

where $\mu > 0$ is the penalty parameter. When $\mu \to \infty$, the relaxed problem (4.23) reduces to the original problem (4.22). Problem (4.23) can be solved by alternately minimizing over $\mathbf{x}$ and $\mathbf{z}$:

$$\mathbf{z}_{k+1} \leftarrow \arg\min_{\mathbf{z}} \mathcal{R}(\mathbf{z}) + \mu \|\mathbf{x}_k - \mathbf{z}\|_2^2, \tag{4.24}$$

$$\begin{aligned}\mathbf{x}_{k+1} &\leftarrow \arg\min_{\mathbf{x}} \lambda \|\mathbf{y} - \mathbf{SFx}\|_2^2 + \mu \|\mathbf{x} - \mathbf{z}_{k+1}\|_2^2 \\ &= \mathbf{F}^H \mathbf{\Lambda} \mathbf{F} \mathbf{z}_k + \frac{\lambda_0}{1+\lambda_0} \mathbf{SF}^H \mathbf{y} \\ &:= \mathcal{DC}(\mathbf{z}_k; \mathbf{y}, \lambda_0, \Omega), \end{aligned} \tag{4.25}$$

where $\lambda_0 = \lambda/\mu$, $\mathbf{\Lambda}$ is a diagonal matrix accounting for the undersampling scheme, and Ω is the index set of the acquired k-space samples. Qin et al. [37] proposed replacing (4.24) with a CRNN, providing the following iterations:

$$\mathbf{z}_{k+1} \leftarrow \mathbf{x}_k + \text{CRNN}(\mathbf{x}_k),$$
$$\mathbf{x}_{k+1} \leftarrow \mathcal{DC}(\mathbf{z}_{k+1}; \mathbf{y}, \lambda_0, \Omega).$$

The CRNN is a learnable RNN which integrates convolutions and evolves both over iterations and time. To this end, the authors introduced bidirectional convolutional recurrent units, evolving both over time and iterations, and convolutional recurrent units, evolving over iterations only. CRNN introduces inter-iteration connections so that information can be propagated across iterations, whereas in conventional unrolling techniques the iterations are typically disconnected. In addition, CRNN exploits temporal dependency and data redundancies of the dynamic MRI data by introducing temporal progressions. In contrast with typical RNNs, the weights for CRNN are independent across layers. The network is trained end-to-end by minimizing the MSE loss between the predicted MR images and the ground-truth data. Experimentally, CRNN achieves nearly 1 dB improvement in PSNR for dynamic MRI, compared with state-of-the-art approaches including both iterative reconstruction and deep learning. In terms of running time, it is also significantly faster than both classes of approaches.

On a different algorithmic front, Pramanik et al. [38] unrolled the structured low-rank (SLR) algorithm for calibration less parallel MRI and multishot MRI applications. The SLR algorithm performs MR image reconstruction by lifting the signal into a structured matrix and solving the following rank-minimization problem:

$$\min_{\Gamma} \quad \text{rank}\left[\mathcal{T}(\mathcal{G}(\widehat{\Gamma}))\right]$$
$$\text{s.t.} \quad \mathbf{B} = \mathcal{A}(\Gamma) + \mathbf{P}, \quad (4.26)$$

where Γ is the matrix representing multi-channel images on different coils, $\widehat{\Gamma}$ denotes the discrete Fourier transform (DFT), $\mathbf{B}$ is the corresponding noisy undersampled multi-channel Fourier measurement, $\mathcal{A}$ is the linear operator formed by composing the Fourier transform with subsampling, and $\mathbf{P}$ is the multi-channel noise matrix; $\mathcal{G}$ is the mapping from Fourier samples to their gradients, and $\mathcal{T}$ is the lifting operator, which lifts the signal into a higher-dimensional structured matrix. The low-rank constraint comes from the fact that the matrix $\mathcal{T}(\mathcal{G}(\widehat{\Gamma}))$ has a high-dimensional null space, which in turns originates from the sparsity of edges in natural images. For the technical details, see [38] and [39].

Problem (4.26) is difficult to tackle directly owing to its nonconvex nature, and the following convex relaxation usually serves as a tractable surrogate:

$$\min_{\Gamma} \|\mathcal{A}(\Gamma) - \mathbf{B}\|_2^2 + \lambda \left\|\mathcal{T}(\mathcal{G}(\widehat{\Gamma}))\right\|_*, \quad (4.27)$$

where λ is the regularization coefficient and $\|\cdot\|_*$ is the nuclear norm whose minimization promotes low-rank structures of the solution. Problem (4.27) can be solved by the iteratively reweighted least squares (IRLS) algorithm, which majorizes the nuclear norm with a weighted Frobenius norm and solves the problem:

$$\min_{\Gamma, \mathbf{Q}} \|\mathcal{A}(\Gamma) - \mathbf{B}\|_2^2 + \lambda \left\|\mathcal{T}(\mathcal{G}(\widehat{\Gamma}))\mathbf{Q}\right\|_F^2. \quad (4.28)$$

By introducing an auxiliary variable $\widehat{\mathbf{Z}}$, problem (4.28) can be further relaxed into

$$\min_{\Gamma, \mathbf{Q}, \widehat{\mathbf{Z}}} \|\mathcal{A}(\Gamma) - \mathbf{B}\|_2^2 + \lambda \left\|\mathcal{T}(\widehat{\mathbf{Z}})\mathbf{Q}\right\|_F^2 + \beta \left\|\mathcal{G}(\widehat{\Gamma}) - \widehat{\mathbf{Z}}\right\|_2^2, \quad (4.29)$$

where $\beta > 0$ is the penalty coefficient. Problem (4.29) can be solved by alternately minimizing over Γ, $\widehat{\mathbf{Z}}$, and $\mathbf{Q}$:

$$\Gamma_{k+1} \leftarrow \arg\min_{\Gamma} \|\mathcal{A}(\Gamma) - \mathbf{B}\|_2^2 + \beta \left\|\mathcal{G}(\widehat{\Gamma}) - \widehat{\mathbf{Z}}_k\right\|_2^2$$
$$= \left(\mathcal{A}^H \mathcal{A} + \beta \mathcal{G}^H \mathcal{G}\right)^{-1} \left(\mathcal{A}^H \mathbf{B} + \beta \mathcal{G}^H (\widehat{\mathbf{Z}}_k)\right)$$
$$:= \mathcal{M}\left(\mathcal{A}^H \mathbf{B}, \mathbf{Z}_k\right), \quad (4.30)$$

$$\widehat{\mathbf{Z}}_{k+1} \leftarrow \arg\min_{\widehat{\mathbf{Z}}} \beta \left\|\mathcal{G}(\widehat{\Gamma}_{k+1}) - \widehat{\mathbf{Z}}\right\|_2^2 + \lambda \left\|\mathcal{T}(\widehat{\mathbf{Z}})\mathbf{Q}\right\|_F^2$$
$$= \left[\mathbf{I} + \frac{\lambda}{\beta} \mathcal{J}(\mathbf{Q}_k)^H \mathcal{J}(\mathbf{Q}_k)\right]^{-1} \mathcal{G}(\widehat{\Gamma}_{k+1}), \quad (4.31)$$

$$\mathbf{Q}_{k+1} = \left[\mathcal{T}(\mathcal{G}(\widehat{\Gamma}_{k+1}))^H \mathcal{T}(\mathcal{G}(\widehat{\Gamma}_{k+1})) + \epsilon_{k+1}\mathbf{I}\right]^{-1/4},$$

where $\mathcal{J}$ is a lifting operator such that $\mathcal{J}(\mathbf{Q})$ represents a filter bank. Pramanik et al. [38] proposed to replace the update step (4.31) with a denoising CNN, and thus

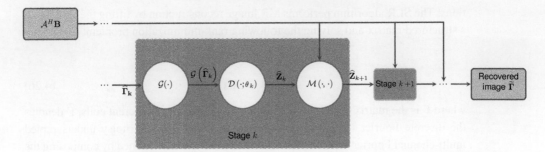

Figure 4.11 Diagrammatic representation of DSLR [38]: at each stage, the input $\mathbf{\Gamma}_k$ passes through a gradient operator, a CNN defined in (4.32), and a data-consistency layer defined by (4.30).

adopted the following iteration procedures:

$$\widehat{\mathbf{Z}}_k \leftarrow \mathcal{D}\left[\mathcal{G}\left(\widehat{\mathbf{\Gamma}}_k\right); \theta_k\right], \tag{4.32}$$

$$\widehat{\mathbf{\Gamma}}_{k+1} \leftarrow \mathcal{M}\left(\mathcal{A}^H \mathbf{B}, \mathbf{Z}_k\right),$$

where $\mathcal{D}(\cdot; \theta_k)$ is a CNN with parameter θ_k. The unrolled network, called deep SLR (DSLR), is illustrated in Fig. 4.11. At implementation, DSLR adopts a hybrid regularization scheme which incorporates an additional prior to exploit image-domain redundancies. Training is performed using the MSE loss.

Experimentally, DSLR demonstrates superior performance on several evaluation metrics over state-of-the-art techniques: SNR, PSNR, and SSIM. In addition, DSLR runs significantly faster compared with SLR methods as it is free from expensive operations such as singular-value decomposition (SVD), bringing down the running time from tens of minutes to less than a second. Figure 4.12 provides a visual comparison between DSLR and state-of-the-art MRI algorithms, where it is clearly observed that DSLR achieves much more accurate recovery, with lower errors and higher quantitative scores.

4.3.5 Unrolling Techniques across Multiple Biomedical Imaging Modalities

In addition to application-specific approaches, there are numerous techniques applicable to multiple medical imaging modalities. Different medical imaging problems may share a similar forward model and have a common prior structure, which enables the development of an abstract imaging algorithm that can be adapted towards each particular problem. An example that unrolls the well-known alternating direction method of multipliers (ADMM) algorithm is the ADMM-CSNet proposed in [4], where Yang et al. developed a compressive sensing (CS) technique for a linear imaging model. Specifically, the measurement vector $\mathbf{y} \in \mathbb{C}^m$ is collected through $\mathbf{y} \approx \mathbf{W}\mathbf{x}$, where $\mathbf{W} \in \mathbb{C}^{n \times m}$ is a measurement matrix with $m > n$ and $\mathbf{x} \in \mathbb{R}^m$ is the signal to be recovered.

The CS problem is typically solved by exploiting the underlying sparse structure of the signal $\mathbf{x}$ in a certain transformation domain [1]. In [4], a slightly generalized

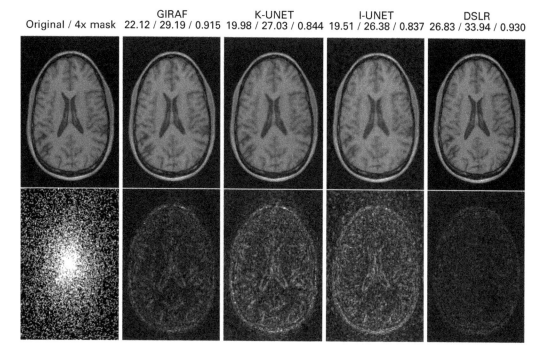

Figure 4.12 Experimental results on the reconstruction of 4× accelerated single-channel brain data. The data were undersampled using a Cartesian 2D nonuniform variable-density mask. The SNR (dB) / PSNR (dB) / SSIM values are also included for quantitative comparison. The upper row and the lower row show the reconstructed images (in magnitude) and the corresponding error images, respectively. DSLR was proposed in [38], while the SLR algorithm (GIRAF) [40], k-space U-net (K-UNET) [41], and image domain UNET (I-UNET) are state-of-the-art techniques for comparison. Images are courtesy of Mathews Jacob at University of Iowa [38].

CS model was employed, which amounts to solving the following optimization problem [4]:

$$\min_{\mathbf{x}} \frac{1}{2}\|\mathbf{W}\mathbf{x} - \mathbf{y}\|_2^2 + \sum_{i=1}^{C} \lambda_i g(\mathbf{D}_i \mathbf{x}), \tag{4.33}$$

where the λ_i are positive regularization coefficients, $g(\cdot)$ is a sparsity-inducing function, and $\{\mathbf{D}_i\}_{i=1}^{C}$ is a sequence of C operators which effectively perform linear filtering operations. Concretely, $\mathbf{D}_i$ can be taken as a wavelet transform and g can be chosen as the ℓ^1-norm. However, instead of resorting to analytical filters and prior functions, in [4] they are learned from training data through unrolling.

In general, there are numerous techniques for solving (4.33) such as the ISTA algorithm discussed in Section 4.2.1. Among them a simple yet efficient algorithm is ADMM [42], which has found wide application in various imaging domains. To employ it, problem (4.33) is first recast into a constrained minimization problem through variable splitting, by introducing variables $\{z_i\}_{i=1}^C$:

$$\min_{x,\{z\}_{i=1}^C} \frac{1}{2}\|Wx - y\|_2^2 + \sum_{i=1}^C \lambda_i g(z_i),$$

$$\text{s.t. } z_i = D_i x, \quad \forall i. \tag{4.34}$$

The corresponding augmented Lagrangian is then formed:

$$\mathcal{L}_\rho(x, z; \alpha_i) = \frac{1}{2}\|Wx - y\|_2^2 + \sum_{i=1}^C \lambda_i g(z_i) + \frac{\rho_i}{2}\|D_i x - z_i + \alpha_i\|_2^2, \tag{4.35}$$

where $\{\alpha_i\}_{i=1}^C$ are the dual variables and $\{\rho_i\}_{i=1}^C$ are positive penalty coefficients. The ADMM algorithm then alternately minimizes (4.35) and uses a dual-variable update, leading to the following iterations:

$$\begin{aligned}
x^l &= \left(W^H W + \sum_{i=1}^C \rho_i D_i^T D_i\right)^{-1} \left[W^H y + \sum_{i=1}^C \rho_i D_i^T \left(z_i^{l-1} - \alpha_i^{l-1}\right)\right] \\
&:= \mathcal{U}^1\left\{y, \alpha_i^{l-1}, z_i^{l-1}; \rho_i, D_i\right\}, \\
z_i^l &= \mathcal{P}_g\left\{D_i x^l + \alpha_i^{l-1}; \frac{\lambda_i}{\rho_i}\right\} \\
&:= \mathcal{U}^2\left\{\alpha_i^{l-1}, x^l; \lambda_i, \rho_i, D_i\right\}, \\
\alpha_i^l &= \alpha_i^{l-1} + \eta_i(D_i x^l - z_i^l) \\
&:= \mathcal{U}^3\left\{\alpha_i^{l-1}, x^l, z_i^l; \eta_i, D_i\right\}, \quad \forall i,
\end{aligned} \tag{4.36}$$

where the η_i are constant parameters and $\mathcal{P}_g\{\cdot; \lambda\}$ is the proximal mapping for g with parameter λ. The unrolled network can thus be constructed by concatenating these operations. The parameters $\lambda_i, \rho_i, \eta_i, D_i$ in each layer are learned from real datasets. Figure 4.13 depicts the resulting network architecture. In [4] the authors discuss several implementation issues, including efficient matrix inversion and the

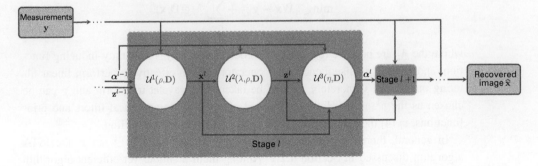

Figure 4.13 Diagrammatic representation of ADMM-CSNet [4]: each stage comprises a series of inter-related operations, whose analytic forms are given in Eq. (4.36).

analytic backpropagation rules. The network is trained by minimizing a normalized version of the root mean square error (RMSE).

ADMM-CSNet has demonstrated its efficacy through various experiments. In particular, for MRI applications, ADMM-CSNet achieves the same reconstruction accuracy using 10% less sampled data and speeds up recovery by around 40 times compared with conventional iterative methods. When compared with state-of-the art deep networks, it exceeds their performance by a margin of around 3 dB PSNR under a 20% sampling rate on brain data.

In a similar fashion, Aggrawal et al. [43] unrolled an alternating minimization algorithm over a regularized image reconstruction problem. By integrating a CNN into the image prior, they started with the following optimization problem:

$$\min_{\mathbf{x}} \|\mathbf{W}\mathbf{x} - \mathbf{y}\|_2^2 + \lambda \|\mathbf{x} - \mathcal{D}_\theta(\mathbf{x})\|_2^2, \qquad (4.37)$$

where $\mathbf{x} \in \mathbb{C}^m$ is the underlying image to be recovered, $\mathbf{y} \in \mathbb{C}^n$ is the measurement vector, $\mathbf{W} \in \mathbb{C}^{n \times m}$ is the measurement model matrix, and $\mathcal{D}_\theta$ is a denoising CNN carrying parameters θ. Aggrawal et al. proposed to solve problem (4.37) via the following iterations:

$$\mathbf{z}_k \leftarrow \mathcal{D}_\theta(\mathbf{x}_k),$$

$$\mathbf{x}_{k+1} \leftarrow \arg\min_{\mathbf{x}} \|\mathbf{W}\mathbf{x} - \mathbf{y}\|_2^2 + \lambda \|\mathbf{x} - \mathbf{z}_k\|_2^2$$

$$= \left(\mathbf{W}^H \mathbf{W} + \lambda \mathbf{I}\right)^{-1} \left(\mathbf{W}^H \mathbf{y} + \lambda \mathbf{z}_k\right), \qquad (4.38)$$

where $\mathbf{I} \in \mathbb{R}^{m \times m}$ is the identity matrix. The network parameters θ are shared across different iterations. The unrolled network, dubbed MoDL, has a very similar structure to DSLR in Fig. 4.11. Training of MoDL is conducted by minimizing the MSE loss.

For applications such as MRI, $\mathbf{W}$ corresponds to a composition of the sampling operator with the Fourier matrix, and thus the iteration step (4.38) admits a simple analytical formula. However, in more complex cases, such as multi-channel MRI, the operator $\mathbf{W}^H \mathbf{W} + \lambda \mathbf{I}$ is not analytically invertible and iterative numerical methods such as conjugate gradient (CG) need to be plugged into the network. Aggrawal et al. [43] verified that, although CG is an iterative technique, the intermediate results need not be stored in order to perform backpropagation. Instead, another CG step simply needs to be applied in the back propagation step.

Numerous experiments verify the effectiveness of MoDL. When compared with other deep-learning frameworks, MoDL achieves nearly 3 dB higher PSNR values for 6× acceleration on the dataset of Aggrawal et al. acquired through 3D T2 CUBE. Furthermore, as MoDL incorporates the forward model $\mathbf{y} = \mathbf{W}\mathbf{x}$, it is relatively insensitive to the undersampling patterns in MRI. In particular, when trained on a 10× acceleration setting, MoDL is capable of faithfully recovering images under 12× and 14× acceleration settings.

4.4 Perspectives and Recent Trends

We have so far reviewed several successes of algorithm unrolling, including both conceptual and practical breakthroughs in many biomedical imaging topics. In this section we will put algorithm unrolling into perspective, in order to understand why it is beneficial compared with traditional iterative algorithms and contemporary DNNs. Furthermore, we summarize some recent trends that we observe in many biomedical imaging works.

4.4.1 Why is Unrolling So Effective for Biomedical Imaging?

In recent years, algorithm unrolling has proved highly effective in achieving superior performance and higher efficiency in many practical domains, while inheriting interpretability from the underlying iterative algorithms. In addition, it frequently demonstrates improved robustness under deviations of training data and forward models compared with generic DNNs. A question that naturally arises is, why is it so powerful?

From a machine learning perspective, a fundamental trade-off for learning-based models is the *bias–variance trade-off* [44]. Models that have low capacity, such as linear models, typically generalize better but may prove inadequate to capture complicated data patterns. Such models generally exhibit high bias and low variance. In contrast, models that have abundant capacity, such as DNNs, are capable of fitting sophisticated functions but have a high risk of overfitting.

Unrolled networks achieve a more favorable bias–variance trade-off than alternative techniques [4], in particular, generic neural networks and iterative algorithms. Figure 4.14 provides a high-level illustration of the representation capability of these classes of techniques, from a functional approximation perspective. The areas corresponding to various classes of methods represent their capability in fitting functions, i.e., their modeling capacity as functional approximators. A traditional iterative algorithm spans a relatively small subset of functions compared with deep-learning techniques, and thus has low variance but high bias. Indeed, in practice, iterative algorithms typically underperform deep-learning techniques, which provides strong evidence of their low modeling capacity. Therefore, there is generally an irreducible gap between the spanned set and the target function. Nevertheless, in implementing an iterative method, typically the user tweaks the architecture and fine-tunes the parameters, rendering it capable of approximating a given target function reasonably well and effectively reducing the gap. On the other hand, iterative algorithms generalize relatively well in limited training scenarios thanks to their low variance. In biomedical imaging, the data is generally difficult to gather owing to the high costs of imaging devices and the requirements for patients. Therefore, iterative algorithms have played a critical role for many years.

However, a generic neural network is capable of more accurately approximating the target function thanks to its universal approximation capability [45]. As shown in

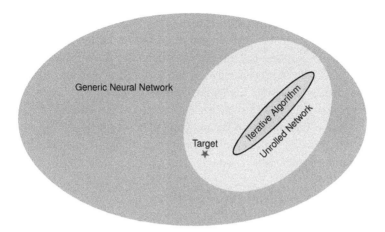

Figure 4.14 A high-level unified interpretation of algorithm unrolling from a functional approximation perspective: the ellipses depict the scope of the functions that can be approximated by each category of methods. Compared with iterative algorithms, which have limited representational power and usually underfit the target, unrolled networks often approximate the target better thanks to their higher representational power. Although unrolled networks have lower representational power than generic neural networks they usually generalize better in practice, hence providing an attractive balance.

Fig. 4.14, it constitutes a large subset in function space. In practice, the fact that DNNs can approximate complicated mappings as long as the data is sufficient is empirically supported. Nowadays, DNNs typically consists of an enormous number of parameters and have low bias but high variance. The high dimensionality of parameters requires abundant training samples, otherwise generalization becomes a serious issue. Furthermore, searching for the target function out of its huge spanned space is rather difficult, which poses great challenges in training.

In contrast, the unrolled networks are typically constructed by expanding the capacity of iterative algorithms, through generalizations of the underlying iterations. Therefore, such networks integrate domain knowledge from iterative algorithms and can approximate the target function more accurately, while spanning a relatively small subset in the function space. The reduced size of the search space alleviates the burden of training and the requirement of large-scale training datasets. In addition, unrolled networks have better generalizability thanks to their lower variance and are less prone to overfitting. As an intermediate model between generic networks and iterative algorithms, unrolled networks typically have relatively low bias and variance simultaneously.

4.4.2 Emerging Unrolling Trends for Biomedical Imaging

As we have seen through various case studies, there are a few common trends shared by many methods, which are summarized below.

Using DNNs as Learnable Operators

As discussed in Section 4.2, an important procedure in constructing unrolled networks is to generalize the functional form at each iteration. A commonly employed technique is to substitute certain operators by DNNs. In particular, the proximal operators which are manifest in various regularized inverse problems are frequently replaced by parameterized deep networks such as CNNs [5, 17, 20] or RNNs [37]. This substitution technique is reminiscent of the plug-and-play scheme [46], where the proximal operators are replaced with an off-the-shelf image denoiser such as block-matching and 3D filtering (BM3D) [47].

Exploring Broader Classes of Algorithms

In the early stage of unrolling research the focus was primarily in unrolling ISTA-like algorithms [3, 18]. Nowadays, researchers are pursuing other alternatives such as PGD [20], ADMM [4], and IRLS [38], to name a few. This trend has contributed greatly to the novelty and variety of unrolling approaches.

Balancing Performance and Efficiency

In practice, when designing unrolled deep networks, there is a fundamental trade-off between performance and efficiency: wider and deeper networks with numerous parameters generally achieve better performance but are usually less efficient and more data-demanding. Therefore, under practical constraints important design choices have to be made. One question is whether to use shared parameters across all the layers or adopt layer-specific parameters. Another issue is how to determine the depth of the networks, i.e., the number of iterations in the algorithm.

4.5 Conclusions

In this chapter we have discussed algorithm unrolling, a systematic approach to constructing deep networks out of iterative algorithms, and its applications to biomedical imaging. We first provided a comprehensive tutorial on algorithm unrolling and how to leverage it to construct deep networks. We then showcased how algorithm unrolling can be applied in biomedical imaging by discussing concrete examples in different medical imaging modalities. Next, we put algorithm unrolling into perspective, illustrated why it can be so powerful from a bias–variance trade-off standpoint, and summarized several recent trends in unrolling research.

To facilitate future research, we conclude this chapter by briefly discussing open challenges and suggesting possible future research directions. First, although algorithm unrolling has inherited the merits of iterative algorithms to a large extent, customizations of the iteration procedures might undermine some of them. In particular, convergence guarantees may no longer hold. Therefore, it is of interest to extend the theories around iterative algorithms to incorporate promising unrolled networks. Second, as algorithm unrolling serves as a valuable bridge between iterative algorithms and deep networks, it can be regarded as a tool to understand why deep networks are so effective in practical imaging applications, in order to complement their lack of interpretability. Finally, in biomedical imaging, training data is relatively deficient. Since algorithm unrolling has already demonstrated superior generalizability in many previous works, it should be exploited as a viable alternative under limited-training scenarios, in addition to and/or integrated with semisupervised learning.

References

[1] Y. C. Eldar and G. Kutyniok, *Compressed Sensing: Theory and Applications*. Cambridge University Press, 2012.

[2] S. Min, B. Lee, and S. Yoon, "Deep learning in bioinformatics," *Briefings in Bioinformatics*, vol. 18, no. 5, pp. 851–869, 2017.

[3] K. Gregor and Y. LeCun, "Learning fast approximations of sparse coding," in *Proc. International Conference on Machine Learning*, 2010.

[4] Y. Yang, J. Sun, H. Li, and Z. Xu, "ADMM-CSNet: A deep learning approach for image compressive sensing," *IEEE Transactions on Pattern Analysis and Machine Intelligence*, 2018.

[5] A. Hauptmann, F. Lucka, M. Betcke, N. Huynh, J. Adler, B. Cox, P. Beard, S. Ourselin, and S. Arridge, "Model-based learning for accelerated, limited-view 3D photoacoustic tomography," *IEEE Transactions on Medical Imaging*, vol. 37, no. 6, pp. 1382–1393, 2018.

[6] O. Solomon, R. Cohen, Y. Zhang, Y. Yang, Q. He, J. Luo, R. J. G. van Sloun, and Y. C. Eldar, "Deep unfolded robust PCA with application to clutter suppression in ultrasound," *IEEE Transactions on Medical Imaging*, vol. 39, no. 4, pp. 1051–1063, 2020.

[7] M. Mischi, M. A. L. Bell, R. J. van Sloun, and Y. C. Eldar, "Deep learning in medical ultrasound – from image formation to image analysis," *IEEE Transactions on Ultrasonics, Ferroelectrics, and Frequency Control.*, vol. 67, no. 12, pp. 2477–2480, 2020.

[8] V. Monga, Y. Li, and Y. C. Eldar, "Algorithm unrolling: Interpretable, efficient deep learning for signal and image processing," *arXiv:1912.10557*, 2019.

[9] I. Daubechies, M. Defrise, and C. De Mol, "An iterative thresholding algorithm for linear inverse problems with a sparsity constraint," *Communications on Pure and Applied Mathematics*, vol. 57, no. 11, pp. 1413–1457, 2004.

[10] S. S. Chen, D. L. Donoho, and M. A. Saunders, "Atomic decomposition by basis pursuit," *SIAM Review*, vol. 43, no. 1, pp. 129–159, 2001.

[11] Y. A. LeCun, L. Bottou, G. B. Orr, and K. Müller, "Efficient backprop," in *Neural Networks: Tricks of the Trade*, Springer, 2012, pp. 9–48.

[12] B. Xin, Y. Wang, W. Gao, D. Wipf, and B. Wang, "Maximal sparsity with deep networks?" in *Advances in Neural Information Processing Systems*, 2016, pp. 4340–4348.

[13] X. Chen, J. Liu, Z. Wang, and W. Yin, "Theoretical linear convergence of unfolded ISTA and its practical weights and thresholds," in *Advances in Neural Information Processing Systems*, 2018.

[14] J. Liu, X. Chen, Z. Wang, and W. Yin, "ALISTA: Analytic weights are as good as learned weights in LISTA," in *Proc. International Conference on Learning Representation*, 2019.

[15] O. Solomon, Y. C. Eldar, M. Mutzafi, and M. Segev, "SPARCOM: Sparsity based super-resolution correlation microscopy," *SIAM Journal of Imaging Science*, vol. 12, no. 1, pp. 392–419, 2019.

[16] Z. Wang, D. Liu, J. Yang, W. Han, and T. Huang, "Deep networks for image super-resolution with sparse prior," in *Proc. IEEE International Conference on Computer Vision*, 2015, pp. 370–378.

[17] J. Adler and O. Öktem, "Learned primal–dual reconstruction," *IEEE Transactions on Medical Imaging*, vol. 37, no. 6, pp. 1322–1332, 2018.

[18] K. H. Jin, M. T. McCann, E. Froustey, and M. Unser, "Deep convolutional neural network for inverse problems in imaging," *IEEE Transactions on Image Processing*, vol. 26, no. 9, pp. 4509–4522, 2017.

[19] O. Ronneberger, P. Fischer, and T. Brox, "U-net: Convolutional networks for biomedical image segmentation," in *Proc. International Conference on Medical Image Computation and Computer Assisted Interventions*, 2015, pp. 234–241.

[20] H. Gupta, K. H. Jin, H. Q. Nguyen, M. T. McCann, and M. Unser, "CNN-based projected gradient descent for consistent CT image reconstruction," *IEEE Transactions on Medical Imaging*, vol. 37, no. 6, pp. 1440–1453, 2018.

[21] E. Kang, W. Chang, J. Yoo, and J. C. Ye, "Deep convolutional framelet denoising for low-dose ct via wavelet residual Network," *IEEE Transactions on Medical Imaging*, vol. 37, no. 6, pp. 1358–1369, 2018.

[22] M. Li, Z. Fan, H. Ji, and Z. Shen, "Wavelet frame based algorithm for 3D reconstruction in electron microscopy," *SIAM Journal on Scientific Computing*, vol. 36, no. 1, pp. B45–B69, 2014.

[23] B. Dong, Q. Jiang, and Z. Shen, "Image restoration: Wavelet frame shrinkage, nonlinear evolution pdes, and beyond," *Multiscale Modeling & Simulation*, vol. 15, no. 1, pp. 606–660, 2017.

[24] J. C. Ye, Y. Han, and E. Cha, "Deep convolutional framelets: A general deep learning framework for inverse problems," *SIAM Journal of Imaging Science*, vol. 11, no. 2, pp. 991–1048, 2018.

[25] R. T. Rockafellar, *Convex Analysis*. Princeton University Press, 1970.

[26] Y. Chun and J. A. Fessler, "Deep BCD-net using identical encoding-decoding CNN structures for iterative image recovery," in *Proc. IEEE Image, Video, and Multidimensional Signal Processing Workshop*. IEEE, 2018, pp. 1–5.

[27] I. Y. Chun, X. Zheng, Y. Long, and J. A. Fessler, "Bcd-net for low-dose CT reconstruction: Acceleration, convergence, and generalization," in *Proc. International Conference on Medical Image Computation and Computer Assisted Interventions*, Springer, 2019, pp. 31–40.

[28] D. Wu, K. Kim, B. Dong, G. E. Fakhri, and Q. Li, "End-to-end Lung nodule detection in computed tomography," in *Machine Learning in Medical Imaging*, Springer, 2018, pp. 37–45.

[29] B. Huang, "Super-resolution optical microscopy: Multiple choices," *Current Opinion in Chemistry and Biology*, vol. 14, no. 1, pp. 10–14, 2010.

[30] G. Dardikman-Yoffe and Y. C. Eldar, "Learned SPARCOM: Unfolded deep super-resolution microscopy," *Optical Express*, vol. 28, no. 19, pp. 27 736–27 763, 2020.

[31] E. Nehme, L. E. Weiss, T. Michaeli, and Y. Shechtman, "Deep-STORM: Super-resolution single-molecule microscopy by deep learning," *Optica*, vol. 5, no. 4, pp. 458–464, 2018.

[32] B. Furlow, "Contrast-enhanced ultrasound," *Radiologic Technology*, vol. 80, no. 6, pp. 547S–561S, 2009.

[33] R. J. Van Sloun, R. Cohen, and Y. C. Eldar, "Deep learning in ultrasound imaging," *Proceedings of the IEEE*, vol. 108, no. 1, pp. 11–29, 2019.

[34] O. Bar-Shira, A. Grubstein, Y. Rapson, D. Suhami, E. Atar, K. Peri-Hanania, R. Rosen, and Y. C. Eldar, "Learned super resolution ultrasound for improved breast lesion characterization," *arXiv:2107.05270*, 2021.

[35] X.-P. Zhang, "Thresholding neural network for adaptive noise reduction," *IEEE Transactions on Neural Networks*, vol. 12, no. 3, pp. 567–584, 2001.

[36] M. Doneva, "Mathematical models for magnetic resonance imaging reconstruction: An overview of the approaches, problems, and future research areas," *IEEE Signal Processing Magazine*, vol. 37, no. 1, pp. 24–32, 2020.

[37] C. Qin, J. Schlemper, J. Caballero, A. N. Price, J. V. Hajnal, and D. Rueckert, "Convolutional recurrent neural networks for dynamic MR image reconstruction," *IEEE Transactions on Medical Imaging*, vol. 38, no. 1, pp. 280–290, 2018.

[38] A. Pramanik, H. Aggarwal, and M. Jacob, "Deep generalization of structured low-rank algorithms (Deep-SLR)," *IEEE Transactions on Medical Imaging*, vol. 39, no. 12, pp. 4186–4197, 2020.

[39] G. Ongie, S. Biswas, and M. Jacob, "Convex recovery of continuous domain piecewise constant images from nonuniform Fourier samples," *IEEE Transactions on Signal Processing*, vol. 66, no. 1, pp. 236–250, 2017.

[40] G. Ongie and M. Jacob, "A fast algorithm for convolutional structured low-rank matrix recovery," *IEEE Transactions on Computational Imaging*, vol. 3, no. 4, pp. 535–550, 2017.

[41] Y. Han, L. Sunwoo, and J. C. Ye, "k-Space deep learning for accelerated MRI," *IEEE Transactions on Medical Imaging*, vol. 39, no. 2, pp. 377–386, 2019.

[42] S. Boyd, N. Parikh, E. Chu, B. Peleato, J. Eckstein *et al.*, "Distributed optimization and statistical learning via the alternating direction method of multipliers," *Foundations and Trends® in Machine Learning*, vol. 3, no. 1, pp. 1–122, 2011.

[43] H. K. Aggrawal, M. P. Mani, and M. Jacob, "MoDL: Model-based deep learning architecture for inverse problems," *IEEE Transactions on Medical Imaging*, vol. 38, no. 2, pp. 394–405, Feb 2019.

[44] J. Friedman, T. Hastie, R. Tibshirani *et al.*, *The Elements of Statistical Learning*. Springer Series in Statistics, 2001, vol. 1, no. 10.

[45] S. Sonoda and N. Murata, "Neural network with unbounded activation functions is universal approximator," *Applied and Computational Harmonic Analysis*, vol. 43, no. 2, pp. 233–268, 2017.

[46] R. Ahmad, C. A. Bouman, G. T. Buzzard, S. Chan, S. Liu, E. T. Reehorst, and P. Schniter, "Plug-and-play methods for magnetic resonance Imaging: Using denoisers for image recovery," *IEEE Signal Processing Magazine*, vol. 37, no. 1, pp. 105–116, 2020.

[47] K. Dabov, A. Foi, V. Katkovnik, and K. Egiazarian, "Image denoising by sparse 3D transform-domain collaborative filtering," *IEEE Transactions on Image Processing*, vol. 16, no. 8, pp. 2080–2095, 2007.

Part II

Deep-Learning Architecture for Various Imaging Architectures

Part II

Deep-Learning Architecture for Various Imaging Architectures

5 Deep Learning for CT Image Reconstruction

Haimiao Zhang, Bin Dong, Ge Wang, and Baodong Liu

5.1 General Background

X-ray computed tomography (CT) is widely used in image-guided surgery and noninvasive diagnosis. It has various imaging modalities such as limited-angle, sparse-view, interior-CT, and low-dose CT. In the past few decades, research works on high-quality CT image reconstruction techniques have developed rapidly. Manually designed (or handcrafted) and deep-learning models are two mainstream approaches. In this chapter, we adopt X-ray CT imaging as the background task for reviewing the development of image reconstruction techniques.

The CT imaging problem can be written as

$$y = Pu + \eta, \tag{5.1}$$

where y is the measured projection data, P is the imaging system, with its specific form determined by the scanning geometry, and η is the noise or measurement error in the projection data. The mathematical difficulties in solving this image reconstruction problem have several aspects. On the one hand, when the dimension of the measured data is less than the number of pixels in the unknown image, resolving u from y is an ill-posed inverse problem. Therefore, the solution u cannot be uniquely and stably determined. When the discrete imaging operator P is a nonsingular, high-dimensional, and sparse matrix, in exceptional cases, there is still no fast algorithm for computing the matrix inversion P^{-1}. On the other hand, an algebraic formulation provides the flexibility to design a hybrid imaging system that compensates for incomplete measurement, irregular scanning geometry, detector sensitivities, and reweighting for inconsistencies.

For the ill-posed inverse problem, priors and regularization are commonly adopted to constrain the reconstructed image u. Some well-known examples are transformed-domain sparsity and smoothness, the low-rank property, and structure similarity. However, the image prior is data dependent and challenging to devise.

Given a prior assumption on the statistics of η, the image-reconstruction problem can be solved by formulating an optimization problem with numerical algorithms. The complex imaging system produces combined measurement noise and error. Also, it is tricky to describe mathematically and accurately the imaging system P. These features will induce structured artifacts in the reconstructed CT images and disturb the diagnosis.

Data-driven modeling is appealing for the design of new image-reconstruction methods. Dictionary learning was an early attempt to utilize the data so as to assist handcrafted modeling. However, handcrafted and data-driven modeling are not time efficient during practical usage. In recent years, the paradigm has shifted to the use of deep learning to improve the quality of the reconstructed image. Deep-learning models have a noticeable advantage over handcrafted models in a nonideal data measurement condition. However, we notice that purely artificial neural network models are sensitive to noise in the data and to the hyperparameter setting. In addition, they have a decline in a performance on out-of-distribution (OOD) data. Fusing handcrafted modeling and data-driven modeling is an appealing methodology to benefit from both handcrafted and deep models [1].

Nowadays, deep-learning models have been widely developed for CT image processing (i.e., denoising, deblurring, and super-resolution) and analysis (i.e., segmentation, region of interest (ROI) detection, and classification). New data processing techniques and impressive image quality improvement are obtained from the paradigm shift from handcrafted modeling to deep modeling. Several research groups have provided new perspectives and have pointed out the challenge and opportunities (or new research directions) in the deep modeling of medical imaging [2–9]. The promising developments in data-driven CT image-enhancement methods reveal the applicability of deep learning for image reconstruction.

5.2 Major Problems and Deep Solutions

Radiation dose reduction is a public concern connection with CT image-based diagnosis. Different image reconstruction or enhancement methods were investigated for decades with the aim of improving image quality while keeping the radiation dose as low as reasonably achievable (the ALARA principle). Incomplete data and a low tube current are feasible approaches in low-dose CT. For incomplete data CT, there are different imaging formats that include sparse-view, limited angle, interior tomography, and exterior tomography. These projection data measurement schemes can not only accelerate the scanning speed but also reduce the radiation dose. For low-tube-current CT, the measured projection has a lower signal-to-noise ratio (SNR). However, these features will degrade the reconstructed images whenever classic methods such as filtered backprojection (FBP) and algebraic reconstruction techniques (ART) were adopted. The enormous success of deep-learning-based models (deep models for short) in computer vision, audio, and natural language processing hopefully improve the quality of CT images while complying with the ALARA principle.

In the following, we discuss the major problems in CT imaging that are solved by deep models. Each specific task is considered separately, to clarify the primary technique. Some cases might combine two or more tasks, such as the limited angle + interior tomography in cardiac CT [10] and sparse-view + interior tomography for fast scanning and low-dose X-rays [11]. We will refer to these cases appropriately during the following discussion.

5.2.1 Low-Dose CT Denoising

Reducing the radiation dose may inevitably lead to increased noise and artifacts in CT images. In low-dose CT (LDCT), the most prominent artifacts are induced by quantum noise and readout noise in the projection domain. Figure 5.1 shows the noise patterns at different radiation doses. It can be observed that the structured noise (or error) is distinct from the normally distributed random noise. A deep neural network (DNN) can enhance the quality of LDCT images by approximating the mapping between LDCT and normal-dose CT (NDCT). The traditional image denoising methods are based on the local or nonlocal structures of the noisy input image and do not require a ground truth. However, manually tuned models and hyperparameters prohibit their wide use in practice. In recent years, data-driven models, especially DNNs, have been introduced into LDCT denoising and have gained promising quality improvement. These methods can be categorized into two classes: (1) image-domain post-processing and (2) sinogram-domain filtering and reconstruction. The following paragraphs will review the details.

Image Post-Processing

Since CT images can be more easily accessed from the picture-archive-and-communication system (PACS) than raw projection data, to develop the image-domain post-processing approach to enhance the quality of the LDCT image is straightforward.

When deep-learning models are adopted to denoise the LDCT, it is necessary to prepare a large-scale task-specific dataset to optimize the neural network's model param-

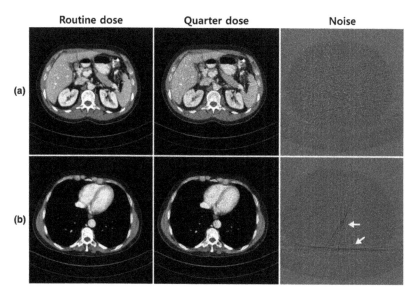

Figure 5.1 Various noise patterns in low-dose CT images: (a) Gaussian noise and (b) streaking artifacts. © 2017 John Wiley and Sons. Reprinted, with permission, from [12]

eters. Various neural network architectures have been used to reduce the noise and artifacts in LDCT. Ge Wang's group was the earliest to learn the mapping between LDCT to NDCT by a vanilla convolutional neural network (CNN) [13, 14]. Since vanilla CNN is sensitive to the parameter initialization, Chen et al. [15] proposed RED-CNN, which embeds shortcut connections into the encoder–decoder convolutional neural network. This residual learning architecture helps to alleviate the gradient vanishing and exploding problem during the training of deep models. Then, Kang et al. [12, 16] combined the multiscale wavelet decomposition of LDCT and residual structures to design the NN architecture. Fan et al. [17] introduced quadratic neurons to build an encoder–decoder structured NN for LDCT denoising. The generative adversarial networks (GAN) model [18] is capable of resembling the distribution of real data. Wolterink et al. [19] adopted a GAN model to suppress the artifacts in LDCT images. Even though a mean squared error (MSE) loss is commonly adopted in deep models for LDCT denoising, it is compromised in such a way as to preserve tiny structures in the denoised images. The authors of [20, 21] introduced a hybrid loss function that is composed of the Wasserstein distance and the perceptual similarity between the recovered image and the ground-truth NDCT. A self-attention mechanism was introduced in the CNN architecture to utilize the intra-slice and inter-slice dependencies in low-dose three-dimensional (3D) CT denoising [22]. We should conclude from these works that a proper NN architecture is crucial to overcoming the difficulties in optimizing deep models' parameters and improving their performance in LDCT denoising.

Raw-to-Image

Whenever low-dose projection data is available, it is reasonable to directly reconstruct CT images from the raw data. However, the reconstruction algorithm is sensitive to the noise from the projection domain and is easily influenced by manually chosen hyperparameters. On the one hand, the noise in the projection domain can induce severe artifacts in the reconstructed image. Therefore, it is desirable to enhance the raw data before reconstruction [23, 24] or design more data-adaptive image priors to regularize the reconstructed CT image [25]. For example, the authors of [23] adopted the CNN to progressively denoise the raw data in the projection domain and then enhanced the subsequently reconstructed image from FBP. On the other hand, semi-supervised learning was introduced by [24] to restore the sinogram and was then followed by FBP reconstruction. Wu et al. [25] adopted the K-sparse autoencoder (KSAE) to learn an image prior over the NDCT dataset through unsupervised learning; they then plugged this image prior into the iterative reconstruction algorithm. On the other hand, adequately selected hyperparameters guarantee the algorithm to be convergent and produce high-quality reconstruction. He et al. [26] proposed optimizing the hyperparameters in the unrolled alternative-directional method of multipliers (ADMM) algorithm and learned the image prior from the training dataset through a convolutional neural network for LDCT image reconstruction. By incorporating the physical-imaging-process knowledge and prior information into the raw-to-image model, deep models enhance the reconstructed LDCT image quality and are robust to the measurement noise. However, increased computational complexity also follows.

5.2.2 CT Image Super-Resolution

The resolution of a CT image is commonly constrained by the X-ray focal spot size, detector element pitch, and reconstruction algorithms. These limitations are equivalent to sampling a lower-resolution projection or thicker slices along the z-axis of the scanned patients. Even though hardware improvement can provide high-resolution projection data, the scanning time and reconstruction algorithm's computational budget increase along with the dimensions of the projection data and CT image. The conventional image reconstruction algorithms will smear the radiomic features. In recent years, deep learning has introduced to deblur low-resolution CT images by super-resolving the sinogram or post-processing the CT images from conventional algorithms. For example, Park et al. [27] adopted a U-Net to model the CT image super-resolution problem. The thin-slice CT image can be enhanced by this model from the thicker-slice data sequence. You et al. [28] adopted a cycle-consistency GAN to approximate the mapping between low-resolution CT images and the denoised and deblurred high-resolution images. Tang et al. [29] proposed super-resolving the low-resolution projection data in the raw-data domain by training a GAN model and reconstructing the CT image by filtered back projection (FBP). These deep models defined either in the sinogram domain or image domain for CT super-resolution are challenged by the requirement to preserve the fine details in the super-resolved CT images.

5.2.3 Limited Angle, Sparse-View, Interior CT

Image reconstruction problems such as limited angle, sparse-view, and interior CT are closely related tasks. Figure 5.2 shows the scanning modality of these tasks. The imaging model can be mathematically written into a uniform formula as in Eq. (5.1). Based on this fact, the CT image reconstruction algorithm can be derived from the following optimization model:

$$\min_{u} \frac{1}{2} \|Pu - Y\|^2 + \lambda R(u), \qquad (5.2)$$

where the first term constraints the reconstructed image u to be consistent with the measured projection data Y. The second term $R(\cdot)$ is chosen to incorporate the hypothesis on the reconstructed image u such as transform-domain sparse representation, smoothness, and global–local structure similarity.

Classical CT image reconstruction methods are typically handcrafted with expert knowledge on image and noise gained from theoretical analysis or experience. This class of modeling methodology is suitable for solving a wide range of CT image-reconstruction problems. In recent years, deep models have been introduced into CT imaging to mitigate the difficulties encountered in traditional modeling methods, for example parameter selection, regularization design, and data consistency constraints. Pure neural network models such as the automated-transform-by-manifold approximation (AUTOMAP) [30] approximate the mapping between the measured

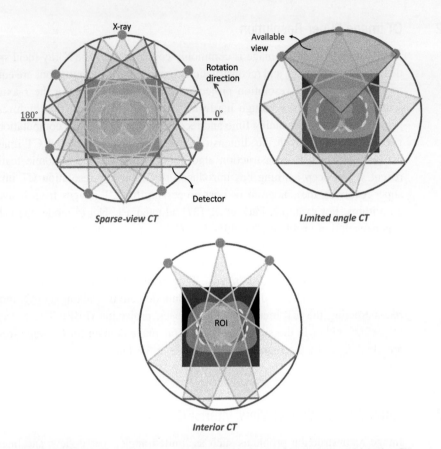

Figure 5.2 Illustration of object scanning: sparse-view, limited angle, and interior CT.

projection data and the reconstructed image. Knowledge of the imaging system is not needed during the modeling process. The authors of [31] mimicked filtered back-projection to design a neural network for directly reconstructing CT images from measured projection data. However, DNNs must be trained on the new dataset with task-specific scanning geometry. Another popular way to build a deep model is to unroll a handcrafted iterative algorithm as the basic building block of a DNN. As a result, the hyperparameters, image priors, and data consistency constraints are regressed concurrently with the parameterized mathematical operators in the backbone on a task-specific dataset. For example, Yang *et al.* [32] proposed learning the regularization parameters and image priors in the ADMM algorithm. Adler and Öktem [33] parameterized the proximal operators in the primal–dual hybrid gradient (PDHG) algorithm by a CNN. Zhang *et al.* [34] adopted a hypernetwork to predict the initialization of the conjugate gradient algorithm involved in the unrolled half-quadratic splitting algorithm. A hybrid loss function was designed in [35] to boost the performance of the unrolled iterative algorithm of the joint spatial-Radon-domain image-reconstruction model [36]. These deep models provide a general scheme to solve each inverse problem in limited angle, sparse-view, and interior CT.

Deep models with task-specific knowledge can improve the quality of reconstructed CT images. Below, we review task-specific deep models for limited angle CT, sparse-view CT, and interior CT.

Limited Angle CT

In limited angle CT (LACT), the acquired data covers a short angular range over the object. This scanning geometry is commonly encountered in radiation dose reduction, C-arm CT, and dental CT. The incomplete projection data brings challenges to image reconstruction. Models and algorithms for LDCT can be categorized as handcrafted and data-driven methods. The handcrafted models are conventionally addressed by an extrapolation technique and solved by an iterative reconstruction algorithm. When the image-reconstruction problem is formulated as an optimization–variational model, an image prior is necessary to regularize the feasible solution. Examples of image priors are the total variation, dictionary learning, and data-driven tight wavelet frame [37–42]. Handcrafted models and those using manually tuned hyperparameters or prior hypothesis are not data-efficiently adaptable to various data distributions.

Deep models provide more general image regularization than the traditional handcrafted image priors on a dataset. Significant image quality improvements are obtained from the deep models in LACT. Extrapolation of the missing projection in the Radon domain is another big challenge in LACT. Anirudh *et al.* [43] proposed completing the limited angle projection to full view data with a fully convolutional one-dimensional CNN. Analytical and iterative algorithms were adopted to reconstruct the image. Würfl *et al.* [44] proposed learning the compensation weights and apodization windows in limited angle cone-beam CT (CBCT) reconstruction. Huang *et al.* [45] introduced the data fidelity constraint to assist the data completion in LACT. Bubba *et al.* [46] fused the conventional sparse regularization model with deep learning to learn the invisible singularities data in LACT. Wang *et al.* [47] trained an extrapolation network to predict the projection and reconstructed the CT image via an unrolled-dynamics neural network model named MetaInv-Net [34]. Despite the excellent performance of these deep models on LACT, they are difficult to generalize on an OOD-dataset and scanning setting. For LACT, the challenging case occurs when the measured projection has a much narrow scanning range and sparsely sampled view. Thus, deep models will struggle to produce a reliable reconstruction.

Sparse-View CT

Sparse-view CT (SVCT) is implemented in clinical applications to reduce the radiation dose and speed up the scanning. Unfortunately, analytical methods lead to severe streaking artifacts in the reconstructed image owing to the undersampled projection data. Various iterative approaches have been developed to mitigate the challenge of solving the ill-posed inverse problem for SVCT image reconstruction and suppressing the artifacts. However, the heavy computation and the burden of parameter selection prevent the use of iterative algorithms in practice. Deep-learning models can partially overcome these challenges while producing images with significantly enhanced quality. Current deep neural network (DNN) models for SVCT can be categorized into three groups: pre-processing, post-processing, and raw-to-image.

In the pre-processing approach, the sparsely sampled projection is interpolated by a CNN and then reconstructed by FBP [48, 49] or the Feldkamp, Davis, and Kress (FDK) algorithm [50]. Since the analytical reconstruction algorithm is sensitive to errors in the sinogram, new artifacts are inevitably introduced by the pre-processing approach.

A DNN was commonly adopted in the post-processing approach to reduce the streak artifacts in an image initially reconstructed from conventional methods such as FBP and iterative reconstruction techniques. Examples of deep-learning-based post-processing models are U-Net [51, 52], DenseNet [53], and deep models inspired by the projected gradient descent algorithm [54]. Xie et al. [55] adopted a Wasserstein generative adversarial network (WGAN) to enhance the quality of sparse-view breast CT. Restricted by the limited information in the pre-reconstructed CT image, the post-processing approach cannot improve the image's overall quality, especially the local anatomical details. When the number of scanning views is changed, deep models need to be retrained in the target scanning geometry to produce satisfactory CT images.

The raw-to-image approach is more appealing for optimizing the sparse-view CT image reconstruction workflow systematically. The backbone architecture of the SVCT image reconstruction model can be built by unrolling the iterative reconstruction algorithm of optimization models such as the fields of experts (FoE) model [56], the iterative joint spatial-Radon-domain reconstruction model (JSR) [35], and the proximal forward–backward splitting algorithm [57]. Zhang et al. [34] introduced the meta inversion network (MetaInv-Net) with a hypernetwork module in the unrolled half-quadratic splitting (HQS) algorithm to predict the initialization for one of the subproblems. This type of image reconstruction model is known as an unrolled-dynamics model.

Deep models for SVCT are currently widely studied at more than tens of scanning views through different modeling methodologies. The digital breast tomosynthesis (DBT) task is a challenging example of SVCT. Less than ten projection views are measured, while tiny structures and accurate geometric shapes are needed in diagnosis and volume computation. Deep modeling is a potential tool for solving the SVCT problem in DBT.

Interior CT

The CT image reconstruction problem reduces to interior tomography when the detector length is not large enough to cover the scanned object. The measured projection covers a subregion (the region of interest, ROI) of the object as shown in Fig. 5.2. Classical analytical methods (i.e., FBP and ART) for reconstructing the image suffer from severe cupping artifacts (see Fig. 5.3) due to the existence of null space in the projection domain (i.e., the Radon transform in parallel-beam CT scanning). Whenever the property of a subregion inside the ROI is known, we can obtain an exact solution of the interior tomography problem [58–62]. Another way to improve the quality of interior CT images is to develop new data-driven models such as DNN-based image processing. For example, in [63], a U-Net was adopted to remove cupping artifacts in CT images from FBP reconstruction. Li et al. [64] proposed a raw-to-image model that

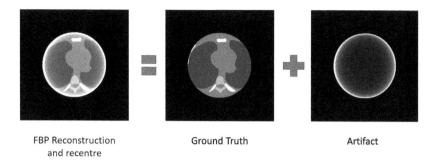

FBP Reconstruction and recentre Ground Truth Artifact

Figure 5.3 Interior CT. From left to right: FBP reconstruction, ground-truth image, and artifact image.

mimicked the FBP-based image reconstruction procedure to design an intelligent-CT neural network (iCT-Net) for sparse-view interior tomography. This model removed the cupping artifacts and preserved the fine structures in sparse-view interior tomographic images. When designing new deep-learning models for interior CT, we should consider image-boundary-preservation mechanisms to avoid introducing new artifacts.

5.2.4 Spectral CT

Spectral and photon-counting CT is a novel imaging modality that enhances the monochromatic X-ray CT with an energy-dependent attenuation characteristic [65]. It resolves not only the spatial structure but also the spectral property of the scanned object. There are various potential applications of spectral CT, such as in reducing metal artifacts, enhancing image contrast, and material decomposition (i.e., soft tissues, bones, or contrast agents). However, there still exist some challenges in spectral CT imaging, such as scattering, energy dependency, and complex noise. The scattered signal will degrade the reconstructed image. To modeling the scattering effect, Monte Carlo simulation (MCS) is commonly adopted in practice. However, the use of MCS to estimate the scatter background is time consuming. A photon-counting detector in spectral CT will lead to serious noise in projection at each energy bin. Owing to the energy-dependent attenuation and data correlation, material decomposition is a challenging ill-posed inverse problem. In addition to the issues just mentioned, ring artifacts and beam-hardening effects commonly appear in material decomposition tasks.

Recently, deep learning has been introduced in spectral CT imaging for scatter correction in the projection domain [66], material decomposition in the spectral CT image [67–69], spectral CT image reconstruction [70], and monochromatic image synthesis [71]. These deep models accelerate image reconstruction and improve the quality. However, it is challenging to prepare the training data in practice. Numerical simulation is a possible solution, but it requires heavy computation and might be biased to the simulated spectral CT system.

5.2.5 Metal-Artifacts Reduction

In clinics, metal implants frequently appear in the patient's body. However, owing to the high attenuation of metal components compared with tissue, metal-induced artifacts (metal artifacts for short) will degrade the fine structure in reconstructed CT images [72]. Hardware development (i.e., dual-energy and spectral CT) and software innovations are practical ways to solve the metal-artifact reduction (MAR) task. The currently developed MAR models can be divided into three categories: (1) model-based iterative algorithms [73–76], (2) image domain post-processing, and (3) projection completion methods. However, iterative reconstruction is time consuming and needs tedious hyperparameter tuning. Thus, it is inefficient for processing a large-scale dataset.

In image-domain post-processing, Gjesteby *et al.* [77, 78] trained a CNN to learn the mapping between the normalized MAR reconstructed image and the simulated metal-artifact-free image. A perceptual loss function was introduced in [79] to improve further the detail-preservation of the learned CNN model. Generative adversarial networks were adopted in [80, 81] to remove metal artifacts in the image domain. The post-processing approach has some drawbacks during the MAR process. On the one hand, its performance is restricted by limited knowledge of the initially reconstructed image. On the other hand, the post-processing approach can induce new artifacts and blurry patterns in the enhanced image.

In projection-domain processing methods, the metal trace (the support set of the projection of metal components in the sinogram domain) was completed using different strategies based on the learned prior image [82–85] and the metal-artifacts-free CT image was reconstructed from the complete projection data by FBP. In addition, Wang *et al.* [86] proposed an unrolled dynamics model, named an interpretable dual-domain network (InDuDoNet), to reconstruct a metal-artifacts-free image from the projection data. The projection completion approach provides a flexible scheme for fully utilizing the hypothesis on the artifact induction mechanism in both the projection and image domains. The challenge in MAR is how to preserve the structure of tissue or bone around the metal region and keep the global contrast during image reconstruction.

5.2.6 Motion-Artifacts Reduction

Motion artifacts commonly appear in dental CT and cone beam (CB)CT, wherein the protocol requires a long scanning time. The reconstructed CT image is degraded by blurring and double edge artifacts induced by both rigid and non-rigid motion. Motion-artifacts reduction can be realized by image registration, fiducial marker estimation, fast scanning, multi-source-detector systems [87, 88], and patient motion estimation and compensation [89–91]. Even though image registration showed good motion-compensation performance, it is computationally expensive. To save the time consumption on placing the external markers, the authors of [92] adapted a CNN to predict the anatomical landmarks for motion compensation. The recently proposed deep-learning models for motion-artifact reduction were equipped with new DNN architectures such as deep residual with attention module networks [93] and Wasserstein generative adversarial networks [94]. Since deep-learning models are sensitive to weight initialization and data distribution, to reduce the motion artifacts induced by large position shifts and dataset distribution shift is challenging.

5.3 Deep-Learning-Empowered Dedicated Systems

A dedicated CT imaging system is more promising if empowered with deep learning than the general CT image-reconstruction methods. Owing to the powerful representation-learning property, deep-learning models can be utilized to optimize the whole workflow for a specific imaging system. This section reviews the representative CT imaging systems such as C-arm CT, dental CT, cabin CT, CT of Covid-19, and computed laminography (CL).

5.3.1 C-Arm CT

C-arm CT is a useful high-contrast imaging modality. Nowadays, C-arm CBCT has entered the clinical routine for 3D angiography to visualize low-contrast matter such as the complex cerebral vasculature. The main challenge in C-arm CT is the system geometric-parameter calibration and motion compensation. The authors of [95, 96] adopted a DNN to generate 3D cerebral angiograms from a single contrast-enhanced C-arm CBCT. This model has the potential to reduce misregistration and also to reduce the radiation dose in 3D angiography. To reduce the artifacts during the in-planar scanning trajectory in C-arm CBCT, the authors of [97] proposed autonomously adjusting the out-of-plane scanning trajectory to that predicted by a DNN. When the scanning trajectory is complex or irregular, one must train a new deep model to match the changed imaging system.

5.3.2 Dental CT

The CBCT is a conventional method to determine the 3D structure of the teeth for treatment planning. The challenge of adopting dental CT in the clinic is the poor spatial resolution and the metal artifacts. While micro-CT or μCT provides a more precise measurement of the root canal morphology, it brings high radiation dose, long acquisition time, and size limitation. There are several ways to improve the quality of dental CT images, for example, deep learning for dental CT super-resolution [98] and for metal-artifacts reduction in dental CT [99], and the low-dose dental CT image enhancement by WGAN [100]. Currently, deep models are mainly focused on a specific stage involved in the overall dental CT system. The challenge of dental CT is how to solve the hybrid imaging tasks such as metal-artifacts reduction, super-resolution, and low-dose CT image reconstruction. Each image processing stage should be carefully devised to balance the efficiency and efficacy of the adopted models and algorithms. In the ideal case, deep-learning-based models for dental CT should be built in an end-to-end format so that the imaging workflow can be flexibly and thoroughly optimized.

5.3.3 Cabin CT

The pandemic of the novel coronavirus disease 2019 (Covid-19) has infected people around the world. Cabin hospitals played a vital role in effectively screening infection and cutting off the transmission channels of Covid-19. The equipped cabin CT in a mobile emergency hospital is more flexible, fast, and convenient than the conventional CT room in community hospitals. Cabin CT can be used to evaluate the severity and the effects of therapeutic interventions of Covid-19 and prevent cross-infection between patients and medical staff. The authors of [101] introduced the workflow of the standardized cabin CT scan in module hospitals. With the equipped 5G network communication system in cabin CT, data post-processing and diagnosis can be conveniently implemented remotely [102]. For cabin CT, the challenge is efficiently transferring a massive amount of patient data with low cost and reliable privacy preservation. Cabin CT will benefit from the development of new techniques in data compression and privacy-preservation.

5.3.4 CT for Covid-19

Covid-19 is a threat to public health. X-ray-based chest CT scanning and chest radiographs provide helpful imaging clues for the control or diagnosis of Covid-19; CT screening provides a 3D pulmonary view while chest X-rays are more affordable and widely available. However, the manual delineation of lung infections is tedious and time consuming. Deep learning holds great promise of mitigating the heavy workload of radiologists in diagnosing Covid-19 from different perspectives. For example, deep-learning techniques have been introduced efficiently into diagnosis, data analysis, and screening using chest X-ray images [103–106] and CT images [107–113]. Recently, a "square cabin" hospital in China equipped with a deep-learning system was used in the management and discharge of Covid-19 patients [114].

The rapid development of deep-learning techniques for Covid-19 has provided new tools of defense against the pandemic [115–117]. However, physicians and researchers should be careful to implement deep-learning models for diagnosis and prognostication that are based on automatic radiological data analysis [118]. Deep-learning models provide flexible schemes to quantify the uncertainty of reconstructed images and the image analysis results.

5.3.5 Computed Laminography

Computed laminography (CL) is a noninvasive imaging method for planar objects with large aspect ratios such as printed circuit boards (PCB), fossils, and paintings [119].

Conventional reconstruction methods such as FBP and FDK are sensitive to measurement error, which easily leads to artifacts in the CT image. It is necessary to correct the projection data [120, 121] and geometric error [122] during CL image reconstruction; CL imaging can adopt a sparse-view trajectory to accelerate the object scanning. An incomplete measurement will lead to streak artifacts in the reconstructed image slices [123]. Recently, the authors of [124] utilized a 3D CNN architecture to enhance the quality of CL reconstruction that uses the conventional SIRT method. For CL data analysis, the authors of [125] adopted DeepLabv3+ [126] to segment the solder region and voids in insulated-gate bipolar transistors from the CL images. The complex and flexible scanning trajectory challenges deep models for CL reconstruction. It requires the trained deep model to be easily transferred or stably generalized to new scanning geometry. Since CL has a lower z-axis resolution than on the xy planes, a deep model for CL image analysis should be adaptable to different slices.

5.4 Data Synthesis and Transfer

Data synthesis and transfer is essentially an image-to-image translation model. It aims at learning a mapping from the reference domain (reference modality) to the target domain (target modality) while preserving the high-level semantic contents. There are various applications of data synthesis in medical imaging such as image registration [127, 128] and fusion [129–131]. A deep-learning model can be adopted to align the target image to the reference image for image registration. Furthermore, owing to the data processing efficiency of deep learning, the computational demands are reduced in volume data processing and in real-time image guidance for surgery [132, 133].

Data transfer can also be termed data synthesis or data generation, e.g., MRI to CT [134]. Cross-domain image translation models were adopted in [135, 136] to map label-rich data to label-poor data, e.g., spleen MRI to CT and brain CT to MRI. Regarding the cycle-consistent generative adversarial network (cycleGAN) [137], there is a list of works on synthesizing novel modality images. For example, the authors of [138] proposed synthesizing cardiac MR images from CT images, and in [139] mapping the brain MR images to CT was proposed. To explicitly preserve anatomical structure, the authors of [140] combined the image synthesis and segmentation tasks into one deep model and trained it end-to-end.

The above deep-learning-based data-transfer models are commonly trained from scratch with different network architectures to represent the mapping between the reference and target domains. Images from different imaging modalities and parts of the body were processed separately by a specific deep model. It is interesting to clarify whether a single deep model can process the multi-organ and multi-modality data-transfer task in a unified framework. A more general challenge in transfer learning is how to transfer a trained deep model to new tasks or novelty imaging modalities with a small batch of training data.

5.5 Important Topics

In addition to the topics on deep-learning techniques for CT imaging tasks (i.e., image reconstruction and image analysis), some other general issues are closely related to the development of deep-learning models.

The publicly available large-scale dataset is an indispensable source for rapidly developing deep-learning models for CT imaging [141]. A widely used public CT datasets are the Mayo Low-Dose X-ray CT Grand Challenge dataset [142] and the DeepLesion dataset [143]. Free software accelerates the devising of more advanced image processing algorithms. The operator discretization library (ODL) is one of the commonly used Python packages that defined the trainable operators needed in deep-learning models, i.e., the Radon transform in CT imaging [144]. The ASTRA toolbox [145] boosts computation of the forward and backward Radon transforms with a graphics processing unit (GPU).

Important concerns in using deep-learning models in clinical applications include but are not limited to generalization, OOD detection [146], stability, interpretability, and ethical artificial intelligence (AI).

Generalization and the question of out-of-distribution data are closely related topics in deep modeling. Let P_A and P_B be two distinct data distributions. Assume the neural network model $\mathcal{F}_\Theta$ is trained on a dataset sampled from P_A. Then, P_A and P_B are called in-distribution and out-of-distribution, respectively. Modern deep-learning models generalize well on the in-distribution data at the test phase. It is necessary to investigate the generalization property to check that deep models can be confidently deployed in clinical practice [147, 148]. There are different perspectives for understanding the generalization of deep-learning models, such as experimental analysis [147], theoretical analysis [149, 150], and domain generalization through data augmentation and domain alignment [151]. The OOD sample detection is a helpful technique to build reliable deep-learning models; OOD detectors can be built with modern deep modeling techniques such as GAN [152], convolutional neural networks [153], and transfer-representation learning [154]. These models require modifying or retraining the DNNs. The authors of [146] provided a method to detect OOD data without retraining the deep model. However, a unified mathematical analysis framework is needed to understand and solve the OOD transfer problem in deep learning.

As verified by the numerical experiments in [155], the stability of the deep-learning model has the following aspects: (1) a tiny structure is challenging to reconstruct, (2) each failure of a deep-learning model leads to different structure changes, (3) deep models should be trained for each specific imaging setting. The deep-learning and medical imaging communities make significant efforts to mitigate these challenges. Coordinate encoding through Fourier feature mapping enables the deep-learning model to efficiently represent tiny structures in the image [156]. In addition, in [157] a new deep-learning framework to stabilize the tomographic reconstruction was proposed. The stabilization of medical imaging models will benefit from the rapid development of new deep modeling techniques in computer vision and graphics and classical mathematical imaging theory such as compressed sensing.

Deep models have achieved remarkable progress in the field of medical image processing and analysis. However, these advanced models lack interpretability when employed in medical diagnosis and therapy planning. In [158, 159] the authors presented a very detailed survey on recent developments in the interpretability of artificial neural networks. The mathematics of causal inference and causal analysis [160] provide a potential tool to understand the interpretability problem in deep modeling. Uncertainty quantification can assist understanding of the output of deep models in medical imaging.

Artificial intelligence has a great potential to improve the efficiency and accuracy of diagnoses based on medical images. However, ethical and responsibility issues should be carefully considered during clinical use of an autonomous AI system [161, 162]. Even though the current deep-learning models have several weaknesses, the situation will change with the rapid development of deep-learning theory. Soon, AI systems will play an essential role in improving the quality of public health care [163] and increasing human well-being.

References

[1] G. Wang, J. C. Ye, K. Mueller, and J. A. Fessler, "Image reconstruction is a new frontier of machine learning," *IEEE Transactions on Medical Imaging*, vol. 37, no. 6, pp. 1289–1296, 2018.

[2] G. Wang, "A perspective on deep imaging," *IEEE Access*, vol. 4, pp. 8914–8924, 2016.

[3] G. Wang, M. D. M. Kalra, and C. G. Orton, "Machine learning will transform radiology significantly within the next 5 years," *Medical Physics*, vol. 44, no. 6, pp. 2041–2044, 2017.

[4] M. T. McCann, K. H. Jin, and M. Unser, "Convolutional neural networks for inverse problems in imaging: A review," *IEEE Signal Processing Magazine*, vol. 34, no. 6, pp. 85–95, 2017.

[5] V. Monga, Y. Li, and Y. C. Eldar, "Algorithm unrolling: Interpretable, efficient deep learning for signal and image processing," *IEEE Signal Processing Magazine*, vol. 38, no. 2, pp. 18–44, 2021.

[6] G. Wang, Y. Zhang, X. Ye, and X. Mou, *Machine Learning for Tomographic Imaging*. IOP Publishing, 2019.

[7] H. Zhang and B. Dong, "A review on deep learning in medical image reconstruction," *Journal of the Operational Research Society of China*, vol. 8, pp. 311–340, 2020.

[8] G. Wang, J. C. Ye, and B. D. Man, "Deep learning for tomographic image reconstruction," *Nature Machine Intelligence*, vol. 2, p. 737–748, 2020.

[9] S. K. Zhou, H. Greenspan, C. Davatzikos, J. S. Duncan, B. van Ginneken, A. Madabhushi, J. L. Prince, D. Rueckert, and R. M. Summers, "A review of deep learning in medical imaging: Image traits, technology trends, case studies with progress highlights, and future promises," *arXiv:2008.09104*, 2020.

[10] B. Liu, G. Wang, E. L. Ritman, G. Cao, J. Lu, O. Zhou, L. Zeng, and H. Yu, "Image reconstruction from limited angle projections collected by multisource interior X-ray imaging systems," *Physics in Medicine and Biology*, vol. 56, no. 19, p. 6337, 2011.

[11] G. Wang, H. Yu, and Y. Ye, "A scheme for multisource interior tomography," *Medical Physics*, vol. 36, no. 8, pp. 3575–3581, 2009.

[12] E. Kang, J. Min, and J. C. Ye, "A deep convolutional neural network using directional wavelets for low-dose x-ray CT reconstruction," *Medical Physics*, vol. 44, no. 10, pp. e360–e375, 2017.

[13] H. Chen, Y. Zhang, W. Zhang, P. Liao, K. Li, J. Zhou, and G. Wang, "Low-dose CT via convolutional neural network," *Biomedical Optical Express*, vol. 8, no. 2, pp. 679–694, 2017.

[14] ——, "Low-dose CT denoising with convolutional neural network," in *Proc. International Symposium on Biomedical Imaging*. IEEE, 2017, pp. 143–146.

[15] H. Chen, Y. Zhang, M. K. Kalra, F. Lin, Y. Chen, P. Liao, J. Zhou, and G. Wang, "Low-dose CT with a residual encoder-decoder convolutional neural network," *IEEE Transactions on Medical Imaging*, vol. 36, no. 12, pp. 2524–2535, 2017.

[16] E. Kang, W. Chang, J. Yoo, and J. C. Ye, "Deep Convolutional Framelet Denosing for Low-dose CT via Wavelet Residual Network," *IEEE Transactions on Medical Imaging*, vol. 37, no. 6, pp. 1358–1369, 2018.

[17] F. Fan, H. Shan, M. K. Kalra, R. Singh, G. Qian, M. Getzin, Y. Teng, J. Hahn, and G. Wang, "Quadratic autoencoder (Q-AE) for low-dose CT denoising," *IEEE Transactions on Medical Imaging*, vol. 39, no. 6, pp. 2035–2050, 2019.

[18] I. Goodfellow, J. Pouget-Abadie, M. Mirza, B. Xu, D. Warde-Farley, S. Ozair, A. Courville, and Y. Bengio, "Generative adversarial nets," in *Advances in Neural Information Processing Systems*, 2014, pp. 2672–2680.

[19] J. M. Wolterink, T. Leiner, M. A. Viergever, and I. Išgum, "Generative adversarial networks for noise reduction in low-dose CT," *IEEE Transactions on Medical Imaging*, vol. 36, no. 12, pp. 2536–2545, 2017.

[20] H. Shan, Y. Zhang, Q. Yang, U. Kruger, M. K. Kalra, L. Sun, W. Cong, and G. Wang, "3-D convolutional encoder-decoder network for low-dose CT via transfer learning from a 2-D trained network," *IEEE Transactions on Medical Imaging*, vol. 37, no. 6, pp. 1522–1534, 2018.

[21] Q. Yang, P. Yan, Y. Zhang, H. Yu, Y. Shi, X. Mou, M. K. Kalra, Y. Zhang, L. Sun, and G. Wang, "Low-dose CT image denoising using a generative adversarial network with wasserstein distance and perceptual loss," *IEEE Transactions on Medical Imaging*, vol. 37, no. 6, pp. 1348–1357, 2018.

[22] M. Li, W. Hsu, X. Xie, J. Cong, and W. Gao, "SACNN: Self-attention convolutional neural network for low-dose CT denoising with self-supervised perceptual loss network," *IEEE Transactions on Medical Imaging*, vol. 39, no. 7, pp. 2289–2301, 2020.

[23] X. Yin, Q. Zhao, J. Liu, W. Yang, J. Yang, G. Quan, Y. Chen, H. Shu, L. Luo, and J.-L. Coatrieux, "Domain progressive 3D residual convolution network to improve low-dose CT imaging," *IEEE Transactions on Medical Imaging*, vol. 38, no. 12, pp. 2903–2913, 2019.

[24] M. Meng, S. Li, L. Yao, D. Li, M. Zhu, Q. Gao, Q. Xie, Q. Zhao, Z. Bian, J. Huang et al., "Semi-supervised learned sinogram restoration network for low-dose CT image reconstruction," in *Medical Imaging 2020: Physics of Medical Imaging*, vol. 11 312. International Society for Optics and Photonics, 2020, p. 113 120B.

[25] D. Wu, K. Kim, G. El Fakhri, and Q. Li, "Iterative low-dose CT reconstruction with priors trained by artificial neural network," *IEEE Transactions on Medical Imaging*, vol. 36, no. 12, pp. 2479–2486, 2017.

[26] J. He, Y. Yang, Y. Wang, D. Zeng, Z. Bian, H. Zhang, J. Sun, Z. Xu, and J. Ma, "Optimizing a parameterized plug-and-play ADMM for iterative low-dose CT reconstruction," *IEEE Transactions on Medical Imaging*, vol. 38, no. 2, pp. 371–382, 2019.

[27] J. Park, D. Hwang, K. Y. Kim, S. K. Kang, Y. K. Kim, and J. S. Lee, "Computed tomography super-resolution using deep convolutional neural network," *Physics in Medicine and Biology*, vol. 63, no. 14, p. 145011, 2018.

[28] C. You, G. Li, Y. Zhang, X. Zhang, H. Shan, M. Li, S. Ju, Z. Zhao, Z. Zhang, W. Cong et al., "CT super-resolution GAN constrained by the identical, residual, and cycle learning ensemble (GAN-Circle)," *IEEE Transactions on Medical Imaging*, vol. 39, no. 1, pp. 188–203, 2019.

[29] C. Tang, W. Zhang, L. Wang, A. Cai, N. Liang, L. Li, and B. Yan, "Generative adversarial network-based sinogram super-resolution for computed tomography imaging," *Physics in Medicine and Biology*, vol. 65, no. 23, p. 235006, 2020.

[30] B. Zhu, J. Z. Liu, S. F. Cauley, B. R. Rosen, and M. S. Rosen, "Image reconstruction by domain-transform manifold learning," *Nature*, vol. 555, no. 7697, p. 487, 2018.

[31] J. He, Y. Wang, and J. Ma, "Radon inversion via deep learning," *IEEE Transactions on Medical Imaging*, vol. 39, no. 6, pp. 2076–2087, 2020.

[32] Y. Yang, J. Sun, H. Li, and Z. Xu, "ADMM-Net: A deep learning approach for compressive sensing MRI," *arXiv:1705.06869*, 2017.

[33] J. Adler and O. Öktem, "Learned Primal-dual Reconstruction," *IEEE Transactions on Medical Imaging*, vol. 37, no. 6, pp. 1322–1332, 2018.

[34] H. Zhang, B. Liu, H. Yu, and B. Dong, "MetaInv-Net: Meta inversion network for sparse view CT image reconstruction," *IEEE Transactions on Medical Imaging*, vol. 40, no. 2, pp. 621–634, 2021.

[35] H. Zhang, B. Dong, and B. Liu, "JSR-Net: A deep network for joint spatial-Radon domain CT reconstruction from incomplete data," in *Proc. International Conference on Acoustics, Speech, and Signal Processing*. IEEE, 2019, pp. 1–5.

[36] B. Dong, J. Li, and Z. Shen, "X-ray CT image reconstruction via wavelet frame based regularization and Radon domain inpainting," *Journal of Scientific Computing*, vol. 54, no. 2, pp. 333–349, 2013.

[37] L. I. Rudin, S. Osher, and E. Fatemi, "Nonlinear total variation based noise removal algorithms," *Physica D*, vol. 60, no. 1, pp. 259–268, 1992.

[38] E. Y. Sidky and X. Pan, "Image reconstruction in circular cone-beam computed tomography by constrained, total-variation minimization," *Physics in Medicine and Biology*, vol. 53, no. 17, p. 4777–4807, 2008.

[39] J.-F. Cai, H. Ji, Z. Shen, and G.-B. Ye, "Data-driven tight frame construction and image denoising," *Applied and Computational Harmonic Analysis*, vol. 37, no. 1, pp. 89–105, 2014.

[40] R. Zhan and B. Dong, "CT image reconstruction by spatial-Radon domain data-driven tight frame regularization," *SIAM Journal of Imaging Science*, vol. 9, no. 3, pp. 1063–1083, 2016.

[41] T. Wang, K. Nakamoto, H. Zhang, and H. Liu, "Reweighted anisotropic total variation minimization for limited-angle CT reconstruction," *IEEE Transactions on Nuclear Science*, vol. 64, no. 10, pp. 2742–2760, 2017.

[42] F. Mahmood, N. Shahid, U. Skoglund, and P. Vandergheynst, "Adaptive graph-based total variation for tomographic reconstructions," *IEEE Signal Processing Letters*, vol. 25, no. 5, pp. 700–704, 2018.

[43] R. Anirudh, H. Kim, J. J. Thiagarajan, K. Aditya Mohan, K. Champley, and T. Bremer, "Lose the views: Limited angle CT reconstruction via implicit sinogram completion," in *Proc. IEEE Conference on Computer Vision and Pattern Recognition*, 2018, pp. 6343–6352.

[44] T. Würfl, M. Hoffmann, V. Christlein, K. Breininger, Y. Huang, M. Unberath, and A. K. Maier, "Deep learning computed tomography: Learning projection-domain weights from image domain in limited angle problems," *IEEE Transactions on Medical Imaging*, vol. 37, no. 6, pp. 1454–1463, 2018.

[45] Y. Huang, A. Preuhs, G. Lauritsch, M. Manhart, X. Huang, and A. Maier, "Data consistent artifact reduction for limited angle tomography with deep learning prior," in *Proc. International Workshop on Machine Learning for Medical Image Reconstruction*. Springer, 2019, pp. 101–112.

[46] T. A. Bubba, G. Kutyniok, M. Lassas, M. März, W. Samek, S. Siltanen, and V. Srinivasan, "Learning the invisible: A hybrid deep learning-shearlet framework for limited angle computed tomography," *Inverse Problems*, vol. 35, no. 6, p. 064002, 2019.

[47] C. Wang, H. Zhang, Q. Li, K. Shang, Y. Lyu, B. Dong, and S. K. Zhou, "Improving generalizability in limited-angle CT reconstruction with sinogram extrapolation," in *Proc. International Conference on Medical Image Computing and Computer-Assisted Intervention*. Springer, 2021.

[48] H. Lee, J. Lee, H. Kim, B. Cho, and S. Cho, "Deep-neural-network-based sinogram synthesis for sparse-view CT image reconstruction," *IEEE Transactions on Radiation and Plasma Medical Sciences*, vol. 3, no. 2, pp. 109–119, 2018.

[49] W. Cheng, Y. Wang, H. Li, and Y. Duan, "Learned full-sampling reconstruction from incomplete data," *IEEE Transactions on Computational Imaging*, vol. 6, pp. 945–957, 2020.

[50] D. Hu, J. Liu, T. Lv, Q. Zhao, Y. Zhang, G. Quan, J. Feng, Y. Chen, and L. Luo, "Hybrid-domain neural network processing for sparse-view CT reconstruction," *IEEE Transactions on Radiation and Plasma Medical Sciences*, vol. 5, no. 1, pp. 88–98, 2021.

[51] K. H. Jin, M. T. Mccann, E. Froustey, and M. Unser, "Deep convolutional neural network for inverse problems in imaging," *IEEE Transactions on Image Processing*, vol. 26, no. 9, pp. 4509–4522, 2017.

[52] Y. Han and J. C. Ye, "Framing U-Net via deep convolutional framelets: Application to sparse-view CT," *IEEE Transactions on Medical Imaging*, vol. 37, no. 6, pp. 1418–1429, 2018.

[53] Z. Zhang, X. Liang, X. Dong, Y. Xie, and G. Cao, "A sparse-view CT reconstruction method based on combination of densenet and deconvolution," *IEEE Transactions on Medical Imaging*, vol. 37, no. 6, pp. 1407–1417, 2018.

[54] H. Gupta, K. H. Jin, H. Q. Nguyen, M. T. McCann, and M. Unser, "CNN-based projected gradient descent for consistent CT image reconstruction," *IEEE Transactions on Medical Imaging*, vol. 37, no. 6, pp. 1440–1453, 2018.

[55] H. Xie, H. Shan, W. Cong, C. Liu, X. Zhang, S. Liu, R. Ning, and G. Wang, "Deep efficient end-to-end reconstruction (DEER) network for few-view breast CT image reconstruction," *IEEE Access*, vol. 8, pp. 196 633–196 646, 2020.

[56] H. Chen, Y. Zhang, Y. Chen, J. Zhang, W. Zhang, H. Sun, Y. Lv, P. Liao, J. Zhou, and G. Wang, "LEARN: Learned experts' assessment-based reconstruction network for sparse-data CT," *IEEE Transactions on Medical Imaging*, vol. 37, no. 6, pp. 1333–1347, 2018.

[57] Q. Ding, G. Chen, X. Zhang, Q. Huang, H. Ji, and H. Gao, "Low-dose CT with deep learning regularization via proximal forward backward splitting," *Physics in Medicine and Biology*, 2020.

[58] H. Yu and G. Wang, "Compressed sensing based interior tomography," *Physics in Medicine and Biology*, vol. 54, no. 9, p. 2791, 2009.

[59] J. Yang, H. Yu, M. Jiang, and G. Wang, "High-order total variation minimization for interior tomography," *Inverse Problems*, vol. 26, no. 3, p. 035013, 2010.

[60] H. Yu, J. Yang, M. Jiang, and G. Wang, "Supplemental analysis on compressed sensing based interior tomography," *Physics in Medicine and Biology*, vol. 54, no. 18, p. N425, 2009.

[61] K. Taguchi, J. Xu, S. Srivastava, B. M. Tsui, J. Cammin, and Q. Tang, "Interior region-of-interest reconstruction using a small, nearly piecewise constant subregion," *Medical Physics*, vol. 38, no. 3, pp. 1307–1312, 2011.

[62] G. Wang and H. Yu, "The meaning of interior tomography," *Physics in Medicine and Biology*, vol. 58, no. 16, p. R161, 2013.

[63] Y. Han, J. Gu, and J. C. Ye, "Deep learning interior tomography for region-of-interest reconstruction," *arXiv:1712.10248*, 2017.

[64] Y. Li, K. Li, C. Zhang, J. Montoya, and G.-H. Chen, "Learning to reconstruct computed tomography images directly from sinogram data under a variety of data acquisition conditions," *IEEE Transactions on Medical Imaging*, vol. 38, no. 10, pp. 2469–2481, 2019.

[65] L. Li, Z. Chen, W. Cong, and G. Wang, "Spectral CT modeling and reconstruction with hybrid detectors in dynamic-threshold-based counting and integrating modes," *IEEE Transactions on Medical Imaging*, vol. 34, no. 3, pp. 716–728, 2015.

[66] S. Xu, P. Prinsen, J. Wiegert, and R. Manjeshwar, "Deep residual learning in CT physics: scatter correction for spectral CT," in *Proc. 2017 IEEE Nuclear Science Symposium and Medical Imaging Conference*. IEEE, 2017.

[67] Z. Chen and L. Li, "Application of deep learning in multi-material decomposition of spectral CT," in *Proc. 2017 IEEE Nuclear Science Symposium and Medical Imaging Conference*. IEEE, 2017.

[68] X. Wu, P. He, Z. Long, P. Li, B. Wei, and P. Feng, "Study on spectral CT material decomposition via deep learning," in *Proc. 15th International Meeting on Fully Three-Dimensional Image Reconstruction in Radiology and Nuclear Medicine*, vol. 11072. International Society for Optics and Photonics, 2019, p. 110723C.

[69] Z. Chen and L. Li, "Robust multimaterial decomposition of spectral CT using convolutional neural networks," *Optical Engineering*, vol. 58, no. 1, p. 013104, 2019.

[70] D. Li, S. Li, M. Zhu, Q. Gao, Z. Bian, H. Huang, S. Zhang, J. Huang, D. Zeng, and J. Ma, "Unsupervised data fidelity enhancement network for spectral CT reconstruction," in *Medical Imaging 2020: Physics of Medical Imaging*, vol. 11312. International Society for Optics and Photonics, 2020, p. 113124D.

[71] A. Zheng, H. Yang, L. Zhang, and Y. Xing, "A novel deep learning-based method for monochromatic image synthesis from spectral CT using photon-counting detectors," *arXiv:2007.09870*, 2020.

[72] L. Gjesteby, B. De Man, Y. Jin, H. Paganetti, J. Verburg, D. Giantsoudi, and G. Wang, "Metal artifact reduction in CT: Where are we after four decades?" *IEEE Access*, vol. 4, pp. 5826–5849, 2016.

[73] G. Wang, D. L. Snyder, J. A. O'Sullivan, and M. W. Vannier, "Iterative deblurring for CT metal artifact reduction," *IEEE Transactions on Medical Imaging*, vol. 15, no. 5, pp. 657–664, 1996.

[74] H. Zhang, B. Dong, and B. Liu, "A reweighted joint spatial-Radon domain CT image reconstruction model for metal artifact reduction," *SIAM Journal of Imaging Sciences*, vol. 11, no. 1, pp. 707–733, 2018.

[75] H. S. Park, D. Hwang, and J. K. Seo, "Metal artifact reduction for polychromatic X-ray CT based on a beam-hardening corrector," *IEEE Transactions on Medical Imaging*, vol. 35, no. 2, pp. 480–487, 2015.

[76] H. S. Park, J. K. Choi, and J. K. Seo, "Characterization of metal artifacts in X-ray computed tomography," *Communications in Pure and Applied Mathematics*, vol. 70, no. 11, pp. 2191–2217, 2017.

[77] L. Gjesteby, Q. Yang, Y. Xi, H. Shan, B. Claus, Y. Jin, B. De Man, and G. Wang, "Deep learning methods for CT image-domain metal artifact reduction," in *Developments in X-Ray Tomography XI*, vol. 10 391. International Society for Optics and Photonics, 2017, p. 103 910W.

[78] L. Gjesteby, Q. Yang, Y. Xi, B. Claus, Y. Jin, B. De Man, and G. Wang, "Reducing metal streak artifacts in CT images via deep learning: Pilot results," in *Proc. 14th International Meeting on Fully Three-Dimensional Image Reconstruction in Radiology and Nuclear Medicine*, vol. 14, no. 6, 2017, pp. 611–614.

[79] L. Gjesteby, H. Shan, Q. Yang, Y. Xi, B. Claus, Y. Jin, B. De Man, and G. Wang, "Deep neural network for CT metal artifact reduction with a perceptual loss function," in *Proc. 5th International Conference on Image Formation in X-ray Computed Tomography*, vol. 1, 2018.

[80] J. Wang, Y. Zhao, J. H. Noble, and B. M. Dawant, "Conditional generative adversarial networks for metal artifact reduction in CT images of the ear," in *MICCAI*. Springer, 2018, pp. 3–11.

[81] H. Liao, W.-A. Lin, S. K. Zhou, and J. Luo, "ADN: Artifact disentanglement network for unsupervised metal artifact reduction," *IEEE Transactions on Medical Imaging*, vol. 39, no. 3, pp. 634–643, 2019.

[82] Y. Zhang and H. Yu, "Convolutional neural network based metal artifact reduction in X-ray computed tomography," *IEEE Transactions on Medical Imaging*, vol. 37, no. 6, pp. 1370–1381, 2018.

[83] M. U. Ghani and W. C. Karl, "Fast enhanced CT metal artifact reduction using data domain deep learning," *IEEE Transactions on Computational Imaging*, vol. 6, pp. 181–193, 2019.

[84] W.-A. Lin, H. Liao, C. Peng, X. Sun, J. Zhang, J. Luo, R. Chellappa, and S. K. Zhou, "DuDoNet: Dual domain network for CT metal artifact reduction," in *CVPR*, 2019, pp. 10 512–10 521.

[85] L. Yu, Z. Zhang, X. Li, and L. Xing, "Deep sinogram completion with image prior for metal artifact reduction in CT images," *IEEE Transactions on Medical Imaging*, vol. 40, no. 1, pp. 228–238, 2020.

[86] H. Wang, Y. Li, H. Zhang, J. Chen, K. Ma, D. Meng, and Y. Zheng, "InDuDoNet: An interpretable dual domain network for CT metal artifact reduction," in *MICCAI*. Springer, 2021.

[87] B. De Man, S. Basu, D. Bequé, B. Claus, P. Edic, M. Iatrou, J. LeBlanc, B. Senzig, R. Thompson, M. Vermilyea *et al.*, "Multi-source inverse geometry CT: a new system

concept for X-ray computed tomography," in *Medical Imaging 2007: Physics of Medical Imaging*, vol. 6510. International Society for Optics and Photonics, 2007, p. 651 00H.

[88] Y. Liu, H. Liu, Y. Wang, and G. Wang, "Half-scan cone-beam CT fluoroscopy with multiple X-ray sources," *Medical Physics*, vol. 28, no. 7, pp. 1466–1471, 2001.

[89] G. Wang and M. W. Vannier, "Preliminary study on helical CT algorithms for patient motion estimation and compensation," *IEEE Transactions on Medical Imaging*, vol. 14, no. 2, pp. 205–211, 1995.

[90] W. Lu and T. R. Mackie, "Tomographic motion detection and correction directly in sinogram space," *Physics in Medicine and Biology*, vol. 47, no. 8, p. 1267, 2002.

[91] H. Yu and G. Wang, "Data consistency based rigid motion artifact reduction in fan-beam CT," *IEEE Transactions on Medical Imaging*, vol. 26, no. 2, pp. 249–260, 2007.

[92] B. Bier, K. Aschoff, C. Syben, M. Unberath, M. Levenston, G. Gold, R. Fahrig, and A. Maier, "Detecting anatomical landmarks for motion estimation in weight-bearing imaging of knees," in *Proc. International Workshop on Machine Learning for Medical Image Reconstruction*. Springer, 2018, pp. 83–90.

[93] Y. Ko, S. Moon, J. Baek, and H. Shim, "Rigid and non-rigid motion artifact reduction in X-ray CT using attention module," *Medical Image Analysis*, p. 101 883, 2020.

[94] C. Jiang, Q. Zhang, Y. Ge, D. Liang, Y. Yang, X. Liu, H. Zheng, and Z. Hu, "Wasserstein generative adversarial networks for motion artifact removal in dental CT imaging," in *Medical Imaging 2019: Physics of Medical Imaging*, vol. 10 948. International Society for Optics and Photonics, 2019, p. 1 094 836.

[95] J. C. Montoya, Y. Li, C. Strother, and G.-H. Chen, "3D deep learning angiography (3D-DLA) from C-arm conebeam CT," *American Journal of Neuroradiology*, vol. 39, no. 5, pp. 916–922, 2018.

[96] ——, "Deep learning angiography (DLA): Three-dimensional C-arm cone beam CT angiography generated from deep learning method using a convolutional neural network," in *Medical Imaging 2018: Physics of Medical Imaging*, vol. 10 573. International Society for Optics and Photonics, 2018, p. 105 731N.

[97] J.-N. Zaech, C. Gao, B. Bier, R. Taylor, A. Maier, N. Navab, and M. Unberath, "Learning to avoid poor images: Towards task-aware C-arm cone-beam CT trajectories," in *Proc. International Conference on Medical Image Computing and Computer-Assisted Intervention*. Springer, 2019, pp. 11–19.

[98] J. Hatvani, A. Horváth, J. Michetti, A. Basarab, D. Kouamé, and M. Gyöngy, "Deep learning-based super-resolution applied to dental computed tomography," *IEEE Transactions on Radiation and Plasma Medical Sciences*, vol. 3, no. 2, pp. 120–128, 2018.

[99] K. Liang, L. Zhang, H. Yang, Y. Yang, Z. Chen, and Y. Xing, "Metal artifact reduction for practical dental computed tomography by improving interpolation-based reconstruction with deep learning," *Medical Physics*, vol. 46, no. 12, pp. e823–e834, 2019.

[100] Z. Hu, C. Jiang, F. Sun, Q. Zhang, Y. Ge, Y. Yang, X. Liu, H. Zheng, and D. Liang, "Artifact correction in low-dose dental CT imaging using Wasserstein generative adversarial networks," *Medical Physics*, vol. 46, no. 4, pp. 1686–1696, 2019.

[101] H. Wang, B. Wu, L. Xu, and H. Xu, "Experience about workflow of standardized cabin CT scan in module hospital during the outbreak of Covid-19," *Chinese Journal of Radiological Medicine and Protection*, vol. 40, no. 4, 2020.

[102] Q. Yang, H. Xu, X. Tang, C. Hu, P. Wang, Y. X. J. Wáng, Y. Wang, G. Ma, and B. Zhang, "Medical imaging engineering and technology branch of the Chinese society

of biomedical engineering expert consensus on the application of emergency mobile cabin CT," *Quantitative Imaging in Medicine and Surgery*, vol. 10, no. 11, p. 2191, 2020.

[103] L. Wang, Z. Q. Lin, and A. Wong, "Covid-net: A tailored deep convolutional neural network design for detection of Covid-19 cases from chest X-ray images," *Science Reports*, vol. 10, no. 1, pp. 1–12, 2020.

[104] A. Abbas, M. M. Abdelsamea, and M. M. Gaber, "Classification of Covid-19 in chest X-ray images using detrac deep convolutional neural network," *arXiv:2003.13815*, 2020.

[105] J. Zhang, Y. Xie, Y. Li, C. Shen, and Y. Xia, "Covid-19 screening on chest X-ray images using deep learning based anomaly detection," *arXiv:2003.12338*, 2020.

[106] A. I. Khan, J. L. Shah, and M. M. Bhat, "Coronet: A deep neural network for detection and diagnosis of Covid-19 from chest X-ray images," *Computer Methods and Programs in Biomedicine*, vol. 96, p. 105 581, 2020.

[107] J. Chen, L. Wu, J. Zhang, L. Zhang, D. Gong, Y. Zhao, Q. Chen, S. Huang, M. Yang, X. Yang *et al.*, "Deep learning-based model for detecting 2019 novel coronavirus pneumonia on high-resolution computed tomography," *Science Reports*, vol. 10, no. 1, pp. 1–11, 2020.

[108] B. Wang, S. Jin, H. Xu, C. Luo, L. Wei, W. Zhao, Q. Yan, X. Hou, W. Ma, Z. Xu *et al.*, "AI-assisted CT imaging analysis for Covid-19 screening: Building and deploying a medical AI system," *Applied Soft Computing*, vol. 98, p. 106 897, 2020.

[109] X. Wu, H. Hui, M. Niu, L. Li, L. Wang, B. He, X. Yang, L. Li, H. Li, J. Tian *et al.*, "Deep learning-based multi-view fusion model for screening 2019 novel coronavirus pneumonia: a multicentre study," *European Journal of Radiology*, vol. 128, p. 109 041, 2020.

[110] X. Xu, X. Jiang, C. Ma, P. Du, X. Li, S. Lv, L. Yu, Q. Ni, Y. Chen, J. Su *et al.*, "A deep learning system to screen novel coronavirus disease 2019 pneumonia," *Engineering*, vol. 6, no. 10, pp. 1122–1129, 2020.

[111] F. Shan, Y. Gao, J. Wang, W. Shi, N. Shi, M. Han, Z. Xue, and Y. Shi, "Lung infection quantification of Covid-19 in CT images with deep learning," *arXiv:2003.04655*, 2020.

[112] Q. Yan, B. Wang, D. Gong, C. Luo, W. Zhao, J. Shen, Q. Shi, S. Jin, L. Zhang, and Z. You, "Covid-19 chest CT image segmentation – a deep convolutional neural network solution," *arXiv:2004.10987*, 2020.

[113] D.-P. Fan, T. Zhou, G.-P. Ji, Y. Zhou, G. Chen, H. Fu, J. Shen, and L. Shao, "Inf-net: Automatic Covid-19 lung infection segmentation from CT images," *IEEE Transactions on Medical Imaging*, 2020.

[114] Q. Meng, W. Liu, X. Dou, J. Zhang, A. Sun, J. Ding, H. Liu, Z. Lei, and X. Chen, "Role of novel deep-learning-based CT used in management and discharge of Covid-19 patients at a "square cabin" hospital in China," online, 2020.

[115] M. Islam, F. Karray, R. Alhajj, J. Zeng *et al.*, "A review on deep learning techniques for the diagnosis of novel coronavirus (Covid-19)," *arXiv:2008.04815*, 2020.

[116] D. Dong, Z. Tang, S. Wang, H. Hui, L. Gong, Y. Lu, Z. Xue, H. Liao, F. Chen, F. Yang *et al.*, "The role of imaging in the detection and management of Covid-19: a review," *IEEE Reviews in Biomedical Engineering*, vol. 14, pp. 16–29, 2020.

[117] F. Shi, J. Wang, J. Shi, Z. Wu, Q. Wang, Z. Tang, K. He, Y. Shi, and D. Shen, "Review of artificial intelligence techniques in imaging data acquisition, segmentation and diagnosis for Covid-19," *Reviews in Biomedical Engineering*, vol. 14, pp. 4–15, 2020.

[118] M. Roberts, D. Driggs, M. Thorpe, J. Gilbey, M. Yeung, S. Ursprung, A. I. Aviles-Rivero, C. Etmann, C. McCague, L. Beer *et al.*, "Common pitfalls and recommendations for

using machine learning to detect and prognosticate for Covid-19 using chest radiographs and CT scans," *Nature Machine Intelligence*, vol. 3, no. 3, pp. 199–217, 2021.

[119] Z. Wei, L. Yuan, B. Liu, C. Wei, C. Sun, P. Yin, and L. Wei, "A micro-CL system and its applications," *Review of Scientific Instruments*, vol. 88, no. 11, p. 115 107, 2017.

[120] L. Sun, G. Zhou, Z. Qin, S. Yuan, Q. Lin, Z. Gui, and M. Yang, "A reconstruction method for cone-beam computed laminography based on projection transformation," *Measurement Science and Technology*, vol. 32, no. 4, p. 045 403, 2021.

[121] Y. Zhao, J. Xu, H. Li, and P. Zhang, "Edge information diffusion-based reconstruction for cone beam computed laminography," *IEEE Transactions in Image Processing*, vol. 27, no. 9, pp. 4663–4675, 2018.

[122] J. Zhang, L. Shi, C. Wei, B. Liu, and B. Wei, "Research on micro-CL geometric errors," *IEEE Access*, 2021.

[123] M. Park, G. Cho, J. Park, and S. Cho, "Noncircular scanning trajectory for sparse-view computed laminography," in *Proc. 6th Conference on Industrial Computed Tomography*. Research Group Computed Tomography, 2016.

[124] L. F. A. Pereira, J. De Beenhouwer, J. Kastner, and J. Sijbers, "Extreme sparse X-ray computed laminography via convolutional neural networks," in *Proc. IEEE International Conference on Tools with Artificial Intelligence*. IEEE, 2020, pp. 612–616.

[125] Y. Li, S. Liu, C. Li, Y. Zheng, C. Wei, B. Liu, and Y. Yang, "Automated defect detection of insulated gate bipolar transistor based on computed laminography imaging," *Microelectronic Reliability*, vol. 115, p. 113 966, 2020.

[126] L.-C. Chen, Y. Zhu, G. Papandreou, F. Schroff, and H. Adam, "Encoder-decoder with atrous separable convolution for semantic image segmentation," in *Proc. European Conference on Computer Vision*, 2018, pp. 801–818.

[127] Y. Fu, Y. Lei, T. Wang, W. J. Curran, T. Liu, and X. Yang, "Deep learning in medical image registration: a review," *Physics in Medicine and Biology*, vol. 65, no. 20, p. 20TR01, 2020.

[128] G. Haskins, U. Kruger, and P. Yan, "Deep learning in medical image registration: A survey," *Machine Vision and Applications*, vol. 31, no. 1, p. 8, 2020.

[129] Y. Liu, X. Chen, J. Cheng, and H. Peng, "A medical image fusion method based on convolutional neural networks," in *Proc. International Conference on Information Fusion*. IEEE, 2017, pp. 1–7.

[130] J. Du, W. Li, and B. Xiao, "Anatomical–functional image fusion by information of interest in local Laplacian filtering domain," *IEEE Transactions on Image Processing*, vol. 26, no. 12, pp. 5855–5866, 2017.

[131] W. Kong, Q. Miao, and Y. Lei, "Multimodal sensor medical image fusion based on local difference in non-subsampled domain," *IEEE Transactions on Instrumentation and Measurement*, vol. 68, no. 4, pp. 938–951, 2018.

[132] R. Liao, S. Miao, P. de Tournemire, S. Grbic, A. Kamen, T. Mansi, and D. Comaniciu, "An artificial agent for robust image registration," in *Proc. AAAI Conference on Artificial Intelligence*, vol. 31, no. 1, 2017.

[133] J. Krebs, T. Mansi, H. Delingette, L. Zhang, F. C. Ghesu, S. Miao, A. K. Maier, N. Ayache, R. Liao, and A. Kamen, "Robust non-rigid registration through agent-based action learning," in *MICCAI*. Springer, 2017, pp. 344–352.

[134] D. Nie, X. Cao, Y. Gao, L. Wang, and D. Shen, "Estimating CT image from MRI data using 3D fully convolutional networks," in *Proc. Conference on Deep Learning in Medical Image Analysis*. Springer, 2016, pp. 170–178.

[135] Y. Huo, Z. Xu, H. Moon, S. Bao, A. Assad, T. K. Moyo, M. R. Savona, R. G. Abramson, and B. A. Landman, "Synseg-net: Synthetic segmentation without target modality ground truth," *IEEE Transactions on Medical Imaging*, vol. 38, no. 4, pp. 1016–1025, 2018.

[136] Z. Zhang, L. Yang, and Y. Zheng, "Translating and segmenting multimodal medical volumes with cycle and shape-consistency generative adversarial network," in *CVPR*, 2018, pp. 9242–9251.

[137] J.-Y. Zhu, T. Park, P. Isola, and A. A. Efros, "Unpaired image-to-image translation using cycle-consistent adversarial networks," in *Proc. IEEE International Conference on Computer Vision*, 2017, pp. 2223–2232.

[138] A. Chartsias, T. Joyce, R. Dharmakumar, and S. A. Tsaftaris, "Adversarial image synthesis for unpaired multi-modal cardiac data," in *Proc. International Workshop on Simulation and Synthesis in Medical Imaging*. Springer, 2017, pp. 3–13.

[139] H. Yang, J. Sun, A. Carass, C. Zhao, J. Lee, Z. Xu, and J. Prince, "Unpaired brain MR-to-CT synthesis using a structure-constrained cycleGAN," in *Deep Learning in Medical Image Analysis and Multimodal Learning for Clinical Decision Support*. Springer, 2018, pp. 174–182.

[140] X. Chen, C. Lian, L. Wang, H. Deng, T. Kuang, S. Fung, P.-T. Yap, J. J. Xia, and D. Shen, "Anatomy-regularized representation learning for cross-modality medical image segmentation," *IEEE Transactions on Medical Imaging*, vol. 40, no. 1, pp. 274–285, 2020.

[141] L. Shi, B. Liu, H. Yu, C. Wei, L. Wei, L. Zeng, and G. Wang, "Review of CT image reconstruction open source toolkits," *Journal of X-Ray Science and Technology*, vol. 28, no. 4, pp. 619–639, 2020.

[142] "Low Dose CT Grand Challenge," 2016.

[143] K. Yan, X. Wang, L. Lu, and R. M. Summers, "DeepLesion: Automated mining of large-scale lesion annotations and universal lesion detection with deep learning," *Journal of Medical Imaging*, vol. 5, no. 3, p. 036501, 2018.

[144] "Operator discretization library (ODL)," Latest release, ODL 0.7.0, 2018.

[145] "Astra toolbox," Latest release, v1.9.0.dev11, 2021.

[146] S. Liang, Y. Li, and R. Srikant, "Enhancing the reliability of out-of-distribution image detection in neural networks," in *Proc. International Conference on Learning Representations*, 2018, pp. 1–27.

[147] C. Zhang, S. Bengio, M. Hardt, B. Recht, and O. Vinyals, "Understanding deep learning requires rethinking generalization," *arXiv:1611.03530*, 2016.

[148] V. Nagarajan and J. Z. Kolter, "Uniform convergence may be unable to explain generalization in deep learning," in *Advances in Neural Information Processing Systems*, 2019, pp. 11 615–11 626.

[149] B. Neyshabur, S. Bhojanapalli, D. McAllester, and N. Srebro, "Exploring generalization in deep learning," in *Advances in Neural Information Processing Systems*, vol. 30, 2017, pp. 5947–5956.

[150] K. Kawaguchi, L. P. Kaelbling, and Y. Bengio, "Generalization in deep learning," *arXiv:1710.05468*, 2017.

[151] H. Li, Y. Wang, R. Wan, S. Wang, T.-Q. Li, and A. C. Kot, "Domain generalization for medical imaging classification with linear-dependency regularization," in *Advances in Neural Information Processing Systems*, 2020.

[152] T. Schlegl, P. Seeböck, S. M. Waldstein, U. Schmidt-Erfurth, and G. Langs, "Unsupervised anomaly detection with generative adversarial networks to guide marker discovery," in *Proc. International Conference on Image Processing and Machine Intelligence.* Springer, 2017, pp. 146–157.

[153] M. Sabokrou, M. Fayyaz, M. Fathy, and R. Klette, "Fully convolutional neural network for fast anomaly detection in crowded scenes," *arXiv:1609.00866*, 2016.

[154] J. Andrews, T. Tanay, E. J. Morton, and L. D. Griffin, "Transfer representation-learning for anomaly detection," in *Proc. International Conference on Machine Learning.* JMLR, 2016.

[155] V. Antun, F. Renna, C. Poon, B. Adcock, and A. C. Hansen, "On instabilities of deep learning in image reconstruction and the potential costs of AI," in *Proc. National Academy of Science*, vol. 117, no. 48, pp. 30 088–30 095, 2020.

[156] M. Tancik, P. P. Srinivasan, B. Mildenhall, S. Fridovich-Keil, N. Raghavan, U. Singhal, R. Ramamoorthi, J. T. Barron, and R. Ng, "Fourier features let networks learn high frequency functions in low dimensional domains," in *Advances in Neural Information Processing Systems*, 2020.

[157] W. Wu, D. Hu, S. Wang, H. Yu, V. Vardhanabhuti, and G. Wang, "Stabilizing deep tomographic reconstruction networks," *arXiv:2008.01846*, 2020.

[158] F. Fan, J. Xiong, and G. Wang, "On interpretability of artificial neural networks," *arXiv:2001.02522*, 2020.

[159] A. B. Arrieta, N. Díaz-Rodríguez, J. Del Ser, A. Bennetot, S. Tabik, A. Barbado, S. García, S. Gil-López, D. Molina, R. Benjamins *et al.*, "Explainable artificial intelligence (XAI): Concepts, taxonomies, opportunities and challenges toward responsible AI," *Information Fusion*, vol. 58, pp. 82–115, 2020.

[160] P. W. Holland, "Statistics and causal inference," *Journal of the American Statistical Association*, vol. 81, no. 396, pp. 945–960, 1986.

[161] J. R. Geis, A. P. Brady, C. C. Wu, J. Spencer, E. Ranschaert, J. L. Jaremko, S. G. Langer, A. B. Kitts, J. Birch, W. F. Shields *et al.*, "Ethics of artificial intelligence in radiology: Summary of the joint European and North American multisociety statement," *Canadian Association of Radiologists Journal*, vol. 70, no. 4, pp. 329–334, 2019.

[162] E. Neri, F. Coppola, V. Miele, C. Bibbolino, and R. Grassi, "Artificial intelligence: Who is responsible for the diagnosis?" *La Radiologia Medica*, vol. 125, p. 517–521, 2020.

[163] E. Neri, V. Miele, F. Coppola, and R. Grassi, "Use of CT and artificial intelligence in suspected or Covid-19 positive patients: Statement of the Italian Society of Medical and Interventional Radiology," *La Radiologia Medica*, vol. 125, no. 5, pp. 505–508, 2020.

6 Deep Learning in CT Reconstruction: Bringing the Measured Data to Tasks

Guang-Hong Chen, Chengzhu Zhang, Yinsheng Li, Yoseob Han, and Jong Chul Ye

6.1 Introduction

The positive impact that diagnostic computed tomography (CT) imaging has made upon the diagnosis and treatment of symptomatic patients over the past few decades is unprecedented. However, despite the remarkable contribution of CT to modern healthcare, the issue of the small theoretical cancer risk associated with the use of ionizing radiation in CT has recently become a public concern.

To address the public concerns, many strategic discussion sessions have been organized by the National Institutes of Health (NIH) and the American Association of Physicists in Medicine (AAPM) to map out the scientific steps needed to accomplish the goal of sub-mSv CT imaging. A consensus report was published [1] to guide the development of low-dose CT imaging technologies. The research and development efforts from academic, industrial, and clinical societies in the past decade have resulted in the development and implementation of a variety of radiation-dose-reduction strategies for CT exams.

To lower the radiation dose in CT exams, advanced CT hardware technologies have been developed and incorporated into the current clinical CT systems, including but not limited to, newer generations of CT detectors with improved detection quantum efficiency, beam-shaping filters and dynamic z-axis collimators that limit the dose at the periphery of the patient and edge of the scan range, automatic exposure control with the flexibility to modulate tube potential and tube current, and photon-counting CT.

In addition to hardware technology developments, software technologies have also played a major role in this joint effort of radiation dose reduction. Many advanced image-processing techniques working in the sinogram projection-data domain or in the reconstructed-image domain, or in both domains jointly in fully statistical iterative reconstruction (IR) format have been developed and some of these methods have been incorporated into commercial products for routine clinical uses. In the past decade or so, such newly developed low-dose reconstruction or processing software has been rigorously evaluated with large patient cohorts. It has been found that, after a decade-long effort to reduce CT radiation dose with IR software techniques, the radiation dose can be reduced by up to 25 percent with statistical IR techniques *without sacrificing clinical diagnostic quality* [2, 3].

This status quo of the radiation dose reduction factor using statistical IR leaves plenty of room for the development of new techniques to further reduce radiation dose in clinical CT exams, particularly, new techniques that are sufficiently flexible to leverage the intrinsic advantages of analytical reconstruction methods, such as filtered back projection or differentiated back projection filtration-based methods [4] and the statistical learning principles that have been extensively investigated in statistical IR methods. In this regard, the recent advances in deep learning provide us with an ideal computational framework that can be developed to directly reconstruct CT images from sinogram data, or can be combined with analytical reconstruction methods to address the challenges encountered in their application, or can be combined with statistical IR to address reconstruction accuracy and the generalizability concerns associated with deep-learning methods.

The foundational physics principles of CT image reconstruction problems will be presented in Section 6.2. After that, deep learning in CT image reconstruction will be described, with examples on how to perform reconstruction directly from sinogram to image in Section 6.3, how to combine deep learning with analytical reconstruction methods in Section 6.4, and how to combine deep learning with statistical IR methods in Section 6.5.

6.2 CT Imaging Physics and Reconstruction-Problem Formulations

In this section, by following the experimental image-formation processes in CT, image-reconstruction problems are formulated from an imaging-physics standpoint in order to make the scientific foundations of each reconstruction framework transparent. Particularly, the assumptions implicitly or explicitly used in the past will be made transparent in our formulations. This review of problem formulation also aims to put the rapid rising of deep-learning methods into a proper scientific context and to find a niche for deep-learning methods in the big picture of radiation dose reduction in CT. We also hope that this section will inspire cross-pollination between analytical image reconstruction, statistical IR, and data-driven deep-learning-based reconstruction strategies, with the ultimate objective of developing some innovative reconstruction strategies with a high clinical impact.

6.2.1 CT Image Reconstruction Problem Formulation: A Deterministic Approach

The physics of X-ray attenuation by an image object is captured by the empirical Beer–Lambert law, which characterizes the mean photon number before and after the X-ray beam interacts with the image object. The interactions of the X-ray photons and the image object are characterized by the so-called linear attenuation coefficient, $\mu(\vec{x}, \varepsilon)$, at a given spatial location and X-ray energy ε. The mean photon number recorded at the ith detector element is given by the Beer–Lambert law as follows:

$$\bar{N}_i = \bar{N}_{i0} \int_0^{\varepsilon_{max}} d\varepsilon \, \Omega(\varepsilon) \exp\left(-\int_{\ell_i} d\vec{x} \, \mu(\vec{x}, \varepsilon)\right), \tag{6.1}$$

where $\bar{N}_{i0}$ denotes the mean number of photons, at the ith measurement, before the photons enter the image object while $\bar{N}_i$ denotes the mean number of photons detected at the detector plane; $\Omega(\varepsilon)$ denotes the normalized energy spectral function, which includes the effects of both the entrance X-ray energy spectrum and the detector energy response function, and ε_{max} is the highest X-ray photon energy, determined by the X-ray tube potential.

One can use the fundamental mean value theorem for integrals to linearize the above equation as follows:

$$\bar{N}_i = \bar{N}_{i0} \exp\left(-\int_{\ell_i} d\vec{x}\,\mu(\vec{x},\bar{\varepsilon}_i)\right) \int_0^{\varepsilon_{max}} d\varepsilon\,\Omega(\varepsilon) = \bar{N}_{i0} \exp\left(-\int_{\ell_i} d\vec{x}\,\mu(\vec{x},\bar{\varepsilon}_i)\right), \quad (6.2)$$

where $\bar{\varepsilon}_i \in [0, \varepsilon_{max}]$ is the effective energy of the X-rays on the path ℓ_i, and the normalization condition for the energy response function has been used. By taking the log-transform, one obtains the familiar CT image reconstruction model as follows:

$$\int_{\ell_i} d\vec{x}\,\mu(\vec{x},\bar{\varepsilon}_i) = \ln \frac{\bar{N}_{i0}}{\bar{N}_i}. \quad (6.3)$$

The left-hand side is an integral of the unknown variables $\mu(\vec{x},\bar{\varepsilon}_i)$ while the right-hand side is a quantity that, in principle, can be measured if data acquisition can be performed at each detector element.

Using Eq. (6.3), the CT reconstruction problem is formulated as a problem of jointly solving a series of integral equations, which is also an important topic in integral geometry [5, 6]. In two dimensions, a classical solution of the above integral equation was discovered by Radon [7] and this classical solution is the mathematical foundation of the celebrated filtered backprojection (FBP) analytical reconstruction schemes. Besides its use in X-ray CT, image reconstruction from line integrals also plays an important role in other imaging modalities such as single photon emission computed tomography (SPECT), positron emission tomography (PET), and magnetic resonance imaging (MRI).

In the CT community, many different forms of analytical solutions have been developed to reconstruct images from log-transformed data. Besides the well-known FBP methods with a ramp kernel, a new FBP reconstruction with a Hilbert filter and directed derivatives with respect to the view angle variable have been developed in the past 20 years [8–10]. These methods allow the reconstruction of local regions of interest (ROIs) provided that the measured data at each view angle are not transversely truncated. To handle the reconstruction problem with transversely truncated projection data, differentiated back projection (DBP) filtration methods [11–13] have also been developed for a large class of scanning configurations. For more technical details regarding these new algorithms, we highly recommend a very accessible book by Zeng [4].

Although the reconstruction problem formulation in Eq. (6.3) has been widely used in textbooks and publications to illustrate problem formulation for CT reconstruction, it does have one often overlooked issue: *it is not intrinsically compatible with clinical*

CT acquisitions. Note that the projection data in Eq. (6.3) require us to perform repeated measurements to infer the statistical mean values before the log-transforms are performed to obtain the needed projection data. For laboratory phantom studies and some industrial CT applications where neither the radiation dose nor the data acquisition time are a concern, one can repeatedly measure the photon counts N_i at each detector element. However, in clinical CT acquisitions, one cannot repeatedly perform data acquisitions to meet the requirement in Eq. (6.3) since that would deliver an excessive amount of radiation to the patients and thus it would be medically dangerous. Additionally, repeated measurements prolong data acquisition and thus reduce throughput in clinical operations.

This intrinsic incompatibility issue in deterministic CT reconstruction problem formulation and actual data acquisition has been largely ignored, i.e., the quantities on the right-hand side of Eq. (6.3) have simply been replaced by single measurements N_{i0} and N_i. This is the strategy that has been used in almost all CT scanners in the past 50 years. When the radiation dose level is not too low, namely, the photon count level is not too low, this strategy works well, as is witnessed by the success of CT in modern medicine. However, when the radiation dose level is lowered this approximation becomes more and more problematic, with increased noise level and bias in the CT numbers. To address this intrinsic problem that arises with a deterministic formulation of the CT reconstruction problem, the neglected statistical nature in CT data acquisitions must be properly handled, as will be presented in the next subsection.

6.2.2 CT Reconstruction Problem Formulation (II): Statistical Learning Framework

To facilitate the illustration of the intrinsic statistical nature in CT acquisitions, an ideal photon-counting detector is assumed in the data acquisition process. After X-rays are generated from the X-ray tube and collimated to illuminate the image of the object, information about the image in terms of the local distribution of linear attenuation coefficients $\mu(\vec{x}, \bar{\varepsilon}_i)$ is encoded into the photon attenuation properties via physical interaction processes between the X-ray photons and the object. This information is encoded into the measured dataset with a total of M data points, $\{N_1, N_2, \ldots, N_M\}$. Let us keep in mind that each of these numbers has only been measured once.

Owing to the intrinsic statistical nature of photon behavior, the number of photons recorded at a given detector element i is an intrinsic random variable, X, that follows Poisson statistics:

$$P(X = N_i | \bar{N}_i) = \frac{\bar{N}_i^{N_i} e^{-\bar{N}_i}}{N_i!}, \tag{6.4}$$

where $\bar{N}_i$ is the mean number of photon counts received at detector element i; it is a statistical parameter of the Poisson distribution and not a directly measured quantity. Clearly, the desired object information $\mu(\vec{x}, \bar{\varepsilon}_i)$ in X-ray CT is intrinsically encoded (as shown in Eq. (6.3)) into $\bar{N}_i$ in Eq. (6.4) as a statistical parameter of the Poisson statistical distribution, not in terms of the directly measured photon counts N_i.

Therefore, when the repetition of experimental data acquisition is prohibited, as in clinical CT acquisitions, *CT image reconstruction must be formulated as a statistical learning problem: to statistically infer the object information* $\mu(\vec{x}, \bar{\varepsilon})$, *i.e., the statistical parameters given a set of measured data* $\{N_1, N_2, \ldots, N_M\}$.

Since the measurements at different detector elements can be considered to be statistically independent, the joint probability of a measured dataset is then given by

$$P(\{N_1, N_2, \ldots, N_M\} | \mu(\vec{x})) = \prod_{i=1}^{M} P(N_i | \mu(\vec{x})) = \prod_{i=1}^{M} \frac{\bar{N}_i^{N_i} e^{-\bar{N}_i}}{N_i!}, \quad (6.5)$$

where the effective energy dependence $\bar{\varepsilon}$ in $\mu(\vec{x}, \bar{\varepsilon})$ has been omitted, to avoid notational cluttering. When the statistical parameters, i.e., the mean counts, are known for each measurement, the above equation dictates the joint probability for the set of measurements. Now the CT reconstruction problem can be formulated as a standard statistical learning problem: *after the measurements* $\{N_1, N_2, \ldots, N_M\}$ *are given, how does one estimate the statistical parameters* $\{\bar{N}_1, \bar{N}_2, \ldots, \bar{N}_M\}$? *Or, alternatively, how does one estimate* $\mu(\vec{x})$ *since these linear attenuation coefficients are related to the mean counts as shown in Eq. (6.3)?* There are several different approaches in statistics to address this statistical learning problem, as will be discussed below.

Maximum Likelihood Method

The first way of addressing the above statistical learning problem is to use Fisher's maximum likelihood (ML) method. In this method, when the values of $\{N_1, N_2, \ldots, N_M\}$ are obtained from experimental measurements, the formula on the right-hand side of Eq. (6.5) becomes a function of the statistical parameters $\{\bar{N}_1, \bar{N}_2, \ldots, \bar{N}_M\}$ and this function is called the likelihood of the statistical parameters. To find a point estimate of the statistical parameters, what Fisher suggested is to maximize the log-likelihood function as follows:

$$\hat{\mu}(\vec{x}) = \arg\max_{\mu(\vec{x})} \sum_{i=1}^{M} (N_i \ln \bar{N}_i - \bar{N}_i),$$

$$= \arg\max_{\mu(\vec{x})} \sum_{i=1}^{M} \left(-N_i \int_{\ell_i} d\vec{x} \mu(\vec{x}) - \bar{N}_{i0} \exp\left(-\int_{\ell_i} d\vec{x} \mu(\vec{x})\right) \right) \quad (6.6)$$

where $\bar{N}_i$ in Eq. (6.2) has been substituted in the second line to make the target quantity $\mu(\vec{x})$ transparent. The term that is not related to $\bar{N}_i$ has been dropped in the above equation. If we denote $P_i = \int_{\ell_i} d\vec{x} \mu(\vec{x})$, then the above optimization problem enjoys the following stationarity conditions:

$$P_i = \int_{\ell_i} d\vec{x} \mu(\vec{x}) = \ln \frac{\bar{N}_{i0}}{N_i}, \quad i = 1, 2, \ldots, M. \quad (6.7)$$

Therefore, one can solve for $\mu(\vec{x})$ using the measured quantities $\bar{N}_{i0}$ and N_i. In this ML estimate formulation, repeated measurements are required to infer $\bar{N}_{i0}$, but this can be done experimentally since there is no image object involved and thus there is no concern of radiation risks to a patient.

Quadratic Approximation in ML

In the above ML estimate, $\bar{N}_i$ is replaced by the single measurement N_i. Then one can also replace $\bar{N}_{i0}$ in Eq. (6.3) since $\bar{N}_{i0}$ is much larger than $\bar{N}_i$. In the following, we work out the underlying mathematical approximations for this well-known assumption in CT practice. Note that this assumption dictates that

$$\Delta = \ln \frac{N_{i0}}{N_i} - \ln \frac{\bar{N}_{i0}}{\bar{N}_i} = \ln \frac{N_{i0}}{N_i} - \int_{\ell_i} d\vec{x}\,\mu(\vec{x})$$

is a negligibly small quantity. What would be the consequence of this assumption in the statistical framework? To answer this question, we can perform a Taylor expansion of the objective function in Eq. (6.6) as follows:

$$\bar{N}_{i0} \exp\left(-\int_{\ell_i} d\vec{x}\,\mu(\vec{x})\right) = \bar{N}_{i0} \exp\left(\ln \frac{N_{i0}}{N_i} - \int_{\ell_i} d\vec{x}\,\mu(\vec{x}) - \ln \frac{N_{i0}}{N_i}\right) \quad (6.8)$$

$$= \frac{\bar{N}_{i0}}{N_{i0}} N_i e^{\Delta} \approx \frac{\bar{N}_{i0}}{N_{i0}} N_i \left(1 + \Delta + \frac{1}{2}\Delta^2\right).$$

Using this quadratic approximation, the objective function in Eq. (6.6) becomes

$$-N_i \int_{\ell_i} d\vec{x}\,\mu(\vec{x}) - \bar{N}_{i0} \exp\left(-\int_{\ell_i} d\vec{x}\,\mu(\vec{x})\right) \quad (6.9)$$

$$\approx -\frac{\bar{N}_{i0}}{N_{i0}} N_i \left(1 + \ln \frac{N_{i0}}{N_i}\right) - N_i \left(1 - \frac{\bar{N}_{i0}}{N_{i0}}\right) \int_{\ell_i} d\vec{x}\,\mu(\vec{x})$$

$$-\frac{1}{2}\frac{\bar{N}_{i0}}{N_{i0}} N_i \left[\ln \frac{N_{i0}}{N_i} - \int_{\ell_i} d\vec{x}\,\mu(\vec{x})\right]^2.$$

The first term in the above equation does not depend on the target variable $\mu(\vec{x})$ and can be dropped from the objective function. Now we are left with two terms: one linear and one quadratic term in the line integral $\int_{\ell_i} d\vec{x}\,\mu(\vec{x})$. However, a further simplification can be readily made as follows. The photon fluence N_{i0} is on the order of 10^{5-6} photon/mm^2, or 10^{5-6} photon counts per detector element (note that the detector element size is around 1 mm × 1mm in clinical CT scanners). Therefore, $\bar{N}_{i0} \approx N_{i0}$ is a good approximation and this results in the celebrated least squares objective function:

$$-\frac{1}{2}\sum_{i=1}^{M} N_i \left[\ln \frac{N_{i0}}{N_i} - \int_{\ell_i} d\vec{x}\,\mu(\vec{x})\right]^2. \quad (6.10)$$

Owing to the assumed statistical independence among the measured data, the stationarity condition for the above quadratic objective function becomes

$$\int_{\ell_i} d\vec{x}\,\mu(\vec{x}) = \ln \frac{N_{i0}}{N_i} = y_i \quad (6.11)$$

where $y_i = \ln(N_{i0}/N_i)$ is the log-transformed projection data in sinogram space. This is very similar to Eq. (6.3), but the log-transform is performed over the experimentally measured data, not over their statistical mean! Therefore, Eq. (6.11) can be directly applied to the measured data in clinical CT scanners without the intrinsic incompatibility issue. Therefore, we conclude that the FBP reconstruction problem in deterministic form is the direct consequence of two approximations used in the ML method: (1) $\bar{N}_{i0} \approx N_{i0}$ and (2) orders higher than the quadratic term in the Taylor expansion of e^Δ can be safely neglected.

The above quadratic approximation in the ML method, as in Eq. (6.10), also gives us a natural connection to the well-investigated algebraic reconstruction framework. To make this connection transparent, we can digitize the target function $\mu(\vec{x})$ using spatial basis functions $\{b_j(\vec{x}), j = 1, 2, \ldots, N\}$ as follows:

$$\mu(\vec{x}) = \sum_{j=1}^{N} \mu_j b_j(\vec{x}). \tag{6.12}$$

As a result, the line integral can be written in terms of matrix multiplication:

$$\int_{\ell_i} d\vec{x} \left[\sum_{j=1}^{N} \mu_j b_j(\vec{x}) \right] = \sum_{j=1}^{N} \mu_j \left[\int_{\ell_i} d\vec{x} b_j(\vec{x}) \right] = \sum_{j=1}^{N} a_{i,j} \mu_j := [\mathbf{A}\vec{\mu}]_i, \tag{6.13}$$

where $\mathbf{A}$ is the so-called system matrix. Using matrix notation, the objective function in Eq. (6.10) becomes

$$-\frac{1}{2} \sum_{i=1}^{M} N_i \left[\ln \frac{N_{i0}}{N_i} - \int_{\ell_i} d\vec{x} \mu(\vec{x}) \right]^2 = -\frac{1}{2} \|\vec{y} - \mathbf{A}\vec{\mu}\|_D^2, \tag{6.14}$$

where $\vec{y} = (y_1, y_2, \ldots, y_M)^T$ is an M-dimensional column vector, i.e., $\vec{y} \in R^M$, $y_i = \ln(N_{i0}/N_i)$, and $D = \text{diag}(N_1, N_2, \ldots, N_M)$ is a diagonal matrix with the detector raw counts N_i as its entries. It is important to note that it is the measured detector raw counts that appear in the matrix D, not their statistical mean. This statistical weight matrix has a natural interpretation: the lower-count data (i.e., noisy measurements) are given a lower weight while higher counts (less noisy measurements) are given a higher weight. Clearly, the above least squares objective function in Eq. (6.14) is the one to be used in algebraic reconstruction methods.

Maximum A Posteriori (MAP)-Based Statistical IR Methods

The second approach to addressing the statistical learning problem in CT image reconstruction is the Bayes statistical inference framework. In this framework, the statistical parameters are treated as random variables and their *a priori* information is incorporated in the form of a prior statistical probability density function. Using Bayes' theorem, the posterior probability distribution is

$$P(\mu(\vec{x})|\{N_1, N_2, \ldots, N_M\}) \propto P(\{N_1, N_2, \ldots, N_M\}|\mu(\vec{x})) P(\mu(\vec{x})), \tag{6.15}$$

where the proportionality constant, which has no dependence on $\mu(\vec{x})$, has been ignored.

To obtain a point estimate for $\mu(\vec{x})$, the above posterior probability function is maximized for a chosen prior distribution $P(\mu(\vec{x}))$, i.e., this is a maximum a posteriori

(MAP) procedure. This framework is often formulated as the following minimization problem:

$$\hat{\mu}(\vec{x}) = \arg\min_{\mu(\vec{x})} \left[\sum_{i=1}^{M} (-N_i \ln \bar{N}_i + \bar{N}_i) + \lambda R(\mu(\vec{x})) \right], \quad (6.16)$$

where $\lambda R(\mu(\vec{x})) = -\ln P(\mu(\vec{x}))$ is often referred to as a regularizer.

When the quadratic approximation shown in Eq. (6.8) is used, the above optimization problem is recast into the following popular form:

$$\hat{\vec{\mu}} = \arg\min_{\vec{\mu}} \left[\frac{1}{2} \| \vec{y} - \mathbf{A}\vec{\mu} \|_D^2 + \lambda R(\vec{\mu}) \right]. \quad (6.17)$$

Since the intrinsic statistical nature of CT data acquisitions is incorporated in the above statistical learning problem, the solution from solving the above optimization problem is naturally suitable for addressing the low-dose CT reconstruction problem, which becomes an obvious advance when compared with the deterministic formulation. Such CT image reconstruction via solving the above optimization problem has been the mainstream focus of statistical IR development in the past decade or so.

Despite their great success, there are three remaining challenges in the statistical IR algorithms used in low-dose CT. First, for each individual reconstruction problem, the entire optimization problem must be solved from scratch. As a result, the overhead in reconstruction time is still heavy for demanding clinical applications. Second, the prior distribution functions, i.e., regularizers are handcrafted by algorithm designers using a mathematical formulation of some general knowledge about the image object. It is hard to capture the essential distribution of a given cohort of image objects. Third, a specific parametric statistical model, i.e., the Poisson statistical model, has been used in the problem formulation; however, for data acquired from current CT scanners, owing to the use of energy integration detectors and the associated problems such as cross-talk between neighboring detector elements, it is not clear to what extent the Poisson parametric model is still a good approximation. Finally, owing to the nonlinearity of the optimization problem, it is hard to predict the variations in the performance of the algorithm across patient populations. These challenges have become a niche application of the deep-learning method which can naturally incorporate the essence of statistical IR and provides tremendous flexibility to address many challenges in both analytical reconstruction and statistical reconstruction formulations, as will be discussed in the following subsections.

6.2.3 CT Reconstruction Problem Formulation (III): Deep-Learning Framework

Recent advances in deep learning provide a powerful tool to address the first three problems in statistical learning formulation. In the following, we discuss how the deep-learning framework naturally extends point estimates in the above classical statistical learning frameworks that have been used in CT reconstruction in the past several decades.

The key in classical statistical learning is to maximize the posterior probability density function, which is modeled as a product of the likelihood and prior. The specific form of likelihood depends on the chosen statistical model for the data generation process and the prior distribution is handcrafted by algorithm developers.

Modern deep-learning methods offer us new insights that provide sufficient freedom in statistical model selection for the data generation process and to liberate algorithm developers from the challenges of making educated guesses at the prior distribution. The needed statistical models and prior distributions can all be learned from some well-curated training datasets. This is essentially the same as the classical statistical regression in statistical learning processes. More and more powerful computational facilities and backpropagation training strategies allow us to engage deep neural networks to leverage their powerful function approximation capacity [14].

Specifically, in the deep-learning framework, instead of using an explicit assumption on the prior $P(\vec{\mu})$, the posterior distribution $P(\vec{\mu}|\vec{y})$ is directly learned from the training data via a supervised learning process [15]. In this process, a labeled image $\vec{\mu}_i$ is drawn from the output training image dataset and a sample $\vec{y}_i$ is drawn from the input training dataset, $i = 1, 2, \ldots, D$. The data pairs $(\vec{y}_i, \vec{\mu}_i)$ are used to train the deep neural network to model the posterior probability function. Namely, we aim to learn a regression function $f : \mathcal{Y} \mapsto \mathcal{X}$ ($\mathcal{X}$ denotes the image space and $\mathcal{Y}$ denotes the input data space), i.e., a map to directly generate the final image from the measured data space. Using this regression function, we hope that the learned model distribution $Q(\vec{\mu}|\vec{y}; f)$ can best approximate the underlying posterior distribution $P(\vec{\mu}|\vec{y})$. Once the map $f : \mathcal{Y} \mapsto \mathcal{X}$ is learned, it is applied to predict an image output $\vec{\mu}_{new}$ from the input projection data $\vec{y}_{new}$ not used in the training process.

6.2.4 CT Reconstruction Problem Formulation (IV): Combining Deep Learning with either Analytical or Statistical IR Framework

Once the function approximation capacity of the deep neural network is fully recognized, the deep-learning framework can be readily combined with the analytical reconstruction framework to address its intrinsic problem of ignoring the statistical nature of the acquired CT data. Note that the deep model can be used to model and correct the errors caused by the incorrect handling of data statistics. This can be combined with either the FBP type of reconstruction framework or the DBP type of reconstruction framework. Since there have been many works [16] published on the combination of FBP and deep learning, owing to the abundance of FBP-reconstructed images as training data, in this chapter we highlight the combination of DBP with deep learning to solve some of the difficult reconstruction problems associated with either data truncation or missing data in cone-beam acquisitions.

However, we should also bear in mind that the success of the current deep-learning applications is largely due to the impressive regression capacity offered by deep-neural-network architectures. In statistical regression [17, 18], a regression function is learned that captures the statistical features among all the training data. In other words, regression functions do not aim at a perfect fit of the training dataset. In fact, a variety of training strategies have been deliberately introduced to avoid situations where the regression model fits all the training data, since overfitted regression models generalize very poorly to new input data [17–19]. As a result, deep neural networks intend to ignore some patient-specific features in the output images. Although this might not be a serious issue in nonmedical applications such as natural language processing or computer vision studies, these individual patient-specific image features, e.g., lesions, are too important in medical diagnoses to be omitted or modified in the reconstructed medical images. Additionally, this statistical nature in regression also leads to fundamental challenges in current deep-learning research, i.e., the "generalizability issue." Since regression models are derived from training datasets with limited sample size, they only capture the statistical features present within the training cohort. As a result, when the derived regression models are applied to new test data, which may have been collected under different data acquisition conditions, the results of regression models may be less than optimal if not downright incorrect. Therefore, proper measures must be developed in medical imaging to address the patient-specific reconstruction-accuracy issue, and the generalizability issue, in deep-learning methods. Since the patient-specific information is encoded into the measured projection data, a natural way to address the question of patient-specific reconstruction accuracy and generalizability issues is to check whether the output images from the trained neural network are consistent with the corresponding data. If not, then the final image can be iteratively corrected, as has been done in statistical iterative reconstruction.

There are many different schemes that combine the deep-learning reconstruction framework with statistical IR. For example, numerical solvers of any iterative image reconstruction algorithm can be unrolled and incorporated into a deep-neural-network architecture, and the reconstruction parameters can then be learned using training data, or, alternatively, the handcrafted regularizers used in statistical IR algorithms can be learned from the available training data, as has been shown in a large body of literature [20–40].

In this chapter, we present a simple framework to combine analytical reconstruction, statistical IR, and deep learning in so-called prior-image-constrained compressed sensing (PICCS) [41], in order to address the reconstruction accuracy and generalizability issues encountered in deep-learning regressions [42] and also to show the power of a synergistic use of these reconstruction frameworks in some extremely challenging CT image reconstruction problems such as the high-pitch helical CT reconstruction problem [43].

6.3 Deep Learning in CT Reconstruction: From Sinogram to Image Directly

In this section, we review two sinogram-domain deep neural networks, referred to as the intelligent CT network (iCT-Net) [44] and the ScoutCT-Net [45], which can be trained to reconstruct images directly from sinogram-domain projection data. However, these two deep-learning reconstruction methods have totally different image

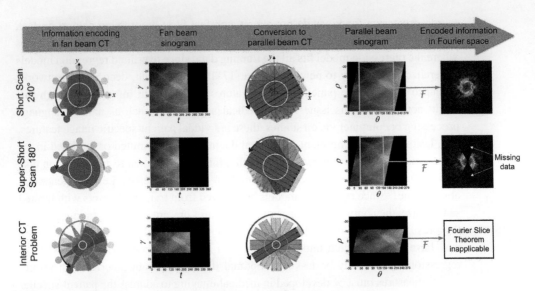

Figure 6.1 Line integral acquisitions and the corresponding reconstruction problems. (1) Short-scan problem (1st row). The tube detector rotates (first column) over an angular range of $180° + \gamma_m$, where γ_m is the fan angle (60° assumed) subtended by the field of view (FOV) marked by the white circle. When the acquired data are not transversely truncated, i.e., the detector size matches the diameter of the FOV, and the view angles (t) are not sparse, the fan-beam sinogram (second column) can be rebinned (third column) into the parallel-beam sinogram (fourth column) through a coordinate transformation $(t, \gamma) \rightarrow (\rho, \theta)$. Using the projection-slice theorem, the acquired line integral data can be transformed into the corresponding Fourier coefficients of the objective function (fifth column). (2) Super-short scan problem (second row), viz., the view angle falls in the range $(180° - \gamma_m, 180° + \gamma_m)$. The acquired data can be converted into a limited region in the corresponding Fourier space provided that the data are not truncated and the view angles are not sparse. When the view angle range is shorter than $180° - \gamma_m$, the problem is referred to as an intrinsic limited-view-angle problem. (3) Interior problem (third row), viz., the sinogram is potentially truncated transversely at every view angle. No portion of the acquired data may be correctly converted to Fourier space. As a further detriment, every measured line integral is contaminated with contributions from both the interior region (inside the FOV) and exterior region (outside the FOV). As a final note, when the view angles are sparse, as in the compressed sensing problem, the attempt to rebin data into the parallel-beam geometry fails in the above three cases, as does the connection with the Fourier space.

reconstruction tasks: in iCT-Net, the overarching objective is to reconstruct diagnostic-quality CT images with high quantitative accuracy under a wide variety of CT data acquisition conditions, as illustrated in Fig. 6.1; the viewing angle sampling conditions can be either dense or sparse. Namely, iCT-Net aims at addressing image reconstruction problems with either complete or incomplete line integral data including problems that have not been solved, or not satisfactorily solved, by human knowledge. In contrast, ScoutCT-Net aims to reconstruct a three-dimensional (3D) tomographic patient model from only two available views of projection data, i.e., so-called scout scans to obtain two radiograph images of the patient anatomy are made before the actual CT scans are executed. The scout radiograph images are often used to (1) position the patient properly on the patient couch, to make sure that the clinical targets are placed in the scanning field of view, and (2) to estimate the necessary 3D tomographic patient model for tube current modulations, either mA modulation or radiation-dose modulation. Current methods are only able to roughly estimate the eccentricity of the elliptical shape of the patient cross section; they do not provide a heterogeneous distribution of the patient anatomy. Therefore, the overarching objective of ScoutCT-Net is to reconstruct accurate 3D tomographic patient models for radiation dose modulation purposes, not for diagnostic purposes.

6.3.1 iCT-Net for Image Reconstruction with Diagnostic Purpose

iCT-Net Reconstruction Pipeline

The design of iCT-Net was inspired by the cascade imaging chain used in the current clinical CT image reconstruction pipeline. This pipeline includes three major cascaded steps: (1) to correct the measured signals to account for erroneous detector counts caused by a variety of physical reasons such as excessive noise and beam hardening; (2) to filter the corrected data with an apodized ramp filter; (3) to backproject the filtered data in order to accomplish the domain transformation from line integral space to tomographic image space. In the iCT-Net architecture (Fig. 6.2), convolutional layers with multi-channels were used to accomplish two objectives: the primary functionality of the above three cascaded steps in the conventional FBP-based image reconstruction, and a new functionality to enable iCT-Net to address difficult image-reconstruction problems such as view angle truncation, view angle undersampling, and interior problems using the same architecture in each case. In the end, iCT-Net comprises the following four components: (1) a data rectification module including five convolutional layers (L1–L5) to suppress excessive noise in line integral data and convert a sparse view sinogram into a dense view sinogram; (2) a data filtration module including four convolutional layers (L6–L9) to learn high-level feature representations from the output data of the L5 layer; (3) a domain transform module with a fully connected layer L10 to transform the extracted data features into an image; (4) an image combination module including two layers (L11 and L12) to learn to combine the partial image from each view angle to generate a final image. These final two components are analogous to the backprojection and summation steps in the conventional FBP-based CT imaging pipeline, but with learnable summation weights to account for potential data redundancy and differences caused by the completely different strategies used in iCT-Net to filter data. The parameters in all layers are directly learned from the input data and training images in the training dataset. The iCT-Net architecture enables us to reconstruct images with a 512×512 matrix since the number of parameters is on the order of $O(N^2 \times N_c)$, which is in contrast with $O(N^4)$ in other architectures [46]. Here, N denotes the image matrix size and N_c denotes the number of detector elements.

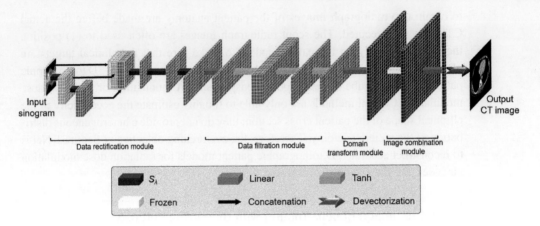

Figure 6.2 Architecture of iCT-Net. The proposed deep neural network consists of a total of 12 layers (L1–L12). The L11 layer is a frozen layer, which means that parameters in this layer are not updated in the training process. Both linear and nonlinear activations are used as indicated in the graphics. S_λ is a hard thresholding activation function. © 2019 IEEE. Reprinted, with permission, from [44]

Training Strategies, Training Data, and Quantitative Image Evaluation Metrics

A two-stage training strategy was used to train iCT-Net: (1) a pre-training stage and (2) a fine-tuning stage. In the pre-training stage, the iCT-Net architecture was trained module by module using numerical simulation data. In the fine-tuning stage, the entire pre-trained iCT-Net architecture was trained from end to end using a carefully curated large experimental dataset acquired from a 64-slice MDCT scanner (Discovery CT750 HD, GE Healthcare, Waukesha, WI). The final end-to-end training using experimental data is crucial for the trained iCT-Net model to be able to handle physical confounding factors such as beam hardening, scatter, the X-ray tube heel effect, and the limited dynamic range of X-ray detectors.

The curated training datasets include: (1) numerical simulation data; (2) experimental phantom data; and (3) clinical human subject data. For all three types of training data, raw counts, log-transformed sinogram projection data, and the corresponding FBP reconstructed images were prepared.

To establish the generalizability of the trained iCT-Net, its reconstruction performance was tested using experimental data acquired using totally different phantoms and different data acquisition conditions, such as different tube potentials and different radiation dose levels. The model training was done using chest CT protocols, but the generalizability tests were also performed for head CT exams and abdominal CT exams.

Reconstruction accuracy is quantified using two standard metrics: the relative root mean square error (rRMSE) and the structural similarity index metric (SSIM), defined as follows:

$$\text{rRMSE} = \frac{\|x - x_0\|_2}{\|x_0\|_2} \times 100\%, \tag{6.18}$$

where x denotes the reconstructed image and x_0 denotes the corresponding reference image,

$$\text{SSIM}(x, x_0) = \frac{(2\mu_x \mu_{x_0} + a_1)(2\sigma_{x,x_0} + a_2)}{(\mu_x^2 + \mu_{x_0}^2 + a_1)(\sigma_x^2 + \sigma_{x_0}^2 + a_2)}, \qquad (6.19)$$

where μ_x denotes the mean value of x, σ_x^2 denotes the variance of x, and similar properties are defined for the reference image x_0. In Eq. (6.19), σ_{x,x_0} is the covariance of x and x_0 and $a_1 = 10^{-6}$ and $a_2 = 3 \times 10^{-6}$ are two constants which are used to stabilize the division with a weak denominator. The size of each of the ROIs used to calculate SSIM was 30 mm × 30 mm.

In addition to the above metrics to assess reconstruction accuracy, line profiles across images were also used to demonstrate the reconstruction accuracy across the images.

Results and Performance Evaluations

Short-scan acquisition with both dense and sparse view samplings Results presented in Fig. 6.3 demonstrate that iCT-Net can be trained to accurately reconstruct images given the acquisition condition that the angular range is large enough to satisfy the Tuy data-sufficiency condition for the entire field of view, but the view angle sampling interval can be both dense, as required in standard clinical conditions, and can also be sparse by a factor 4, i.e., only 25% of the acquired view angles are used in reconstruction.

To demonstrate the generalizability of iCT-Net with quantifiable reconstruction accuracy, numerical simulation data without added noise were generated from human CT images and the trained iCT-Net was directly applied to these sinogram data to reconstruct images. As shown in Fig. 6.3, iCT-Net is able to accurately reconstruct images with lower overall rRMSE values and higher SSIM values compared with the corresponding FBP and CS reconstructions. Note that iCT-Net was trained using both numerical simulation data with added noise and experimental data with both noise and nonideal confounding factors. The prediction results for the clean numerical simulation data provide us with the needed benchmark of reconstruction accuracy. To further demonstrate that the trained iCT-Net is able to reconstruct images directly from experimental data, sinogram data from human subject cases were used as input to reconstruct images. As shown in Fig. 6.3, iCT-Net is able to accurately reconstruct images for sparse-view problems.

Super-short-scan acquisition with both dense and sparse view samplings As was shown in Fig. 6.1, when the view angle range is shorter than the standard short-scan condition, even if the detector is wide enough to cover the entire FOV, the data are not sufficient for the reconstruction of the entire FOV since there will be missing data in the corresponding Fourier space. Under this data acquisition condition, one cannot expect that conventional FBP reconstruction can accurately reconstruct the entire image for either the dense view or sparse view reconstruction problem. In contrast, iCT-Net can be trained to accurately reconstruct images for the regions inside FOV for which the Tuy data-sufficiency condition is satisfied. When the angular range is 180°, the Tuy data-sufficiency condition is satisfied for the image pixels in the upper half of the FOV. As shown in Fig. 6.4, the trained iCT-Net is able to accurately

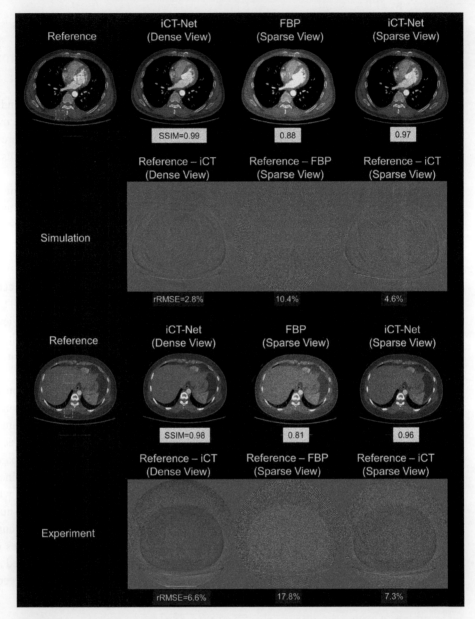

Figure 6.3 The iCT-Net **short-scan** reconstruction results for **dense view** and **sparse view** conditions. (First column) Ground-truth and reconstructed results from (second column) the trained iCT-Net for dense view conditions, (third column) FBP for sparse view conditions, (fourth column) the trained iCT-Net for sparse view conditions. The first and third rows were reconstructed from numerical simulation data without added noise and from experimental human subject data, respectively. The second and fourth rows are difference images obtained by subtracting the reconstructed images from the ground truth. W/L = 1000/100 HU (Houndsfield units) for reconstructed images and 300/0 HU for the difference images. The ROIs inside the images show where the SSIM calculations were performed. © 2019 IEEE. Reprinted, with permission, from [44]

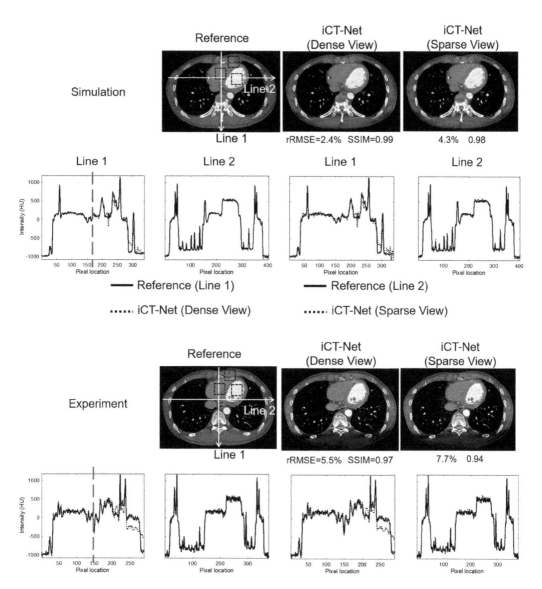

Figure 6.4 The iCT-Net **limited angle** reconstruction results for **dense view** and **sparse view** conditions. (second column) Ground truth and reconstructed results from (third column) the trained iCT-Net for dense view conditions, (fourth column) the trained iCT-Net for sparse view conditions. The first and third rows were reconstructed from numerical simulation data without added noise and from experimental human subject data, respectively; intensity is plotted versus pixel location. The reconstruction accuracy (second and fourth rows) is obtained by comparing plots of the iCT-Net reconstructed image values along a vertical line (Line 1) and a horizontal line (Line 2) crossing the FOV with the corresponding plots from the ground truth. W/L = 1000/100 HU for the reconstructed images. The ROIs inside the images show where the SSIM calculations were performed. © 2019 IEEE. Reprinted, with permission, from [44]

reconstruct the image content in the same upper half of the FOV. This is also true when the view angles are down sampled by a factor 4, as shown in Fig. 6.4.

Interior tomography acquisition with both dense and sparse view samplings
As illustrated in Fig. 6.1, the interior tomography reconstruction problem is a notoriously difficult one. Very limited analytical results are available when view angles are densely sampled [47–49]. When the view angles are down-sampled, analytical results are currently not available. This is one of the most difficult CT image reconstruction problems but the deep-learning framework can provide us with some encouraging results, with some surprises.

As shown in Fig. 6.5, trained iCT-Net can be used to reconstruct interior tomography data with high quantification accuracy for both dense-view-angle and sparse-view-angle sampling conditions.

We will revisit the interior tomography reconstruction problem in the next section using the combined analytical DBP and deep-learning method to leverage the available human knowledge of interior tomography and deep learning.

Generalizability Study of iCT-Net

It is important to fully recognize the potential challenges met when the deep-learning method is being developed. One of the most fundamental challenges is that there is no theoretical guarantee for its generalizability. Therefore, it is not known in advance how well a trained model can be applied to new data, particularly when data acquisition conditions have changed. Therefore, for any deep-learning model, we will have to pay extra attention to testing the applicable conditions for trained deep-learning models. There is no exception for trained iCT-Net.

Since iCT-Net was trained using chest CT data acquired at either 100 or 120 kV tube potentials, it is important to test whether the trained iCT-Net model can be directly applied to CT data acquired at other tube potentials and for other body parts. Our initial generalizability test results are presented in Fig. 6.6; as shown here, the trained iCT-Net is indeed able to accurately reconstruct images directly from experimental data acquired at different tube potentials. As shown in Fig. 6.7, the trained iCT-Net is found to be able to accurately reconstruct abdominal CT cases even with significant anatomical change: this subject only has one kidney instead of the normal two. Note that this generalizability test was performed for all listed data acquisition conditions and for both dense view and sparse view samplings. The results demonstrated that the trained iCT-Net is relatively robust against ordinary clinical scan conditions. However, this should not yet be considered as the end of testing. The model should be constantly tested for even more harsh clinical scanning conditions to pinpoint any potential limitations in generalizability.

6.3.2 ScoutCT-Net: Reconstruction of 3D Tomographic Patient Models from Two-Scout-View of Projections

As shown in the previous subsection, with the FBP-inspired iCT-Net architecture the end-to-end trained deep-learning model is able to reconstruct diagnostic CT images with a wide variety of data acquisition conditions and strong generalizability. How-

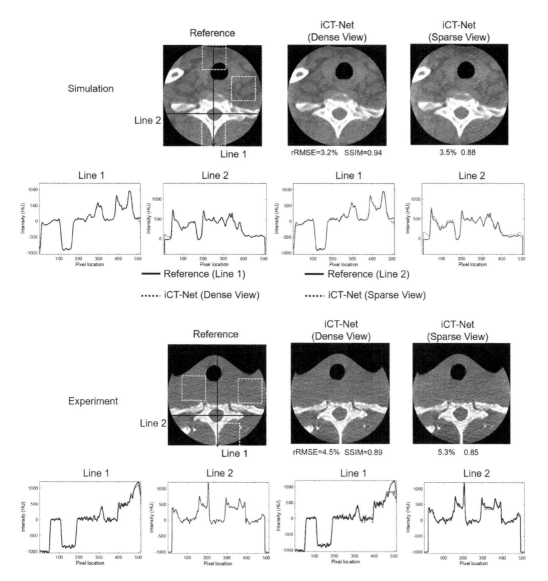

Figure 6.5 The iCT-Net **interior** tomographic reconstruction results for **dense view** and **sparse view** conditions. (second column) Ground truth. After reconstruction, the central portion of the image corresponding to a truncated FOV ($\varnothing = 12.5$ cm) was cropped in this interior problem. Reconstructed results from (third column) the trained iCT-Net for dense view condition and (fourth column) the trained iCT-Net for sparse view conditions. The first and third rows were reconstructed from numerical simulation data without added noise and from anthropomorphic chest phantom data, respectively; intensity is plotted versus pixel number. Reconstruction accuracy (second and fourth rows) is obtained by comparing plots of iCT-Net reconstructed image values along a vertical line (Line 1) and a horizontal line (Line 2) crossing the FOV with the corresponding plots from the ground truth. W/L = 1000/100 HU for the reconstructed images. The ROIs inside the images show where the SSIM calculations were performed.
© 2019 IEEE. Reprinted, with permission, from [44]

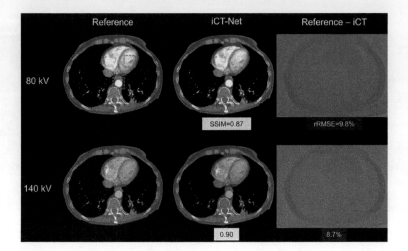

Figure 6.6 The iCT-Net reconstruction results for experimental human subject data acquired at two different X-ray tube potentials with a short-scan angular range. Reference images (first column) were generated with a standard FBP reconstruction with a Ram-Lak filter using data from 644 view angles densely sampled across a short-scan angular range. The trained iCT-Net (second column) was applied to the same input data to reconstruct the corresponding images. Difference images (third column) were generated by subtracting the iCT-Net image from the reference image. W/L = 1000/0 HU for the reconstructed images and 400/0 HU for the difference images. The ROIs inside the images show where the SSIM calculations were performed. © 2019 IEEE. Reprinted, with permission, from [44]

ever, the trained iCT-Net deep model is not capable of reconstructing a 3D tomographic patient model for radiation dose modulation from two scout views. In this subsection, we will show that a dedicated neural network architecture, ScoutCT-Net, shown in Fig. 6.8, can be used for this new task.

Architecture and Data Curation for Training and Testing Purposes

In this architecture, the deep neural network takes the projection data from the two radiograph localizer views with size 888×5 pixels each as input to produce the desired 3D tomographic patient model image slab, with size $288 \times 288 \times 5$ voxels. In order to map the 2D projections onto the 3D image space, the first layer of the network consists of a direct backprojection operation, which is then followed by 24 convolutional layers arranged in a simplified U-Net architecture with four vertical levels [50]. The network was trained using a conditional adversarial loss [51, 52] with a 24-layer convolutional discriminator using a ResNet architecture [53, 54]. The detailed structures of the architecture are described as follows. The U-Net encoder uses the layers C64 – C64 – sc1 – MPs2 – C128 – C128 – sc2 – MPs2 – C256 – C256 – sc3 – MPs2 – C512 – C512 – sc4 – MPs2 – C1024 – C1024, where Ck stands for a convolutional layer with k filters, sc$i$ represents a skip connection at level i and MPs2 represents a max pooling operation with stride 2. Similarly, the U-Net decoder layers are described as follows: TC512 – sc4 – C512 – C512 – TC256 – sc3 – C256 – C256 – TC128 – sc2 – C128 – C128 – TC64 – sc1 – C64 – C64 – C64 – C5, where TCk represents a transpose convolution operation with k filters. All convolutional layers use a 3×3 spatial filter configuration with ReLU activations and no batch normalization.

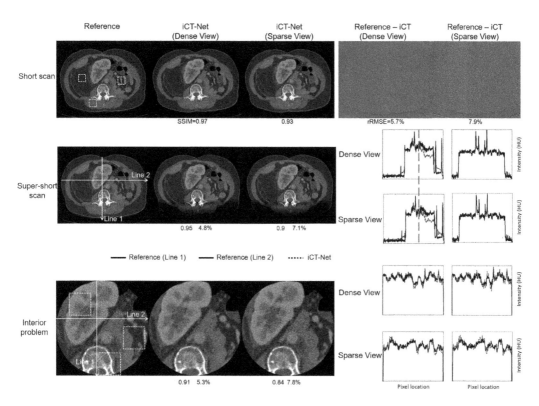

Figure 6.7 The iCT-Net reconstruction results for experimental human subject data acquired in an abdomen–pelvis scan protocol with short-scan angular range, limited angle, and interior tomography problems. The dense view reconstruction results are presented in the second column and sparse view reconstruction results are presented in the third column. The corresponding reference images were generated by applying a standard FBP method with a Ram-Lak filter at full FOV ($\varnothing = 50$ cm) from 644 view angles densely sampled across a short-scan angular range. Note that the central portion of the FBP reconstruction without truncation has been cropped to generate the reference image for the interior problem with the truncated FOV ($\varnothing = 12.5$ cm). The trained iCT-Net was applied to reconstruct images for each of the corresponding data acquisition conditions. W/L = 400/0 HU for the reconstructed images. The ROIs inside the images show where the SSIM calculations were performed. © 2019 IEEE. Reprinted, with permission, from [44]

The backprojection and forward projection layers are implemented as fully connected layers with weights predefined by the system geometry and no trainable parameters. The ResNet configuration for the discriminator network used only during training was the following: C32s2 – C32 – C64 – sc – C64 – C64 – sc – C64 – C64 – sc – C64 – C64 – sc – (C128s2, AP – C128) – C128 – sc – C128 – C128 – sc – C128 – C128 – sc – (C256s2, AP – C256) – C256 – sc – C256 – C256 – sc – C256 – C256 – sc – FC1, where Ck stands for a convolutional layer with k filters and Cks2 represents a convolutional layer with k filters and stride 2. Here, sc represents the characteristic skip connection of the ResNet architecture. Finally, the AP-Ck notation represents an auxiliary layer consisting of a 3×3 average pooling operation with stride 2 followed by a convolutional layer of $1 \times 1 \times k$ filters to match the size and number of channels when spatial down-sampling occurs in the network. The final FC1 layer represents

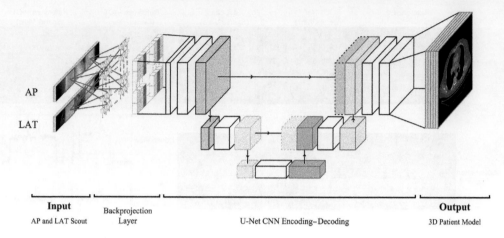

Figure 6.8 ScoutCT-Net deep-neural-network architecture. ScoutCT-Net takes as input two radiograph localizer views with size 888 × 5 pixels each, and it outputs an image slab with size 288 × 288 × 5 voxels. In order to map the 2D projections into the 3D image space, the first layer of the network performs a direct backprojection operation, which is then followed by 24 convolutional layers arranged into a simplified U-Net architecture with four vertical levels (only three vertical levels are shown for simplicity). © 2022 John Wiley and Sons. Reprinted, with permission, from [45]

a fully connected layer with one output. The first layer uses a 5 × 5 receptive field and all other convolutional layers use a 3 × 3 spatial filter configuration with ReLU activations and no batch normalization.

The model was implemented using TensorFlow and the network parameters were initialized using the variance scaling method [55] and trained from scratch using stochastic gradient descent using the adaptive moment estimation (ADAM) optimizer with a learning rate of 2.0×10^{-4}, momentum parameters of $\beta_1 = 0.9$ and $\beta_2 = 0.999$, and batch size of 32. To avoid exploding gradients we adopted gradient clipping in the range $[-1.0, 1.0]$.

To train the ScoutCT-Net architecture, a training dataset with more than 1.1 million CT images and their corresponding projection data from 4214 clinically indicated CT studies were retrospectively collected. Once the ScoutCT-Net model was trained, 3D localizers were reconstructed for a validation cohort and the results were analyzed and compared with the standard multidetector CT (MDCT) images. The data were split into three cohorts consisting of 3790 (90%), 212 (5%), and 212 (5%) CT studies for training, testing, and validation of the deep-learning model, respectively. The total number of unique data samples used was 1 001 496 for training, 55 136 for testing, and 55 136 for validation.

Reconstruction Results and Radiation Dose Modulations from the ScoutCT-Net Reconstructed 3D Patient Model

Figure 6.9 shows axial and coronal images of the ScoutCT-Net results with the reference MDCT images for comparison. As one can observe, the trained ScoutCT-Net is able to reconstruct a 3D tomographic patient model with heterogeneous structures;

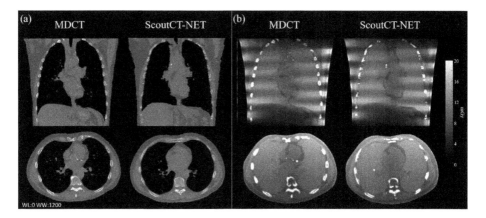

Figure 6.9 Coronal and axial images of the ScoutCT-Net results compared with the reference MDCT images as well as the corresponding radiation dose simulations from a standard helical chest CT. (a) Overall, one notices a similar image impression between the ScoutCT-Net image reconstructed from only two projection views compared with the MDCT images. All major anatomical structures can be clearly delineated in ScoutCT-Net images. (b) Nearly identical radiation dose distributions were calculated from ScoutCT-Net images and MDCT. © 2022 John Wiley and Sons. Reprinted, with permission, from [45]

this cannot be accomplished with the current 3D patient model reconstruction methods used in clinical CT scanners. Actually, the ScoutCT-Net reconstructed images are very similar to the corresponding helical CT images acquired from the same patient. However, we emphasize that the ScoutCT-Net reconstructed CT image should not be expected to have the same quality as that used for clinical diagnosis. Rather, the reconstruction accuracy should be evaluated in terms of radiation dose prescription accuracy. To do so, the radiation dose distribution of a CT acquisition can be calculated using the standard Monte Carlo dose calculation packages and using the standard clinical helical CT patient image volume as the ground-truth 3D patient model. The same radiation dose Monte Carlo simulations were performed using the ScoutCT-Net reconstructed images as the 3D patient model. The results are presented in Fig. 6.9. As one can clearly see, the dose distribution from the ScoutCT-Net patient model closely resembles that from the helical CT patient model. The stripe structure in the Monte Carlo dose simulations is an artifact of the helical CT acquisitions.

To further demonstrate the clinical significance of the ScoutCT-Net reconstructed 3D patient model, radiation dose modulation profiles $\exp(\mu L/2)$ were calculated from the X-ray path length L. The dose modulation profiles calculated from the ground-truth 3D patient model, from the ScoutCT-Net patient model, and from the current clinical standard using the uniform patient model are presented in Fig. 6.10. As one can observe, the current dose modulation profiles using the uniform patient model do not yield optimal dose modulations in CT exams. In contrast, modulations using the ScoutCT-Net reconstructed patient model yield the desired personalized radiation dose modulations in CT exams.

One may wonder how much radiation dose can be saved if radiation dose modulation profiles are prescribed using the accurate 3D patient model when compared with that produced by the ScoutCT-Net deep-learning model? To answer this question,

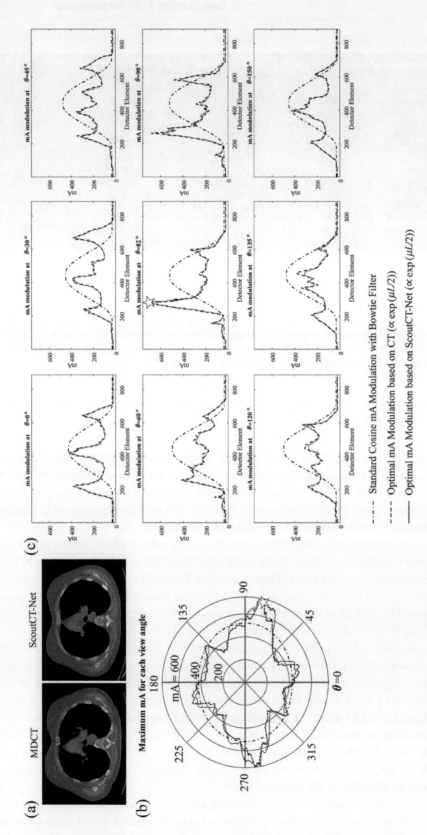

Figure 6.10 For comparison, plots for different mA-modulation techniques in a female patient from our validation cohort, including standard cosine modulation with a modern CT bowtie filter, the optimal mA modulation based on the minimum noise variance criteria using the original CT images for reference and also with the optimal mA modulation based on the reconstructed ScoutCT-Net images: (a) shows images reconstructed in clinical scans and reconstructed ScoutCT-Net and (b) shows a polar plot of the maximum mA value for each projection view; (c) shows mA-modulation profiles from multiple view angles including the acquisition view, with the maximal patient attenuation at 82° labeled by a star. © 2022 John Wiley and Sons. Reprinted, with permission, from [45]

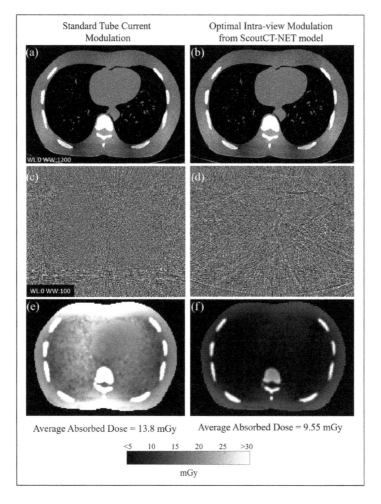

Figure 6.11 Results of reconstructed FBP axial images of an anthropomorphic chest phantom acquired with (a) a standard tube current modulation and (b) the simulated optimal intra-view modulation based on ScoutCT-NET images. The images in panels (c) and (d) show the corresponding noise-only images; in (c) one can see highly the nonuniform noise variance across the image field of view in the standard tube current modulation image. In contrast, (d) shows a relatively uniform noise variance across the field of view with optimal intra-view modulation calculated from ScoutCT-NET images. Images (e) and (f) show the gray-scaled radiation dose distribution from Monte Carlo simulations. © 2022 John Wiley and Sons. Reprinted, with permission, from [45]

two radiation dose modulation schemes were investigated: one was the standard dose modulation scheme used in current clinical CT scans, and the other was the optimal radiation dose modulation profiles for the same phantom. The results are presented in Fig. 6.11. For the same noise variance level in the reconstructed images, the average absorbed dose in the optimal intra-view modulation is 9.55 mGy based on ScoutCT-Net images and 9.84 mGy based on clinical CT images, which may be compared with 13.8 mGy in the standard tube current modulation approach, which represents a 28.6%–30.7% reduction in radiation dose with minimal impact in overall image noise variance.

Before moving on, we want to emphasize that the key point of ScoutCT-Net was not to reconstruct images from two view angles. Our task was to reconstruct a 3D patient model that can provide a better radiation dose modulation prescription. This is an imaging task-driven reconstruction problem. As a matter of fact, without a specific and well-defined imaging task in deep-learning applications, it is not very meaningful to discuss tomographic image reconstruction using projection data with extremely sparse view angle acquisitions. After all, one can also generate tomographic CT images from a noise-only projection data input using the well-known generative adversarial networks (GANs) [51], i.e., one can reconstruct images without projection data from any view angles. Therefore, the problem of image reconstruction from extremely sparse view angles must be pitched in appropriate practical applications such that the success of the reconstruction tasks can be quantitatively evaluated.

6.4 Deep Learning in CT Reconstruction: Hybrid Deep Learning with DBP

In this section, we demonstrate that deep-neural-network training can be done in the DBP domain to address various CT reconstruction problems. However, unlike iCT-Net, applications of the DBP-domain training are problem-specific, as will now be explained.

6.4.1 Region-of-Interest (ROI) Tomography

Interior tomography aims to obtain a region-of-interest (ROI) image by irradiating only within the ROI and is useful when the ROI within a patient's body is small (such as in cardiac and dental imagings). In some applications, such as portable C-arm CT, interior tomography has a cost benefit owing to the small detector size. However, analytical CT reconstruction methods such as filtered backprojection (FBP) generally produce images with severe cupping artifacts owing to the truncation of the projection data along the transverse direction.

A simple method to reduce the cupping artifacts is provided by a projection extrapolation [56], but it is an imprecise approximation method since its performance is sensitively dependent on the fitting process. Other methods have been developed based on statistical IR algorithms with several penalty terms such as the total variation (TV) [49] and generalized L-spline [57, 58].

The ROI tomography system can be regarded as the restriction of the Radon transform $\mathcal{R}f(\theta, s)$ to the region $\{(\theta, s) : |s| < R\}$, where R is the radius of the ROI, and the truncated Radon transform is formulated as $\mathcal{T}_R \mathcal{R} f$, where $\mathcal{T}_R$ is the truncation function in the range $[-R, R]$:

$$\mathcal{T}_R \mathcal{R} f(\theta, s) = \begin{cases} \mathcal{R}f(\theta, s), & |s| < R \\ 0, & \text{otherwise} \end{cases}. \tag{6.20}$$

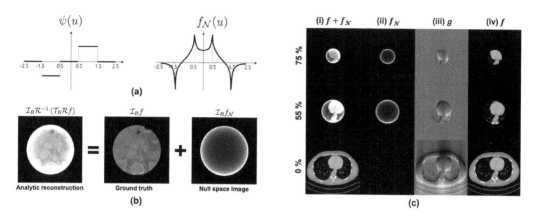

Figure 6.12 (a) 1D space signal, and (b) 2D null space image. (ci) analytic reconstruction image, (cii) null space image, (ciii) DBP measurement, and (civ) ground truth for ROI sizes 75%, 55%, and 0%.

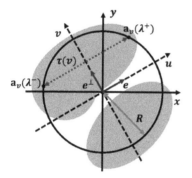

Figure 6.13 A coordinate system for interior tomography.

As shown in Fig. 6.12(b), the FBP method or so-called inverse Radon transform $\mathcal{R}^{-1}$ for the truncated Radon transform $\mathcal{T}_R \mathcal{R} f$ provides a combined result of a ground truth $\mathcal{I}_R f(x)$ and a null space image $\mathcal{I}_R f_{\mathcal{N}}(x)$, where $\mathcal{I}_R$ is the indicator function between $[-R, R]$:

$$\mathcal{I}_R f(x) = \begin{cases} f(x), & |x| < R \\ 0, & \text{otherwise} \end{cases}. \tag{6.21}$$

A 2D interior tomography problem in the FBP domain can be defined to find the unknown image $\mathcal{I}_R f(x)$ from measurements of the truncated Radon transform $\mathcal{T}_R \mathcal{R} f(\theta, s)$. The 2D problem in the FBP domain can be equivalently described in the DBP domain as a 1D interior tomography problem to find the unknown image $f_v(u)$, $|u| < \tau$, on the chord line indexed by v using the DBP measurements $g_v(u)$, $|u| < \tau$, where $\tau := \tau(v)$ denotes the chord line defined by the 1D restriction of the ROI (see Fig. 6.13). Even though the finite Hilbert transform is an analytical reconstruction method for reconstructing the image $f_v(u)$ from the DBP measurement $g_v(u)$, it provides the image $f_v(u)$ as well as the null space image $f_\mathcal{N}(u)$ (see Fig. 6.12(a)) if the DBP measurement $g_v(u)$ is restricted to $[-\tau, \tau]$. That is, since an infinite number of f_v share the same truncated DBP measurement $\mathcal{I}_\tau g_v$, the inverse problem of finding the unknown image $\mathcal{I}_\tau f_v$ from the truncated DBP measurement $\mathcal{I}_\tau g_v$ is strongly ill-posed.

The main technical difficulty for solving the inverse problem of interior tomography is to find the null space of the Hilbert transform,

$$\mathcal{H}_\tau f_\mathcal{N}(u) = 0, \quad |u| < \tau, \tag{6.22}$$

where $\mathcal{H}$ defines the Hilbert transform. Indeed, $f_\mathcal{N}(u)$ is represented by

$$f_\mathcal{N}(u) = -\frac{1}{\pi} \int_{\mathbb{R} \setminus [-\tau, \tau]} \frac{\psi(s)}{u - s} \, ds, \tag{6.23}$$

for any function $\psi(u)$ outside the ROI. Figure 6.12(a) shows a simple example of a 1D null space signal $f_\mathcal{N}$ for a given ψ, and it can be seen that singularities of the null space signal $f_\mathcal{N}$ exist at $u = \pm 0.5$. The characteristics extend to 2D cases as shown in Fig. 6.12(b). In general, the 2D null space images are called cupping artifacts since their shape looks like a cup.

A simple way of using deep learning to find the unknown image f is to remove the null space image $f_\mathcal{N}$ from the image $f + f_\mathcal{N}$ generated using analytical reconstruction methods (Fig. 6.14). The reconstructed images can be simply defined by

$$f + f_\mathcal{N} = \mathcal{M} p, \tag{6.24}$$

where $\mathcal{M}$ denotes the FBP or the BPF method, and p is the zero-padded truncated projection data $\mathcal{T}_R \mathcal{R} f$. Then, an objective function to train the neural network $\mathcal{Q}$ can be formulated as

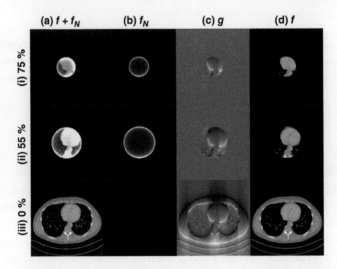

Figure 6.14 (a) Analytic reconstruction image, (b) null space image, (c) DBP measurement, and (d) ground truth, all depending on the ROI sizes: (i) 75%, (ii) 55%, and (iii) 0%.

$$\min_{\mathcal{Q}} \sum_{i=1}^{N} \left\| f^{(i)} - \mathcal{Q}\left(\mathcal{M} p^{(i)}\right) \right\|^2, \qquad (6.25)$$

where $\{(f^{(i)}, p^{(i)})\}_{i=1}^{N}$ denotes the training datasets consisting of the ground truth f and its truncated projection data p. Figure 6.15(a) shows the FBP-domain neural network $\mathcal{Q}$, and this is a simple method to solve the inverse problem of the interior tomography and provides significantly improved performance over the conventional iterative methods.

However, the FBP-domain deep-learning method $\mathcal{Q}$ has a generalizability problem owing to the singularities of the null space images $f_\mathcal{N}$ at the ROI boundaries. Even though this would not matter if we always used a fixed ROI size to measure the truncated projection data, in fact the size of the ROI changes frequently depending on the patient's size and clinical procedures such as interventional imaging, cardiac imaging, etc. Therefore, it is a very important condition to satisfy sufficient generalizability for the ROI size using interior tomography.

In order to avoid singularities caused by the cupping artifacts at the ROI boundaries and to sustain sufficient generalizability for various ROI sizes, the use of the DBP measurement g is one solution. That is, the singularities of the FBP (or BPF) reconstruction results $f + f_\mathcal{N}$ change, depending on the ROI size, as shown in Fig. 6.12(ci), but the DBP measurements g inside the ROI (see Fig. 6.12(ciii)) do not change. Before using the truncated DBP measurements $\mathcal{I}_\tau g$ as inputs to the neural network we need to reformulate the finite Hilbert transform, which reconstructs the image f from the DBP measurement g, to handle the truncated DBP measurement $\mathcal{I}_\tau g_\nu$. If we assume $\tau = 1$, the generalized finite Hilbert transform is represented as

$$\mathcal{I}_{\tau=1} f(u) = \frac{\epsilon(u)}{\pi\sqrt{1-u^2}} - \frac{1}{\pi\sqrt{1-u^2}} \int_{-1}^{1} \frac{\sqrt{1-s^2}g(s)}{u-s} ds, \qquad (6.26)$$

where the offset function $\epsilon(u)$ is defined as

$$\epsilon(u) := c - \int_{-1}^{1} \frac{\sqrt{1-s^2}g_\mathcal{N}(s)}{u-s} ds, \qquad (6.27)$$

with

$$g_\mathcal{N}(u) := -\frac{1}{\pi} \int_{\mathbb{R}\setminus[-1,1]} \frac{f(s)}{u-s} ds. \qquad (6.28)$$

The major problem with this formula is that the offset $\epsilon(u)$ is difficult to calculate analytically. Fortunately, the generalized finite Hilbert transform (6.26) is suitable to use as the working process of the neural network, and the formula (6.26) can be converted to

$$(w_\tau \odot \mathcal{I}_\tau f_\nu)(u) = \epsilon_\nu(u) - w_\tau \odot (h * g)(u), \quad |\nu| < \tau, \qquad (6.29)$$

where $\tau := \tau(u)$ denotes the window size for the chord line index ν (see the dotted chord line in Fig. 6.13), $\odot$ is an element wise product, the star operator denotes convolution, $w_\tau(u)$ is an analytical form of weighting defined as

$$w_\tau = \pi\sqrt{\tau^2 - u^2}, \quad |u| \leq \tau, \tag{6.30}$$

and h and $\epsilon(u)$ are the convolutional kernel for the Hilbert transform and the unknown offset function (or bias function), respectively. That is, the weighted finite Hilbert transform (6.29) is formulated as a convolution operation with a bias function, and the form is similar to a convolutional neural network (CNN). Then, after the reconstruction, the weight w_τ can be removed and the final image f_v can be obtained for all $|v| < \tau$. On the basis of the above mathematical derivation, a DBP-domain neural network $\mathcal{S}$ can be trained as

$$\min_{\mathcal{S}} \sum_{i=1}^{N} \left\| f^{(i)} - \mathcal{S}\left(g^{(i)}\right) \right\|^2, \tag{6.31}$$

where $\{(f^{(i)}, g^{(i)})\}_{i=1}^{N}$ denotes the training datasets consisting of the ground truth f and its DBP measurements g.

To verify the network performance, the FBP-domain and the DBP-domain neural networks are designed as a standard U-Net architecture. Ten subject datasets from the American Association of Physicists in Medicine (AAPM) Low-Dose CT Grand Challenge were used in this experiment. The projection data were numerically generated using a fan-beam projection operator with 1440 detectors and 1200 views. Specifically, out of ten patients, eight patient's datasets were used as training datasets (3720 slices), one patient's dataset was used as a validation dataset (254 slices), and the other dataset was used as the test set (486 slices). The networks were trained twice with different ROI sets. In the first training phase, the truncated projection data were collected from 380 and 1440 detectors. In the next phase, various ROI images from 240, 380, 600, and 1440 detectors were used as the training datasets to improve generalizability.

In contrast with the FBP-domain neural network $\mathcal{Q}$ in Fig. 6.15(a), the DBP-domain neural network $\mathcal{S}$ takes truncated DBP measurements as input and has no singularities in the input data, as shown in Fig. 6.15(b). Because of this, the DBP-domain neural network $\mathcal{S}$ provides much better generalization performance than the FBP-domain neural network $\mathcal{Q}$, so such networks can be used for any ROI size and various acquisition conditions.

6.4.2 Cone-Beam Artifact Removal

Cone-beam X-ray CT with multiple detector rows is often used for interventional imaging, dental CT, etc., because high-resolution projection data can be measured using a relatively simple X-ray source trajectory. Although the Feldkamp, Davis, and Kress (FDK) algorithm [59] has been used as a standard reconstruction method for the cone-beam trajectory, unfortunately, the FDK algorithm suffers from cone-beam artifacts as the cone angle increases. Mathematically, the cone-beam artifact arises from inherent defects arising because the cone-beam CT does not satisfy Tuy's condition [60] on a circular trajectory. In the Fourier domain, the artifact appears as

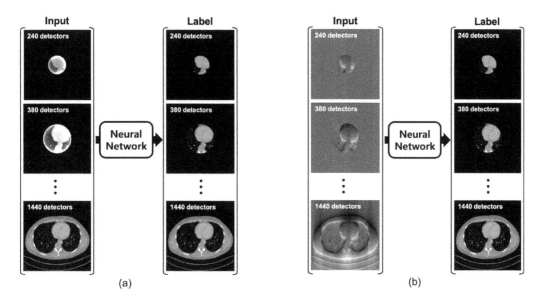

Figure 6.15 Flowcharts of (a) an FBP-domain network $\mathcal{Q}$ and (b) a DBP-domain network $\mathcal{S}$ for various examples of ROI tomography.

missing spectral components at the specific frequencies determined by the scanner geometry [61–63].

To address this issue, some researchers have developed modified FDK algorithms by introducing angle-dependent weighting into the measured projection data [64–66], but these methods only work for small cone angles. Other methods based on model-based iterative reconstruction (MBIR) algorithms [67–70] have been developed by imposing total variation (TV) and other penalties. The role of the penalty functions is to impose constraints to compensate for the missing frequencies. Unfortunately, these algorithms are computationally expensive owing to the iterative routine of 3D projection and backprojection operations.

Using the DBP formulation, Dennerlein *et al.* [71] proposed a very interesting factorized representation of the circular cone-beam reconstruction problem. Specifically, Fig. 6.16 illustrates the geometry of cone-beam CT with circular trajectories. Consider a plane that is parallel to the z-axis and intersects the source trajectory at two locations, $a(\lambda^-)$ and $a(\lambda^+)$. Let $\mathcal{P}$ denote such a plane of interest. Then, the main idea of the factorization methods [71, 72] is to convert the 3D reconstruction problem into a successive 2D problem on the planes of interest. Specifically, on a plane of interest $\mathcal{P}$, the authors in [71] demonstrated the following result:

$$g(t,z) = \pi \int_{-\infty}^{\infty} h_H(t-\tau)\left(f(\tau, z_1(\tau)) + f(\tau, z_2(\tau))\right) d\tau \quad (6.32)$$

where $h_H(t)$ denotes the Hilbert transform, and $z_1(\tau)$ and $z_2(\tau)$ refer to two sources to voxel lines on $\mathcal{P}$. This implies that the cone-beam reconstruction problem can be solved by solving a deconvolution problem in the DBP domain.

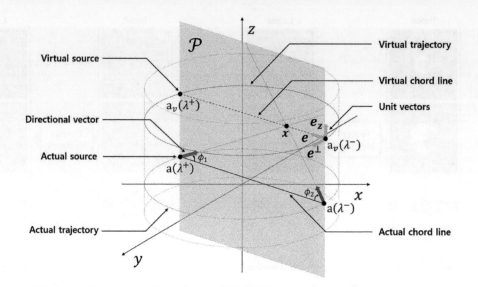

Figure 6.16 Factorization of circular cone-beam geometry.

Although (6.32) seems to suggest that a deconvolution algorithm can recover an artifact-free image $f(t,z)$ from the DBP measurement $g(t,z)$ on each plane of interest $\mathcal{P}$, there are two technical difficulties. First, the grid for the unknown image is not a Cartesian grid based on the $z_1(t)$ and $z_2(t)$ coordinates parameterized by t, because $f(t,z_1(t))$ and $f_2(t,z_2(t))$ can be considered as deformed images of $f(t,z)$ on new coordinate systems. Therefore, standard deconvolution methods are difficult to use satisfactorily. Another technical difficulty is that the deconvolution problem is strongly ill-posed since the DBP measurement $g(t,z)$ consist of two deformed images, $f(t,z_1(t))$ and $f(t,z_2(t))$. To address the above limitations, the authors of [71] proposed a regularized matrix inversion method after discretization with respect to the fixed Cartesian grid for the unknown images, but it is computationally expensive and sensitive to many of the hyperparameters that are used.

Deep learning is an efficient method to address the technical difficulties. A DBP-domain network work $\mathcal{S}$ can be trained as

$$\min_{\mathcal{S}} \sum_{i=1}^{N} \left\| \boldsymbol{f}^{(i)} - \mathcal{S}\left(\boldsymbol{g}^{(i)}\right) \right\|^2, \tag{6.33}$$

where $\{(\boldsymbol{f}^{(i)}, \boldsymbol{g}^{(i)})\}_{i=1}^{N}$ denotes the training datasets composed of the ground truth $\boldsymbol{f}$ and its DBP measurements $\boldsymbol{g}$, and

$$\mathcal{S}: \boldsymbol{g} \mapsto \boldsymbol{f} \tag{6.34}$$

refers to the inverse mapping from the DBP measurement $\boldsymbol{g}$ to the image $\boldsymbol{f}$. Specifically, as shown in Fig. 6.17, the datasets $\{(\boldsymbol{f}^{(i)}, \boldsymbol{g}^{(i)})\}_{i=1}^{N}$ are collected in both the coronal and sagittal directions, and the DBP-domain neural network is trained to solve a deconvolution problem associated with the Hilbert transform. Then, the 3D volumes

6 Deep Learning in CT Reconstruction

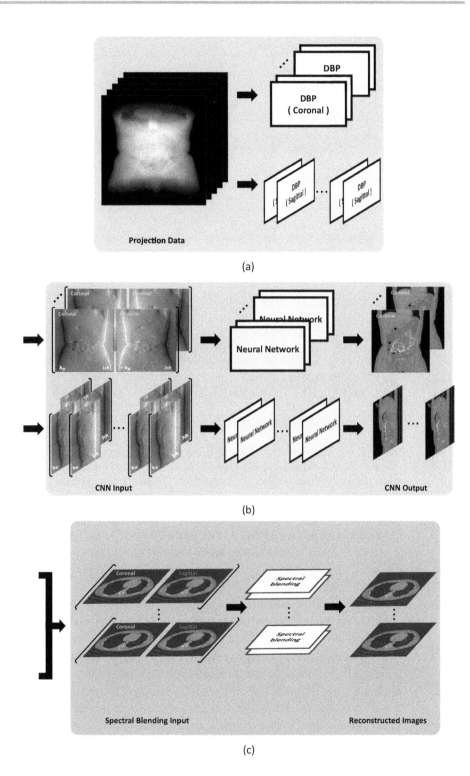

Figure 6.17 Overview of DBP-domain deep learning for cone-beam artifact removal. (a) The module generating DBP measurements along the coronal and sagittal directions, respectively. (b) and (c) The DBP neural network and spectral blending parts.

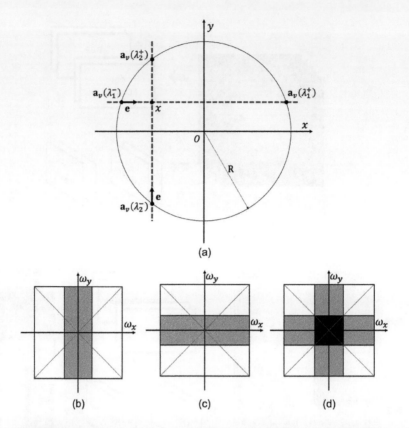

Figure 6.18 The missing frequency region in the cone-beam geometry. (a) The top view of a point $x = (x, y, z)$ and a source trajectory. (b), (c) The missing frequency regions for coronal- and sagittal-direction processing, respectively. (d) The common missing region is shown as the central dark area.

reconstructed from the coronal- and sagittal-direction DBP measurements using the trained DBP-domain network are blended into a 3D volume using a spectral mixing technique.

Although the DBP-domain neural network is programmed to solve the 2D deconvolution problem for each plane of interest $\mathcal{P}$, there are missing frequency regions depending on the orientation of $\mathcal{P}$, as shown in Fig. 6.18. Spectral blending is introduced to minimize artifacts arising from the missing frequency regions. Suppose that we are interested in recovering a voxel at $x = (x, y, z)$ from the DBP measurement in Fig. 6.18(a). Now, consider the image reconstruction along the coronal direction where the plane of interest $\mathcal{P}$ is horizontally aligned between $\mathbf{a}_v(\lambda_1^-)$ and $\mathbf{a}_v(\lambda_1^+)$. According to the spectral analysis in [58], the DBP measurement has missing frequency regions along the sagittal direction, as illustrated in Fig. 6.18(b). In the same manner, the DBP measurements along the sagittal direction (see the vertical broken line between $\mathbf{a}_v(\lambda_2^-)$ and $\mathbf{a}_v(\lambda_2^+)$ in Fig. 6.18(a)) suffer from coronal-direction missing frequency regions, as shown in Fig. 6.18(c). Therefore, the deconvolution processes for the coronal-direction (or sagittal-direction) DBP measurements can amplify noise

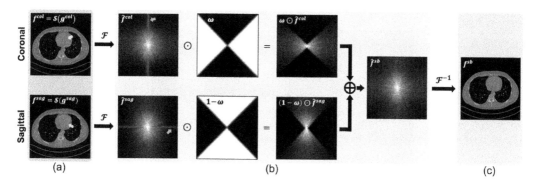

Figure 6.19 Flowchart for the spectral blending. (a) Transverse images of the 3D volume reconstructed by the DBP-domain neural network along the coronal and sagittal directions; the small white arrows indicate the streaking pattern due to the missing frequencies, in (b) these are marked as white arrows in the Fourier domain. (b) The spectral blending process, and (c) the reconstruction image from the spectral blending.

components along the sagittal (or coronal) direction owing to the ill-poseness due to the missing frequency regions. Because for a given voxel at $x = (x, y, z)$ the missing frequency region depends on the orientation of the plane of interest, the missing spectral components can be compensated by combining the reconstruction results from two directions. Specifically, as illustrated in Fig. 6.18(d), the corrected missing frequency region is described by the dark square centered at the origin, which is the intersection of the two orthogonal missing frequency regions in the coronal and sagittal processing.

Figure 6.19 shows a flowchart for the spectral blending used to compensate for the missing frequency regions. The transverse images f^{col} and f^{sag} are the coronal and sagittal 3D volumes reconstructed by the DBP-domain neural network $\mathcal{S}$, and its spectral data $\hat{f}^{col}$ and $\hat{f}^{sag}$ can be calculated by applying the 2D Fourier transform to the transverse images f^{col} and f^{sag}. Streaking artifact patterns are prominently visible in the Fourier domain; these are due to the ill-posedness in the missing frequency regions (see the white arrows in Fig. 6.19(b)). Then, a bow-type spectral weighting ω is applied to suppress the ill-poseness in the missing frequency regions and the weighted spectral data are combined. The spectral blending process can be formulated as

$$f^{sb} = \mathcal{F}^{-1}\left\{\omega \odot \mathcal{F}\left(f^{cor}\right) + (1-\omega) \odot \mathcal{F}\left(f^{sag}\right)\right\}, \tag{6.35}$$

where $\mathcal{F}$ and $\mathcal{F}^{-1}$ are the Fourier and inverse Fourier transforms, respectively, ω is the bow-tie spectral weight, and $\odot$ denotes element wise multiplication.

Figure 6.20 shows the input and target pairs for the FDK-domain and DBP-domain neural networks, and both coronal and sagittal view images are used when the networks are trained. Specifically, for a given plane of interest $\mathcal{P}$, the DBP measurements g can be obtained along the $\vec{e}$ and $-\vec{e}$ directions between two complementary source trajectories $\mathbf{a}_v(\lambda^+)$ and $\mathbf{a}_v(\lambda^-)$, as shown in Fig. 6.18(a). In particular, the two com-

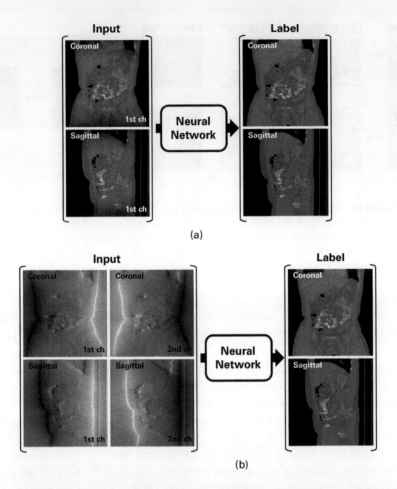

Figure 6.20 (a) FDK-domain network and (b) DBP-domain network for cone-beam artifact removal.

plementary DBP measurements are concatenated along the channel axis and used as the input to the DBP-domain neural network.

To validate the performance of the proposed algorithm, a standard U-Net architecture was used to train the FDK-domain and DBP-domain neural networks. To generate numerical cone-beam datasets, ten subject datasets of the AAPM Low-Dose CT Grand Challenge were used. The following geometric parameters were used for the numerical cone-beam CT. The size of 2D detectors was 1440×1440 elements and the pitch was defined as 1 mm^2. The number of views was set to 1200. The image size was 512×512. Of the ten patients, training and validation datasets were collected from eight patients and one patient, respectively, and the other patient served as the test dataset. The number of datasets in the training set was 8192 (512 slices $\times$ 8 patients $\times$ 2 directions) and the number in the validation and test sets are 1024 (512 slices $\times$ 1 patient $\times$ 2 directions).

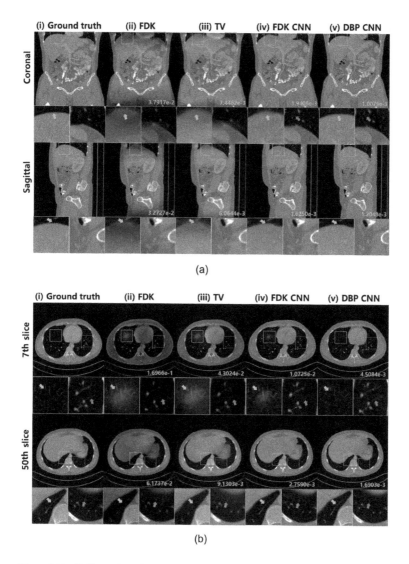

Figure 6.21 (i) Ground truth and reconstructed results from (ii) FDK, (iii) TV, (iv) FDK CNN, and (v) DBP CNN. (a) The transverse results and (b) the coronal and sagittal results. The insets, left and right for each image in the first and third rows, show magnified views. The NMSE values are given at the bottom right-hand corners of the images. The window range is (−1000, 800) [HU]

Figure. 6.21 shows the reconstruction images from FDK, TV, FDK CNN, and DBP CNN. Specifically, in the coronal and sagittal planes (see Fig. 6.21(a)), the FDK results around the midplane ($z = 0$) are similar to the ground truth, but show low intensity and the cone-beam artifacts in the off-midplanes such as the top and bottom regions. The TV and FDK CNN methods show improved image quality, but do not remove the cone-beam artifacts nor preserve the details and textures. On the other hand, the DBP-domain neural network not only removes the artifacts but also

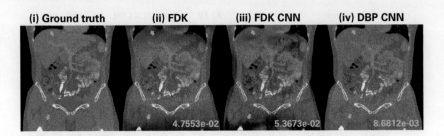

Figure 6.22 Reconstructed results for low-dose cone-beam CT. The initial X-ray intensity is 5×10^5, which results in 15.0 dB of SNR. The NMSE values are given at the bottom right-hand corners of the images. The window range is $(-1000, 800)$ [HU].

retains their sophisticated structures and textures. The transverse views are illustrated in Fig. 6.21(b). The FDK image shows the effect of the missing frequency regions in the Fourier domain. In particular, since a missing frequency region exists in the low-frequency part, the image intensity drops and cone-beam artifacts are generated (see the arrows in Fig. 6.21(b)). Although the TV- and FDK-domain neural network still retain the cone-beam artifacts, the DBP-domain neural network provides reconstructed results with high accuracy.

In addition, the FDK-domain and DBP-domain neural networks were trained with noise-free datasets, but noise simulation was performed to verify the generalizability. Poisson-noise models were applied to the projection data, and various noise levels such as 15.0, 16.5, and 19.2 dB in the signal-to-noise ratio (SNR) were used. In particular, five noise experiments were performed to improve reliability, and the reconstructed results were averaged as shown in Table 6.1. Figure 6.22 shows the reconstructed results for the noise experiments. Specifically, the FDK neural network provided amplified noisy results compared with FDK on its own. In other words, the FDK neural network did not generalize well. However, the DBP-domain neural network showed robust reconstructed results regardless of the noise levels. This confirms that the DBP-domain neural networks offer strong generalizability.

6.5 Deep Learning in CT Reconstruction: Synergy of Deep Learning and Statistical IR

In this section, we discuss how to address the potential challenges when the powerful regression capacity of the deep-learning framework is used to correct images using the analytical reconstruction method directly on the corrupted data, which can be sinogram data acquired at sparsely sampled view angles, or view angle ranges restricted to a limited view angle range, or transversely truncated data in each acquired view angle, or longitudinally truncated data in so-called long object problems. As an example, we outline a strategy in this section to combine deep-learning strategies with the previously published prior-image-constrained compressed-sensing (PICCS) algorithm [41] to improve reconstruction accuracy for individual patients and enhance

Table 6.1. Quantitative comparison with respect to noise levels.

Noise levels	FDK	FDK CNN	DBP CNN
15.0 dB	25.90	26.50	**31.29**
16.5 dB	27.42	30.46	**34.23**
19.2 dB	30.09	36.20	**38.69**

(a) PSNR (dB)

Noise levels	FDK	FDK CNN	DBP CNN
15.0 dB	4.68	1.80	**0.97**
16.5 dB	4.13	1.24	**0.48**
19.2 dB	3.69	0.33	**0.18**

(b) NMSE ($\times 10^{-3}$)

Noise levels	FDK	FDK CNN	DBP CNN
15.0 dB	0.55	0.66	**0.69**
16.5 dB	0.62	0.73	**0.80**
19.2 dB	0.75	0.86	**0.93**

(c) SSIM

generalizability for sparse-view reconstruction problems. The PICCS reconstruction framework uses the following PICCS regularizer:

$$R(\vec{\mu}) = \alpha TV(\vec{\mu} - \vec{\mu}_{DL}) + (1 - \alpha)TV(\vec{\mu}) \tag{6.36}$$

in the general statistical IR framework shown in Eq. (6.17) to form the target convex optimization problem [73]:

$$\hat{\vec{\mu}} = \operatorname{argmin}_{\vec{\mu}} \left[\frac{\lambda}{2} \|\mathbf{A}\vec{\mu} - \vec{y}\|_D^2 + \alpha TV(\vec{\mu} - \vec{\mu}_{DL}) + (1 - \alpha)TV(\vec{\mu}) \right]. \tag{6.37}$$

Here, $\hat{\vec{\mu}}$ denotes the reconstructed CT image, $\vec{\mu}_{DL}$ denotes the output image from a deep-learning network which is introduced as the prior image in PICCS, $\vec{y}$ denotes the sinogram data, $\mathbf{A}$ denotes the system matrix, and D denotes the statistical weighting. Many different numerical solvers can be used to solve the unconstrained problem in Eq. (6.37). The solver used here is the forward–backward proximal splitting scheme, and the alternating direction method of multipliers (ADMM) scheme and pseudo-code can be found in [42].

6.5.1 DL-PICSS Reconstruction Pipeline

In the DL-PICCS framework, the output of the first deep-learning network (DL1) is used as the prior image in the PICCS reconstruction framework to reconstruct the PICCS image using both the prior image produced from the DL1 network and also the

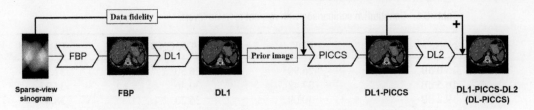

Figure 6.23 Workflow of the DL-PICCS framework. In this flow chart, FBP reconstruction is used as an example. DL1 denotes the first deep-learning neural network architecture and DL2 denotes the second light-duty deep-learning neural network architecture. [74]

sparse-view sinogram data for each individual patient. The output of PICCS reconstruction is denoted DL1-PICCS (Fig. 6.23). The primary purpose of the PICCS reconstruction module is to accomplish accurate reconstruction for each individual patient by invoking the data fidelity term. With the proper choice of reconstruction parameters in PICCS, both high accuracy and streaky artifact reduction can be reached in the DL1-PICCS image to improve reconstruction accuracy in individual patient image reconstruction. However, the noise texture and noise level in DL1-PICCS images may not be ideal for medical diagnoses. To improve image texture and noise level, another deep-neural-network architecture, DL2, can be added to the end of the PICCS output. The final output is referred to as DL1-PICCS-DL2.

All the training data were generated from numerical simulations of human subject CT image volumes by numerical forward-projection operations. A total of 9765 CT image slices from 30 cases of abdomen and contrast-enhanced chest CT exams were used to generate sinogram data for training, validating, and testing purposes. Among the 9765 image slices, 8483 image slices were selected from 26 patient cases to generate data for training and 834 slices from two patient cases for validation purposes, while the remaining 448 images slices from two patient cases were used to generate the dataset for test and performance evaluation purposes.

6.5.2 Results: Reconstruction Accuracy Quantification and Generalizability Tests

Figure 6.24 shows the reconstruction results for a representative sinogram with 123 views at the moderate-dose level. As one can observe in Fig. 6.24, the streaking artifacts present in the stomach (indicated by A3) and spleen (indicated by A4) regions were removed in the DL1-only image, the residual edges in the DL1-only image were reduced in the DL1-PICCS image, and finally the noise level in the DL1-PICCS image is further reduced in the DL-PICCS image. Zoomed-in regions labeled by A1 and A2 highlight the low-contrast hepatic portal veins. Both anatomical regions were precisely reconstructed in DL1-PICCS and DL-PICCS with greatly reduced background streaks and noise. Visual perception also demonstrates that DL-PICCS restores the distorted low-contrast objects in terms of their shape, size, and contrast, eliminates the residual streaks compared with DL1-only, and reduces the uniform noise level in the DL1-PICCS images.

Table 6.2. Quantitative analysis for numerical studies with 123 view moderate-dose data.

Case	Method	rRMSE(%)	SSIM
Liver	**DL-PICCS**	**2.69**	**0.973**
	DL1-PICCS	3.20	0.962
	DL1-Only	3.27	0.965
	FBP	13.40	0.610
Upper GI	**DL-PICCS**	**2.20**	**0.979**
	DL1-PICCS	2.59	0.971
	DL1-Only	2.63	0.972
	FBP	11.60	0.641

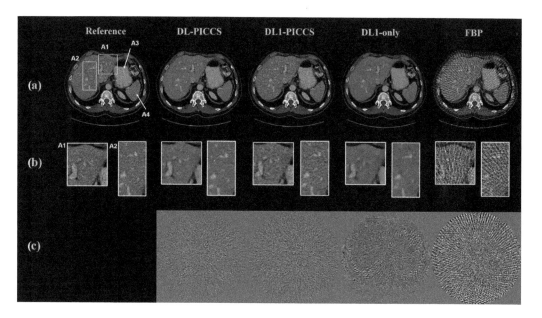

Figure 6.24 Moderate-dose reconstruction results of the liver CT case with 123 view data. Row (a), reconstructed images; (b), zoomed-in images; (c), difference images. $W/L = 400/50$ HU for reconstructed images and $W/L = 200/0$ HU for difference images. [74]

Quantitative reconstruction accuracy results are presented in Table 6.2. As shown in the results, the DL-PICCS shows the best reconstruction accuracy as quantified by both image quality evaluation metrics, rRMSE and SSIM.

To test the generalizability of DL-PICCS, the trained DL1 and DL2 and PICCS in the DL-PICCS workflow were applied directly to a upper gastrointestinal (GI) CT exam, and the results are presented in Fig. 6.25. Compared with DL1-PICCS, DL-PICCS shows improved noise performance. However, the DL1-only method missed vessels in B1 and the muscle details in B2. Quantitative results are presented in Table 6.2 and demonstrates that DL-PICCS can be effective and robust when applied to the reconstruction of different anatomical structures.

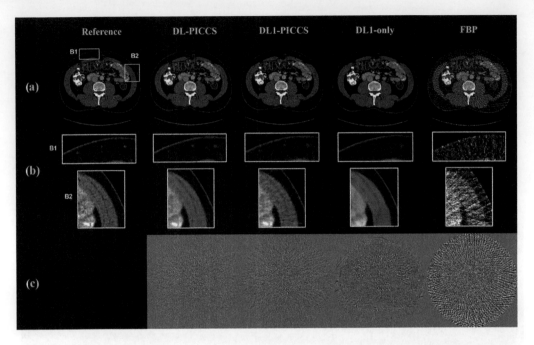

Figure 6.25 DL-PICCS reconstruction results or an upper GI image slice under the moderate-dose condition with 123 view data. Row (a), reconstructed images; (b), zoomed-in images specified by B1 and B2; (c), difference images. $W/L = 350/50$ HU for reconstructed images and $W/L = 200/0$ HU for difference images. [74]

To further test the generalizability of the DL-PICCS workflow, the trained DL1 and DL2 modules using numerical simulation data were used in DL-PICCS to reconstruct three experimental human clinical cases. Reconstructed results from 123 views are presented in Fig. 6.26. As shown in all three clinical cases, while the DL1-only image can effectively eliminate sparse-view streaking artifacts and significantly reduce the noise in FBP images, potential generalizability issues resulting from the statistical regression nature of deep-learning methods are also clearly shown in the DL1-only images. As shown in the zoomed-in images of the first subject, the two hepatic veins of the liver in region C1 are missing from the DL1-only image. In the DL-PICCS image, not only are the hepatic veins restored, but also the noise texture and noise level are significantly improved compared with the DL1-PICCS reconstruction. In the second subject, a false positive lesion is generated in the DL1-only image; distorted and generated structures are present in the zoomed-in uniform regions in the liver as indicated by the arrows. In contrast, the final output from DL-PICCS removed those false positive structures in the D1 and D2 regions. In the third clinical case, the U1-only image failed in reconstructing the liver, as multiple lesion-like objects are present while they are not present in the reference image shown by the zoomed-in areas E1 and E2. In contrast, the DL-PICCS reconstruction successfully identified and removed these false positive lesions.

To further challenge the generalizability of the DL1 network of DL-PICCS, the DL1 segment was replaced by the publicly available trained tight-frame U-Net (called

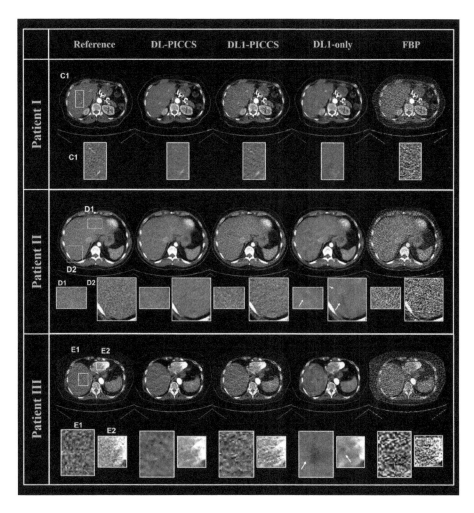

Figure 6.26 Reconstruction results for three experimental human abdomen cases with 123 view data. Four stages of reconstruction are specified and their zoomed-in images, C1, D1, D2, E1, E2 are shown below. $W/L = 400/50$ HU for reconstructed images and $W/L = 200/0$ HU for difference images. [74]

TightU) to process the numerically simulated cases presented in Fig. 6.24. As shown in Fig. 6.27, it is actually amazing to see how well the tight-frame U-Net performed in this extremely challenging test. Most of the sparse-view streaking artifacts in the FBP reconstruction were eliminated by TightU, although the reconstruction accuracy is not ideal, as shown in the difference image. Residual edges for large and small anatomical structures are present in the difference image indicating reduced reconstruction accuracy and degraded spatial resolution. When the TightU-only image is further processed with PICCS to produce a TightU-PICCS image, the reconstruction accuracy is improved and the final results with TightU-PICCS-DL2 are further improved. The quantitative reconstruction accuracy was studied and the results are presented in Table 6.3. The same generalizability test was then applied to the human

Table 6.3. Quantitative analysis for generalizability test (II): DL-PICCS reconstruction with DL1 replaced by tight-frame U-Net (TightU).

Case	Method	rRMSE(%)	SSIM
Liver	**TightU-PICCS-DL2**	**2.81**	**0.971**
	TightU-PICCS	3.95	0.945
	TightU-Only	3.86	0.953
	FBP	13.40	0.610

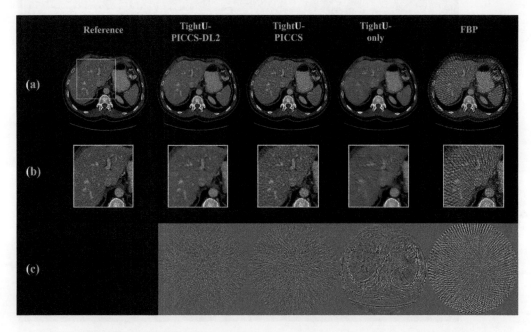

Figure 6.27 DL-PICCS reconstruction results with DL1 replaced by tight-frame U-Net (TightU). The example shows a 123-view result of the liver. Row (a), reconstructed images; (b) zoomed-in images; (c) difference images. $W/L = 400/50$ HU for reconstructed images and $W/L = 200/0$ HU for difference images.

subject cases to test whether DL-PICCS is able to reconstruct clinically acceptable images in clinical cases. As one can observe in Fig. 6.28, TightU-PICCS-DL2 does indeed improve the reconstruction quality for TightU and TightU-PICCS, showing improved reconstruction accuracy and improved generalizability.

6.6 Synergy of FBP, Deep Learning, and Statistical IR

In this short section, as a final example we show that one can synergistically combine the advantages of analytical reconstruction, deep learning, and statistical IR to address some technical challenges encountered in current clinical CT imaging.

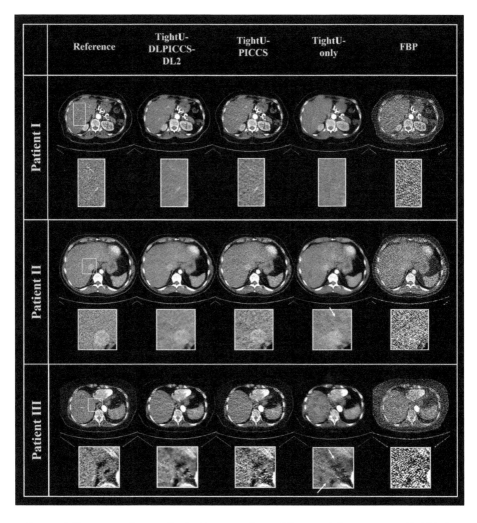

Figure 6.28 Reconstruction results for three experimental human abdomen cases with 123 view data, replacing DL1 with its counterpart tight-frame U-Net (TightU). Four stages of reconstruction are specified. $W/L = 400/50$ HU for reconstructed images and $W/L = 200/0$ HU for difference images.

For current single-source scanners, to ensure high-quality diagnostic CT image reconstruction the helical pitch is often restricted to below 1.5. However, there is a strong desire to use high-pitch helical CT scans in clinical applications for the following reasons: (1) to reduce radiation dose; (2) to improve consistency between different image slabs; (3) to increase throughput in clinical operations; and (4) to potentially reduce the manufacturing cost of CT scanners with the same performance of some higher-end CT scanner with 256-slice or more anatomical coverage.

Unfortunately, when the helical pitch increases beyond 1.5 for single-source scanners, there are two intrinsic scientific challenges in high-pitch helical CT

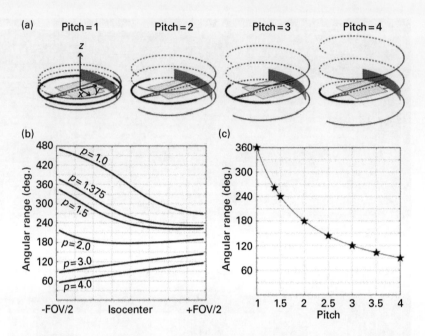

Figure 6.29 (a) Helical trajectories with three revolutions around a given reconstruction plane for each pitch value. The illuminating range for this plane is shown in solid black. (b) The illuminating range of image pixels along the line crossing the isocenter is plotted for different helical pitches. (c) The stars mark the illuminating range for the isocenter as a function of pitch. © 2021 IEEE. Reprinted, with permission, from [43]

reconstruction problems: the angular sample interval is large – as in undersampled view angle reconstruction problems in circular scans – and the view angle range is also truncated – as in limited view angle reconstruction problems. As a result of these scientific challenges, high-pitch helical reconstruction suffers from two main artifacts: limited view artifacts and data inconsistency along the z-direction. These problems are highlighted [43] in Fig. 6.29.

To enable high-pitch helical CT scans for single-source MDCT scanners, a scheme to synergistically leverage the knowledge in analytical reconstruction, deep learning, and statistical IR was recently developed [43]. In this scheme, the limited angular range problem was first approximated by a data extrapolation scheme from the available helical data. This step aims to mitigate limited-view artifacts and also the view angle undersampling artifacts in the reconstructed images. After this step, a deep-learning architecture was used to perform statistical regression between the FBP-reconstructed imaging with data extrapolations and interpolations and the targeted image volumes. Note that the clinically available 64-slice CT image volumes can be used to serve as image labels for 16-slice CT scans at pitch values 2, 3, or 4. Similarly, the state-of-the-art 256-slice CT scanners can, in principle, be used to generate image labels for 64-slice CT scanners at pitch values 2, 3, and 4. After the deep-learning regression model has been trained to clean the artifact-contaminated high-pitch helical

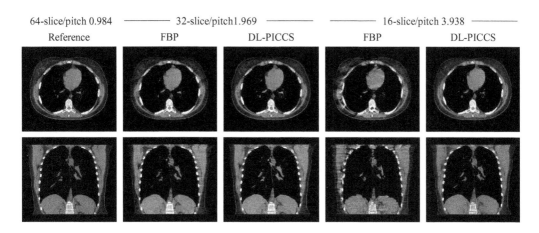

Figure 6.30 Reconstructions of experimental helical scan data acquired on the 64-slice GE Discovery CT750 HD scanner with pitch 0.984. $W/L = 450/50$ HU.

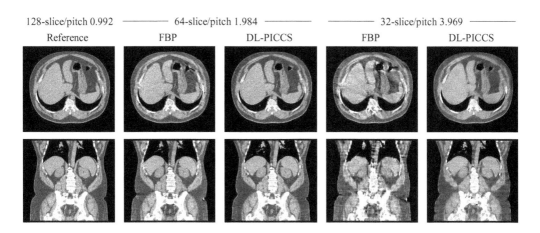

Figure 6.31 Reconstructions of experimental helical scan data acquired with 128 slices on the GE Revolution scanner with pitch 0.992. $W/L = 400/0$ HU.

CT image reconstructed using FBP, nearly artifact-free high-pitch helical CT images can be produced. In the final step, to address the personalized reconstruction accuracy and generalizability challenges associated with the statistical regression nature of deep-learning models, the statistical IR method was introduced to correct the deep-learning output images, as illustrated in the previous section using the DL-PICCS framework [42, 74].

For technical details of the deep model architecture and model training, we refer the readers to [43]. Here we give the results from experimental helical CT data from 64-slice CT scans to show that one can indeed reconstruct high-helical-pitch CT images for 16-slice collimation up to pitch 4.0 (Fig. 6.30) and 32-slice collimations up to pitch 4.0 (Fig. 6.31) with only minor residual artifacts.

6.7 Summary

In this chapter, we have shown that image-domain deep-learning-only reconstruction methods have intrinsic limitations in reconstruction accuracy and in generalizability to individual patients due to the regression nature of these methods. The combination of deep-learning methods with analytic reconstruction methods or statistical IR methods offers a promising opportunity to achieve personalized reconstruction with improved reconstruction accuracy and enhanced generalizability.

Specifically, we showed that the sinogram-domain deep-learning methods such as iCT-Net provide a unified framework to reconstruct CT images for a variety of reconstruction problems under very different conditions but within a unified framework. Such methods have the capability to accurately reconstruct images for those reconstruction problems that have already been completely solved by human efforts, for problems that have been solved only partially by human efforts, and for problems that have not been successfully addressed in any meaningful way using human knowledge (e.g., interior tomography with sparse-view projection data). Additionally, when a clear imaging task is defined in practical applications, one can actually reconstruct a 3D tomographic patient model for radiation modulation purposes using two scout views of projection data. This 3D patient model provides the patient information needed to prescribe dose modulations that will maintain diagnostic image quality while reducing the overall radiation dose to patients. This has been an active research direction in CT for the past several years.

Furthermore, we have also demonstrated that deep neural networks can be trained in the DBP domain. Specifically, for the case of interior tomography for ROI reconstruction, the DBP-domain CNN provided better generalization performance than the FBP-domain CNN, since there are no singularities in the DBP measurements. In addition, inspired by the existing factorization method from the 3D cone-beam reconstruction problem into a set of independent 2D deconvolution problems along the plane of interest, the DBP-domain neural network was designed to learn the mapping function from DBP measurements to images.

We also demonstrated that when deep learning is used in combination with statistical IR, reconstruction accuracy is improved with better noise textures and the elimination of streaks while preserving the low-contrast structure. One can also synergistically use analytical image reconstruction, deep learning, and statistical IR such as PICCS to tackle extremely difficult yet trail-blazing CT reconstruction problems such as the high-pitch helical reconstruction problem, as highlighted in the last example.

To end this chapter, we would like to emphasize that there are still many open questions worth collaborative efforts for the entire community to tackle. Besides the well-recognized generalizability challenge with deep-learning models, the consequences of the intrinsic nonlinearity in deep-learning models remain to be fully investigated. Essentially, the generalizability challenge can be partially attributed to the intrinsic nonlinearity of these models. Owing to nonlinearity, the output of the linearly combined inputs is not equal to the linear combination of the outputs from each individual input. This has been one of the major obstacles in clinical translations of statistical

IR into clinical practice in the past decade, and this nonlinearity problem remains in deep-learning strategies. However, it is hoped that the combination of analytical reconstruction, deep learning, and statistical IR may alleviate this challenge since the reconstruction task can now be shared by several components with different nonlinearity levels. Perhaps there is an opportunity to find a quasi-linear regime in performance such that the imaging performance from phantom studies can be generalized to human subject studies, as is the case for linear FBP reconstruction. If this were possible, clinical translations would become relatively straightforward.

References

[1] C. H. McCollough, G. H. Chen, W. Kalender, S. Leng, E. Samei, K. Taguchi, G. Wang, L. Yu, and R. I. Pettigrew, "Achieving routine submillisievert CT scanning: Report from the summit on management of radiation dose in CT," *Radiology*, vol. 264, no. 2, pp. 567–580, 2012, pMID: 22 692 035.

[2] P. J. Pickhardt, M. G. Lubner, D. H. Kim, J. Tang, J. A. Ruma, A. M. del Rio, and G.-H. Chen, "Abdominal CT with model-based iterative reconstruction (MBIR): Initial results of a prospective trial comparing ultralow-dose with standard-dose imaging," *American Journal of Gerontology*, vol. 199, no. 6, pp. 1266–1274, 2012.

[3] A. Mileto, L. S. Guimaraes, C. H. McCollough, J. G. Fletcher, and L. Yu, "State of the art in abdominal CT: The limits of iterative reconstruction algorithms," *Radiology*, vol. 293, no. 3, pp. 491–503, 2019.

[4] G. L. Zeng, *Medical Image Reconstruction: A Conceptual Tutorial*. Springer, 2010.

[5] I. M. Gel'fand, S. G. Gindikin, and M. I. Graev, *Selected Topics in Integral Geometry*, Translations of Mathematical Monographs, American Mathematical Society, 2003.

[6] S. Helgason, *Integral Geometry and Radon Transforms*. Springer, 2010.

[7] J. Radon, "Über die Bestimmung von Funktionen durch ihre Integralwerte längs gewisser Mannigfaltigkeiten [on the determination of functions from their integral values along certain manifolds]," *Berichte uber die Verhandlungen Gesellshaft der Wissenschaften zu Leipzig. Journal of Mathematical Physics*, vol. 69, pp. 262–277, 1917.

[8] A. Katsevich, "Analysis of an exact inversion algorithm for spiral cone-beam CT," *Physics in Medicine and Biology*, vol. 47, no. 15, p. 2583, 2002.

[9] F. Noo, M. Defrise, R. Clackdoyle, and H. Kudo, "Image reconstruction from fan-beam projections on less than a short scan," *Physics in Medicine and Biology*, vol. 47, no. 14, p. 2525, 2002.

[10] G.-H. Chen, "A new framework of image reconstruction from fan beam projections," *Medical Physics*, vol. 30, no. 6, pp. 1151–61, 2003.

[11] Y. Zou and X. Pan, "Exact image reconstruction on PI-lines from minimum data in helical cone-beam CT," *Physics in Medicine and Biology*, vol. 49, no. 6, p. 941, 2004.

[12] F. Noo, R. Clackdoyle, and J. D. Pack, "A two-step Hilbert transform method for 2D image reconstruction," *Physics in Medicine and Biology*, vol. 49, no. 17, p. 3903, 2004.

[13] T. Zhuang, S. Leng, B. E. Nett, and G. H. Chen, "Fan-beam and cone-beam image reconstruction via filtering the backprojection image of differentiated projection data," *Physics in Medicine and Biology*, vol. 49, no. 24, pp. 5489–503, 2004.

[14] K. Hornik, "Approximation capabilities of multilayer feedforward networks," *Neural Networks*, vol. 4, no. 2, pp. 251–257, 1991.

[15] Z. Ghahramani, "Probabilistic machine learning and artificial intelligence," *Nature*, vol. 521, p. 452, 2015.

[16] G. Wang, J. C. Ye, and B. D. Man, "Deep learning for tomographic image reconstruction," *Nature Machine Intelligence*, vol. 2, p. 737–748, 2020.

[17] C. M. Bishop, *Pattern Recognition and Machine Learning*. Springer, 2006.

[18] J. Friedman, T. Hastie, R. Tibshirani *et al.*, *The Elements of Statistical Learning*. Springer Series in Statistics, 2001.

[19] I. Goodfellow, Y. Bengio, A. Courville, and Y. Bengio, *Deep Learning*. MIT Press, 2016.

[20] Y. Yang, J. Sun, H. Li, and Z. Xu, "Deep ADMM-Net for compressive sensing MRI," in *Advances in Neural Information Processing Systems*, vol. 29, D. D. Lee, M. Sugiyama, U. V. Luxburg, I. Guyon, and R. Garnett, eds. Curran Associates, 2016, pp. 10–18.

[21] A. Hauptmann, F. Lucka, M. Betcke, N. Huynh, J. Adler, B. Cox, P. Beard, S. Ourselin, and S. Arridge, "Model-based learning for accelerated, limited-view 3-D photoacoustic tomography," *IEEE Transactions on Medical Imaging*, vol. 37, no. 6, pp. 1382–1393, 2018.

[22] K. Hammernik, T. Klatzer, E. Kobler, M. P. Recht, D. K. Sodickson, T. Pock, and F. Knoll, "Learning a variational network for reconstruction of accelerated MRI data," *Magnetic Resonance in Medicine*, vol. 79, no. 6, pp. 3055–3071, 2018.

[23] J. Zhang and B. Ghanem, "ISTA-Net: Interpretable optimization-inspired deep network for image compressive sensing," in *Proc. 2018 IEEE/CVF Conference on Computer Vision and Pattern Recognition*, 2018, pp. 1828–1837.

[24] H. Chen, Y. Zhang, Y. Chen, J. Zhang, W. Zhang, H. Sun, Y. Lv, P. Liao, J. Zhou, and G. Wang, "LEARN: Learned experts' assessment-based reconstruction network for sparse-data CT," *IEEE Transactions on Medical Imaging*, vol. 37, no. 6, pp. 1333–1347, 2018.

[25] J. Schlemper, J. Caballero, J. V. Hajnal, A. N. Price, and D. Rueckert, "A deep cascade of convolutional neural networks for dynamic MR image reconstruction," *IEEE Transactions on Medical Imaging*, vol. 37, no. 2, pp. 491–503, 2018.

[26] M. Mardani, E. Gong, J. Y. Cheng, S. S. Vasanawala, G. Zaharchuk, L. Xing, and J. M. Pauly, "Deep generative adversarial neural networks for compressive sensing MRI," *IEEE Transactions on Medical Imaging*, vol. 38, no. 1, pp. 167–179, 2019.

[27] H. K. Aggarwal, M. P. Mani, and M. Jacob, "Model based image reconstruction using deep learned priors (MODL)," in *Proc. International Symposium on Biomedical Imaging*, vol. 2018-April. IEEE Computer Society, 2018, pp. 671–674.

[28] H. Gupta, K. H. Jin, H. Q. Nguyen, M. T. McCann, and M. Unser, "CNN-based projected gradient descent for consistent CT image reconstruction," *IEEE Transactions on Medical Imaging*, vol. 37, no. 6, pp. 1440–1453, 2018.

[29] J. Adler and O. Öktem, "Learned primal–dual reconstruction," *IEEE Transactions on Medical Imaging*, vol. 37, no. 6, pp. 1322–1332, 2018.

[30] D. Wu, K. Kim, G. El Fakhri, and Q. Li, "Iterative low-dose CT reconstruction with priors trained by artificial neural network," *IEEE Transactions on Medical Imaging*, vol. 36, no. 12, pp. 2479–2486, 2017.

[31] B. Chen, K. Xiang, Z. Gong, J. Wang, and S. Tan, "Statistical iterative CBCT reconstruction based on neural network," *IEEE Transactions on Medical Imaging*, vol. 37, no. 6, pp. 1511–1521, 2018.

[32] K. Liang, L. Zhang, H. Yang, Z. Chen, and Y. Xing, "A model-based unsupervised deep learning method for low-dose CT reconstruction," *IEEE Access*, vol. 8, pp. 159 260–159 273, 2020.

[33] Q. Ding, G. Chen, X. Zhang, Q. Huang, H. Ji, and H. Gao, "Low-dose CT with deep learning regularization via proximal forward–backward splitting," *Physics in Medicine and Biology*, vol. 65, no. 12, p. 125 009, 2020.

[34] Y. Ge, T. Su, J. Zhu, X. Deng, Q. Zhang, J. Chen, Z. Hu, H. Zheng, and D. Liang, "ADAPTIVE-NET: Deep computed tomography reconstruction network with analytical domain transformation knowledge," *Quantitative Imaging in Medicine and Surgery*, vol. 10, no. 2, pp. 415–427, 2020.

[35] X. Zheng, S. Ravishankar, Y. Long, and J. A. Fessler, "PWLS-ULTRA: An efficient clustering and learning-based approach for low-dose 3D CT image reconstruction," *IEEE Transactions on Medical Imaging*, vol. 37, no. 6, pp. 1498–1510, 2018.

[36] V. Lempitsky, A. Vedaldi, and D. Ulyanov, "Deep image prior," in *Proc. IEEE Computer Society Conference on Computer Vision and Pattern Recognition*, pp. 9446–9454, 2018.

[37] K. Gong, C. Catana, J. Qi, and Q. Li, "PET image reconstruction using deep image prior," *IEEE Transactions on Medical Imaging*, vol. 38, no. 7, pp. 1655–1665, 2019.

[38] I. Y. Chun, X. Zheng, Y. Long, and J. A. Fessler, "Bcd-net for low-dose CT reconstruction: Acceleration, convergence, and generalization," in *Proc. International Conference on Medical Image Computation – Computer-Assisted Interventions*. Springer, 2019, pp. 31–40.

[39] S. Ye, Y. Long, and I. Y. Chun, "Momentum-net for low-dose CT image reconstruction," *arXiv:2002.12018*, 2020.

[40] I. Y. Chun, Z. Huang, H. Lim, and J. Fessler, "Momentum-net: Fast and convergent iterative neural network for inverse problems," *IEEE Transactions on Pattern Analysis and Machine Intelligence*, vol. 45, no. 4, pp. 4915–4931, 2020.

[41] G.-H. Chen, J. Tang, and S. Leng, "Prior image constrained compressed sensing (PICCS): A method to accurately reconstruct dynamic CT images from highly undersampled projection data sets," *Medical Physics*, vol. 35, no. 2, pp. 660–663, 2008.

[42] C. Zhang, Y. Li, and G.-H. Chen, "Accurate and robust sparse-view angle CT image reconstruction using deep learning and prior image constrained compressed sensing (DL-PICCS)," *Medical Physics*, vol. 48, no. 10, pp. 5765–5781, 2021.

[43] J. W. Hayes, J. Montoya, A. Budde, C. Zhang, Y. Li, K. Li, J. Hsieh, and G.-H. Chen, "High pitch helical CT reconstruction," *IEEE Transactions on Medical Imaging*, vol. 40, no. 11, pp. 3077–3088, 2021.

[44] Y. Li, K. Li, C. Zhang, J. Montoya, and G.-H. Chen, "Learning to reconstruct computed tomography images directly from sinogram data under a variety of data acquisition conditions," *IEEE Transactions on Medical Imaging*, vol. 38, no. 10, pp. 2469–2481, 2019.

[45] J. C. Montoya, Y. Li, C. Zhang, K. Li, and G.-H. Chen, "Reconstruction of three-dimensional tomographic patient model from two-view radiograph localizers for radiation dose prescription in CT," *Medical Physics*, vol. 49, no. 2, pp. 901–916, 2022.

[46] B. Zhu, J. Z. Liu, S. F. Cauley, B. R. Rosen, and M. S. Rosen, "Image reconstruction by domain-transform manifold learning," *Nature*, vol. 555, p. 487, 2018.

[47] Y. Ye, H. Yu, Y. Wei, and G. Wang, "A general local reconstruction approach based on a truncated hilbert transform," *International Journal of Biomedical Imaging*, vol. 2007, p. 63 634, 2007.

[48] H. Kudo, M. Courdurier, F. Noo, and M. Defrise, "Tiny a priori knowledge solves the interior problem in computed tomography," *Physics in Medicine and Biology*, vol. 53, no. 9, pp. 2207–2231, 2008.

[49] H. Yu and G. Wang, "Compressed sensing based interior tomography," *Physics in Medicine and Biology*, vol. 54, no. 9, pp. 2791–2805, 2009.

[50] O. Ronneberger, P. Fischer, and T. Brox, "U-net: Convolutional networks for biomedical image segmentation," in *Proc. International Conference on Medical Image Computing and Computer-Assisted Intervention*. Springer, 2015, pp. 234–241.

[51] I. J. Goodfellow, J. Pouget-Abadie, M. Mirza, B. Xu, D. Warde-Farley, S. Ozair, A. Courville, and Y. Bengio, "Generative adversarial nets," in *Proc. 27th International Conference on Neural Information Processing Systems*, vol. 2, ser. NIPS'14. MIT Press, 2014, pp. 2672–2680.

[52] P. Isola, J. Zhu, T. Zhou, and A. A. Efros, "Image-to-image translation with conditional adversarial networks," *CoRR*, vol. abs/1611.07004, 2016.

[53] K. He, X. Zhang, S. Ren, and J. Sun, "Deep residual learning for image recognition," in *Proc. IEEE Conference on Computer Vision and Pattern Recognition*, 2016.

[54] ——, *Identity Mappings in Deep Residual Networks*. Springer, 2016, pp. 630–645.

[55] ——, "Delving deep into rectifiers: Surpassing human-level performance on imagenet classification," in *Proc. 2015 IEEE International Conference on Computer Vision*, 2015, pp. 1026–1034.

[56] J. Hsieh, E. Chao, J. Thibault, B. Grekowicz, A. Horst, S. McOlash, and T. J. Myers, "Algorithm to extend reconstruction field-of-view," in *Proc. Symposium on Biomedical Imaging: Nano to Macro*. IEEE, 2004, pp. 1404–1407.

[57] J. P. Ward, M. Lee, J. C. Ye, and M. Unser, "Interior tomography using 1D generalized total variation – Part I: mathematical foundation," *SIAM Journal on Imaging Sciences*, vol. 8, no. 1, pp. 226–247, 2015.

[58] M. Lee, Y. Han, J. P. Ward, M. Unser, and J. C. Ye, "Interior tomography using 1D generalized total variation. Part II: Multiscale implementation," *SIAM Journal on Imaging Sciences*, vol. 8, no. 4, pp. 2452–2486, 2015.

[59] L. C. Feldkamp, L. A .and Davis and J. W. Kress, "Practical cone-beam algorithm," *Journal of the Optical Society of America A, Optical Imaging Science*, vol. 1, no. 6, pp. 612–619, 1984.

[60] H. K. Tuy, "An inversion formula for cone-beam reconstruction," *SIAM Journal on Applied Mathematics*, vol. 43, no. 3, pp. 546–552, 1983.

[61] S. Bartolac, R. Clackdoyle, F. Noo, J. Siewerdsen, D. Moseley, and D. Jaffray, "A local shift-variant Fourier model and experimental validation of circular cone-beam computed tomography artifacts," *Medical Physics*, vol. 36, no. 2, pp. 500–512, 2009.

[62] J. D. Pack, Z. Yin, K. Zeng, and B. E. Nett, "Mitigating cone-beam artifacts in short-scan CT imaging for large cone-angle scans," *Fully 3D Image Reconstruction in Radiology and Nuclear Medicine*, pp. 307–310, 2013.

[63] F. Peyrin, M. Amiel, and R. Goutte, "Analysis of a cone beam X-ray tomographic system for different scanning modes," *Journal of the Optical Society of America A*, vol. 9, no. 9, pp. 1554–1563, 1992.

[64] M. Grass, T. Köhler, and R. Proksa, "Angular weighted hybrid cone-beam CT reconstruction for circular trajectories," *Physics in Medicine and Biology*, vol. 46, no. 6, p. 1595, 2001.

[65] X. Tang, J. Hsieh, A. Hagiwara, R. A. Nilsen, J.-B. Thibault, and E. Drapkin, "A three-dimensional weighted cone beam filtered backprojection (CB-FBP) algorithm for image reconstruction in volumetric CT under a circular source trajectory," *Physics in Medicine and Biology*, vol. 50, no. 16, p. 3889, 2005.

[66] S. Mori, M. Endo, S. Komatsu, S. Kandatsu, T. Yashiro, and M. Baba, "A combination-weighted feldkamp-based reconstruction algorithm for cone-beam CT," *Physics in Medicine and Biology*, vol. 51, no. 16, p. 3953, 2006.

[67] E. Y. Sidky and X. Pan, "Image reconstruction in circular cone-beam computed tomography by constrained, total-variation minimization," *Physics in Medicine and Biology*, vol. 53, no. 17, p. 4777–4807, 2008.

[68] E. Y. Sidky, J. H. Jørgensen, and X. Pan, "Convex optimization problem prototyping for image reconstruction in computed tomography with the Chambolle–Pock algorithm," *Physics in Medicine and Biology*, vol. 57, no. 10, p. 3065, 2012.

[69] Z. Zhang, X. Han, E. Pearson, C. Pelizzari, E. Y. Sidky, and X. Pan, "Artifact reduction in short-scan CBCT by use of optimization-based reconstruction," *Physics in Medicine and Biology*, vol. 61, no. 9, p. 3387, 2016.

[70] D. Xia, D. A. Langan, S. B. Solomon, Z. Zhang, B. Chen, H. Lai, E. Y. Sidky, and X. Pan, "Optimization-based image reconstruction with artifact reduction in c-arm CBCT," *Physics in Medicine and Biology*, vol. 61, no. 20, p. 7300, 2016.

[71] F. Dennerlein, F. Noo, H. Schondube, G. Lauritsch, and J. Hornegger, "A factorization approach for cone-beam reconstruction on a circular short-scan," *IEEE Transactions on Medical Imaging*, vol. 27, no. 7, pp. 887–896, 2008.

[72] L. Yu, Y. Zou, E. Y. Sidky, C. A. Pelizzari, P. Munro, and X. Pan, "Region of interest reconstruction from truncated data in circular cone-beam CT," *IEEE Transactions on Medical Imaging*, vol. 25, no. 7, pp. 869–881, 2006.

[73] P. Thèriault-Lauzier, J. Tang, and G.-H. Chen, "Prior image constrained compressed sensing: Implementation and performance evaluation." *Medical Physics*, vol. 39, no. 1, pp. 66–80, 2012.

[74] C. Zhang, Y. Li, and G.-H. Chen, "Deep learning enabled prior image constrained compressed sensing (DL-PICCS) reconstruction framework for sparse-view reconstruction," in *Proc. Conference on Medical Imaging 2020: Physics of Medical Imaging*, vol. 11312. International Society for Optics and Photonics, 2020, p. 1 131 206.

7 Overview of the Deep-Learning Reconstruction of Accelerated MRI

Patricia Johnson and Florian Knoll

7.1 Overview of Image Reconstruction for Accelerated MR Imaging

Image reconstruction is a classic inverse problem. We probe our sample with an imaging device that generates a set of measurements that we then try to relate back to the source inside our sample. Since 2016 [1–3], there has been a substantial amount of research activities related to the use of machine learning methods to solve inverse problems in imaging, and the goal of this chapter is to provide an introduction to these developments for the special case of accelerated magnetic resonance imaging (MRI). To keep the chapter self-contained, we first introduce the core concepts of MR image reconstruction, parallel imaging [4–6] and compressed sensing [7]. This will serve as the foundation for our description of deep-learning methods. The rapid growth of this research area forces us to make compromises in terms of the material that we can present. We have chosen to focus on didactic accessibility at the cost of a comprehensive coverage of a wide range of approaches.

Magnetic resonance scanners employ magnetic gradient fields to encode the spatial positions of magnetized spins of hydrogen atoms via their precession frequency and phase [8]. This corresponds to a Fourier encoding, and an MR acquisition can be described as

$$f_j(k_x, k_y) = \int \int c_j(x, y) u(x, y) e^{-i(k_x x + k_y y)} dx dy. \tag{7.1}$$

Equation (7.1) is generally referred to as the MR-signal equation, where f_j is the measured MR signal in Fourier space with spatial frequencies k_x and k_y; $j = 1, \ldots, n_c$ is the index for the particular element of the radio-frequency coil that is used to receive the MR signal, with sensitivity profile c_j [5]. Receive coils with multiple elements [9] have been the established standard in clinical MR scanners for more than two decades now, and their role in signal encoding and image reconstruction is discussed in more detail in Section 7.2. In Eq. (7.1) u is the MR signal that is measured in a voxel x, y. Equation (7.1) describes a two-dimensional (2D) acquisition. We will limit our description to the 2D case for simplicity of notation, but we would like to point out that Eq. (7.1) can be extended to three dimensions with the introduction of an additional encoding term $e^{-i(k_z z)}$, and all the material that is covered in this chapter can also be generalized accordingly.

In a real-world MR experiment, we always deal with a discretized form of Eq. (7.1):

$$f_j(k_x, k_y) = \sum\sum c_j(x,y) u(x,y) e^{-i(k_x x + k_y y)}. \tag{7.2}$$

Equation (7.2) can then be written in matrix form:

$$f_j = \mathcal{F}_\Omega C_j u = Au. \tag{7.3}$$

The matrix A fully describes the encoding process and is therefore generally referred to as the encoding matrix. In the inverse problem literature it is commonly called the forward operator. It consists of a diagonal matrix C_j that describes the receive coil sensitivity profile and a Fourier transform that maps from image space to Fourier (k-)space coefficients at the coordinates Ω. In order to obtain an MR image u from samples f_j, we need to solve the inverse problem and invert the encoding matrix. This inverse problem can become ill-posed in the case of an accelerated MR acquisition when the number of k-space samples is reduced to speed up the acquisition; this may violate the Nyquist–Shannon sampling theorem. Strategies to deal with this difficulty constitute the topic of the following sections. We will first provide a short review of parallel imaging and compressed sensing, and then focus on deep-learning methods.

7.2 Parallel Imaging

Receive coils with multiple elements [9] have been the established standard in clinical MR scanners for more than two decades now. They were originally designed to increase the signal-to-noise-ratio, and the central idea of parallel imaging [4–6] is to use the multiple elements as a means of an additional spatial encoding strategy on top of the magnetic field gradient encoding. This therefore allows a reduction in the number of gradient encoding steps, which directly reduces the acquisition time of the scan. A trivial reconstruction performed by zero filling of the non-acquired k-space points followed by an inverse Fourier transform is corrupted with aliasing artifacts. Parallel imaging utilizes the encoding capabilities of the multiple receive coils to synthesize these missing k-space points and to recover alias-free images. The first description of parallel imaging, SMASH [4], as well as its refinements, most notably GRAPPA [6], was formulated such that missing coefficients were directly synthesized in k-space. This resulting fully sampled k-space could then be reconstructed with a simple inverse Fourier transform. An alternative approach, SENSE [5] did the inverse Fourier transform as the first step, and the un-aliasing process was then performed in image space. In this chapter we are restricting our discussion of parallel imaging to the SENSE formalism, which will form the basis for the deep-learning methods that are covered in later sections. The data processing pipeline in SENSE is illustrated in Fig. 7.1.

We continue to use the notation from the previous section and so denote f_j^Ω as the undersampled measurement data from $j = 1, \ldots, n_c$ coil receive channels, with estimated receive coil sensitivity profiles c_j and corresponding undersampled images

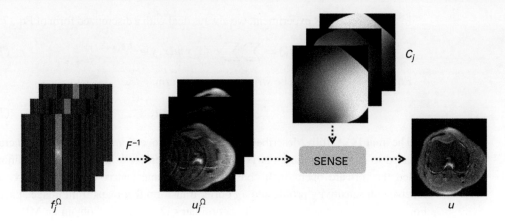

Figure 7.1 Illustration of the SENSE parallel imaging reconstruction process. Undersampled multi-channel k-space data undergoes an initial inverse Fourier transform. A combined alias-free image is then computed with the use of receive coil sensitivity maps. A description of the notation is given in the main body of the text.

u_j^Ω; u is the reconstructed image. It is worth pointing out that the coefficients c_j are only estimations of the true coil sensitivity profiles, and imperfections in estimation can lead to errors in the reconstructions. An equidistant undersampling scheme as used in parallel imaging leads to a superposition of periodic repetitions of the pixels in the accelerated image. In particular, if R is the undersampling factor then the signal of a pixel $u_j^\Omega(x, y)$ is the superposition of R pixels from $u(x, y)c_j(x, y)$. As long as the undersampling factor does not exceed the number of receive coil elements, we can solve the corresponding linear system of equations (7.2) and reconstruct the un-aliased image u.

7.3 Compressed Sensing

The central idea behind the application of compressed sensing (CS) to MR image reconstruction [7] is that MR images are inherently compressible. Owing to this compressibility, they can be represented with a number of nonzero coefficients that is smaller than the number of image pixels after transformation to a particular sparse domain (e.g., the wavelet transform). In combination with parallel imaging, the *a priori* knowledge in the form of the CS sparsity constraint can be seen as a regularization of the inverse problem that arises in the context of an accelerated (undersampled) MR acquisition. Mathematically, the reconstruction problem now takes the form of an optimization problem of the following form (note that we are dropping the indices j and Ω for improved readability from this point on in this chapter, but f should still be considered as undersampled multi-receive-channel raw data):

$$\min \|Au - f\|_2^2 + \lambda \|\Psi(u)\|_1. \tag{7.4}$$

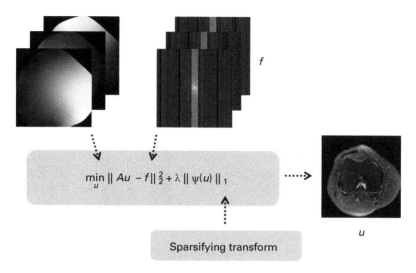

Figure 7.2 Combined parallel imaging and compressed sensing reconstruction. A description of the notation is given in the main body of the text.

Equation (7.4) is illustrated conceptually in Fig. 7.2. It consists of two terms. The first is usually referred to as data fidelity. It enforces consistency of the reconstructed image u with the measured k-space data f. The second is the sparsity constraint in the transform domain Ψ. The two terms are balanced by a hyper parameter λ, commonly referred to as the regularization parameter. It should be noted that in the case when λ is set to zero, Eq. (7.4) is reduced to the parallel imaging matrix inversion, as described in Section 7.2. This optimization problem can then be solved numerically with standard gradient-descent-based optimizers.

7.4 Machine Learning

Since 2016 [1–3], there has been a substantial amount of research activities related to the use of machine learning, in particular deep-learning methods to solve inverse problems in imaging. New research papers on this topic appear every day now, and the goal of this chapter is not to provide an exhaustive overview of the current literature. Instead, we are focusing on methods that can be seen as extensions of CS. In the original CS formulation, the sparsifying transform Ψ was fixed, for example in the wavelet-transform [7], Fourier-transform [10], or Total Variation based total-variation-based methods [11]. Machine learning opens up the possibility of learning a sparsifying transform from a set of training data. Similar ideas were already presented a decade ago in the context of dictionary learning [12], but one of the new contributions of the more recent developments was to map the entire optimization problem from Eq. (7.4) onto a neural network model. This is illustrated conceptually in Fig. 7.3. The input to the neural network is an undersampled k-space f, and the output is the

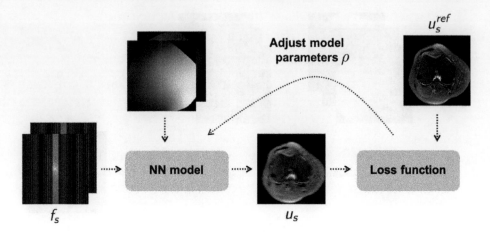

Figure 7.3 Supervised training of a machine learning neural network (NN) model for the reconstruction of accelerated MR images. A description of the notation is given in the main body of the text.

reconstructed image u. The forward operator A can be integrated explicitly in the neural network architecture [2, 13–15], which enforces data consistency in the same way as in CS. In these approaches the neural network can be seen as an unrolled version of a classic gradient descent optimization. Alternatively, the mapping from k-space to the reconstructed image can be fully learned from the training dataset itself in a single step [16]. Similarly, calibration information such as coil sensitivity maps can either be pre-estimated with traditional approaches and then used in the neural network model or estimated from the data as part of the learning procedure [17].

We denote all the parameters of the neural network by ρ and use the index s for k-space data samples f_s from a training set of size S. We also assume that, for each sample, we have a ground-truth reference reconstruction u_s^{ref} available. Situations where a ground truth is not available and self-supervised learning methods need to be employed are active research topics in the field. However, these developments are beyond the scope of this chapter, and we will focus on supervised learning accordingly. The training can be described by solving the following optimization problem:

$$\min \mathcal{L}(\rho) = \frac{1}{S} \sum_{s=1}^{S} \left\| u_s(\rho) - u_s^{ref} \right\|_2^2. \qquad (7.5)$$

The goal of the training that is described by Eq. (7.5) is to minimize the error between the output reconstruction of the network $u_s(\rho)$ with current parameters ρ and the ground truth u_s^{ref}. The cost function $L(\rho)$ that is minimized is defined by a chosen metric, which should be representative of the image quality of the reconstruction. In practice, the mean squared error (MSE) is a simple and popular choice, although it has been criticized as not fully representing the diagnostic quality of medical images [18]. In line with the general machine learning literature, we define one training epoch to be

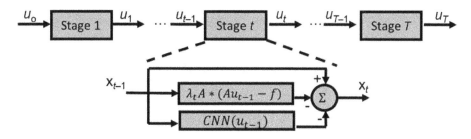

Figure 7.4 A neural network model with T stages viewed as an unrolled gradient descent optimization.

completed when we have looped over all training samples S in our training dataset, and this is repeated for the number of epochs defined by our training convergence criterion.

Deep-learning methods for image reconstruction can be viewed as an unrolled gradient descent optimization. This concept is illustrated in Fig. 7.4, which depicts a neural network model with T stages. Each stage applies regularization with a learned convolutional neural network (CNN) as well as a data consistency operation. One iteration of an iterative reconstruction can be related to one stage in the network. Deep-learning models that are based on this concept vary in complexity and model capacity. One example of such a model – referred to as a variational network (VN) [2] – has unique learned weights in each stage t of the model. Another option is to use the same weights for each layer, which reduces the model capacity of the network [15]. We refer to this model as a shared-weight VN (SW-VN). These methods both require coil sensitivity estimates calculated with conventional methods as input to the model. The model capacities of SW-VN and VN are approximately two hundred thousand and two million, respectively.

7.5 Experimental Results

7.5.1 Data Acquisition and Experimental Design

Here, we present results for the approaches described in previous sections of this chapter for data from a knee scan obtained at the Department of Radiology at NYU School of Medicine. Acquisition of the exam was part of a study approved by our local institutional review board (IRB). Data were acquired on a clinical 3T system (Siemens Magnetom Skyra), using a 15-channel knee coil. A T2-weighted turbo-spin-echo (TSE) sequence was acquired in the axial plane with the following sequence parameters: TR = 4310 ms, TE = 65 ms, in-plane resolution of 0.55×0.55 mm^2 covering a field of view of 140×140 mm^2 (matrix size 256×256), slice thickness 3 mm, echo train length = 9. No acceleration was performed at the data acquisition stage. A total of 41 slices was acquired and a single slice at the center of the volume was used for the examples in this chapter. We obtained the fully sampled ground-truth reconstruction via an inverse Fourier transform and coil-sensitivity-weighted combination of the individual receive channels and obtained coil sensitivity maps with ESPIRiT [19].

We performed an experiment that replicates an accelerated parallel imaging protocol, described in Section 7.2. We simulated an accelerated acquisition with equidistant reduction of k-space lines by a factor $R = 4$ in the left–right direction. A non-undersampled block of 26 lines at the center of k-space was included to estimate the coil sensitivity maps. While this sampling pattern is optimal for parallel imaging, it is not ideal for CS-inspired reconstructions, however. To boost the performance, a sampling pattern that provides a stronger degree of incoherence of the aliasing artifacts, such as Poisson-disk sampling, should be used [20]. However, since sampling pattern optimization is not a topic of this chapter, we have not adapted sampling patterns individually for each reconstruction approach but have used equidistant sampling for all our experiments, owing to its widespread use in clinical practice.

The results of all reconstruction methods, the fully sampled reference, a plot of the sampling pattern, and a trivial zero-filling reconstruction are shown in Fig. 7.5. We have also calculated the MSE with respect to the fully sampled ground truth as a quantitative measure of image quality. The zero-filling reconstruction shows characteristic aliasing artifacts from the undersampling of k-space. Because the k-space center, there is also an over-proportional contribution from low-frequency information, which represents itself as image blurring. Details for the other reconstructions are given in the following subsections.

7.5.2 Parallel Imaging

For the parallel imaging experiment, we performed the matrix inversion of Eq. (7.2) with an iterative conjugate gradient (CG) algorithm according to the CG-SENSE method introduced by Pruessmann et al. [21]. The only hyper-parameter of this reconstruction is the number of CG iterations, which defines the balance between the recovery of resolution and removal of aliasing artifacts and the amplification of noise. We set the number of CG iterations to 4, which minimized the MSE to the ground-truth reference. The MSE (0.038) is reduced substantially in comparison with the zero-filling reconstruction (MSE 0.105), and the image resolution is comparable with the ground-truth reference. However, the reconstruction still shows residual aliasing artifacts and an increase in noise.

7.5.3 Compressed Sensing

We performed a combined parallel imaging and CS reconstruction, as described in Section 7.3, using total generalized variation (TGV) as the sparsity-promoting term in Eq. (7.4). We used 1000 iterations of a first-order primal–dual algorithm [22] to solve the optimization problem of Eq. (7.4) in the same way as described in [23]. The user-defined hyper parameter of this reconstruction is the regularization parameter λ. We again selected it such that the MSE to the ground-truth reference was minimized, which resulted in a value of $\lambda = 10^{-4}$ and an MSE of 0.022. In comparison with

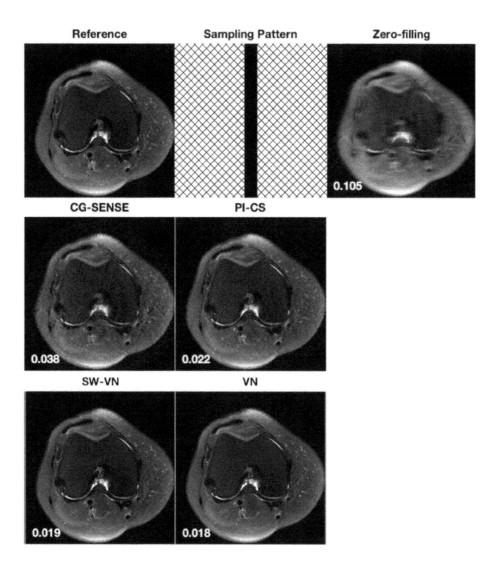

Figure 7.5 Reconstruction results for parallel imaging, compressed sensing, and deep learning together with the fully sampled reference, the sampling pattern, and a trivial zero-filling reconstruction. The reduction factor was $R = 4$, with 26 fully sampled lines at the center of k-space to estimate the coil sensitivity maps. The MSE in comparison to the fully sampled reference is displayed at bottom left for each reconstruction.

the CG-SENSE reconstruction, the CS reconstruction shows reduced aliasing artifacts and reduced noise amplification.

7.5.4 Machine Learning

We trained the deep-learning models discussed in section 4 (SW-VN, VN) with 100 axial knee datasets acquired as part of a study approved by our local IRB. These images had with similar hardware and scan parameters to those described in Section 5.1; however, they were acquired with $R = 2$ under sampling and reconstructed with GRAPPA. The input images were retrospectively undersampled to $R = 4$, with equidistant sampling and 26 center k-space lines. The networks were trained with the adaptive moment estimation (ADAM) optimizer and a learning rate of 1×10^{-4} and the weights of the models were updated after every forward pass through the network (batch size = 1) to minimize the MSE of the prediction relative to the target images. Results of these reconstruction methods along with the calculated MSE relative to the fully sampled ground truth are shown in Fig. 7.5. In comparison with the parallel imaging and compressed sensing reconstructions, the deep-learning reconstructions appear less noisy and have lower MSE. While in this example the two machine learning models perform comparably, with the VN just slightly out-performing the SW-VN, we expect that the VN, as the higher-model-capacity network, may be more prone to over fitting in cases where the training dataset is smaller.

7.6 Summary

The development of deep-learning reconstruction methods for accelerated MR acquisitions has been an ongoing area of research for the last several years. It has been repeatedly demonstrated that deep-learning methods can outperform classic reconstruction approaches in terms of both quantitative image metrics like MSE in relation to ground truth, as well as in qualitative reader studies where radiologists have been asked to rate the image quality in a subjective way [2, 24]. However, the day-to-day diagnostic value of these methods has still to be demonstrated in clinical studies. On the basic science side, their reliability, stability, and robustness is also a topic of increasing research interest [25]. There is also an increasing trend of research-community-based evaluation in the form of publicly shared datasets and, again taking inspiration from the machine learning and computer vision community, publicly open research challenges [26].

References

[1] S. Wang, Z. Su, L. Ying, X. Peng, S. Zhu, F. Liang, D. Feng, and D. Liang, "Accelerating magnetic resonance imaging via deep learning," in *Proc. IEEE International Symposium on Biomedical Imaging*, 2016, pp. 514–517.

[2] K. Hammernik, F. Knoll, D. K. Sodickson, and T. Pock, "Learning a variational model for compressed sensing MRI reconstruction," in *Proc. Proceedings of the International Society of Magnetic Resonance in Medicine*, 2016, p. 1088.

[3] K. H. Jin, M. T. McCann, E. Froustey, and M. Unser, "Deep convolutional neural network for inverse problems in imaging," *IEEE Transactions on Image Processing*, vol. 26, no. 9, pp. 4509–4522, 2017.

[4] D. K. Sodickson and W. J. Manning, "Simultaneous acquisition of spatial harmonics (SMASH): Fast imaging with radiofrequency coil arrays." *Magnetic Resonance in Medicine*, vol. 38, no. 4, pp. 591–603, 1997.

[5] K. P. Pruessmann, M. Weiger, M. B. Scheidegger, and P. Boesiger, "SENSE: Sensitivity encoding for fast MRI," *Magnetic Resonance in Medicine*, vol. 42, no. 5, pp. 952–962, 1999.

[6] M. A. Griswold, P. M. Jakob, R. M. Heidemann, M. Nittka, V. Jellus, J. Wang, B. Kiefer, and A. Haase, "Generalized autocalibrating partially parallel acquisitions (GRAPPA)," *Magnetic Resonance in Medicine*, vol. 47, no. 6, pp. 1202–1210, 2002.

[7] M. Lustig, D. Donoho, and J. M. Pauly, "Sparse MRI: The application of compressed sensing for rapid MR imaging," *Magnetic Resonance in Medicine*, vol. 58, no. 6, pp. 1182–95, 2007.

[8] M. A. Bernstein, K. F. King, and X. J. Zhou, *Handbook of MRI Pulse Sequences*. Academic Press, 2004.

[9] P. B. Roemer, W. A. Edelstein, C. E. Hayes, S. P. Souza, and O. M. Mueller, "The NMR phased array," *Magnetic Resonance in Medicine*, vol. 16, no. 2, pp. 192–225, 1990.

[10] U. Gamper, P. Boesiger, and S. Kozerke, "Compressed sensing in dynamic MRI," *Magnetic Resonance in Medicine*, vol. 59, no. 2, pp. 365–373, 2008.

[11] K. T. Block, M. Uecker, and J. Frahm, "Undersampled radial MRI with multiple coils. Iterative image reconstruction using a total variation constraint," *Magnetic Resonance in Medicine*, vol. 57, no. 6, pp. 1086–1098, 2007.

[12] S. Ravishankar and Y. Bresler, "MR image reconstruction from highly undersampled k-space data by dictionary learning," *IEEE Transactions on Medical Imaging*, vol. 30, no. 5, pp. 1028–1041, 2011.

[13] J. Schlemper, J. Caballero, J. V. Hajnal, A. N. Price, and D. Rueckert, "A deep cascade of convolutional neural networks for dynamic MR image reconstruction," *IEEE Transactions on Medical Imaging*, vol. 37, no. 2, pp. 491–503, 2018.

[14] J. Adler and O. Öktem, "Solving ill-posed inverse problems using iterative deep neural networks," *Inverse Problems*, vol. 33, no. 12, pp. 1–24, 2017.

[15] H. K. Aggarwal, M. P. Mani, and M. Jacob, "MoDL: Model-based deep learning architecture for inverse problems," *IEEE Transactions on Medical Imaging*, vol. 38, no. 2, pp. 394–405, 2019.

[16] J. Y. Zhu, T. Park, P. Isola, and A. A. Efros, "Unpaired image-to-image translation using cycle-consistent adversarial networks," *IEEE International Conference on Computer Vision*, 2017, pp. 2242–2251, 2017.

[17] A. Sriram, J. Zbontar, T. Murrell, A. Defazio, C. L. Zitnick, N. Yakubova, F. Knoll, and P. Johnson, "End-to-end variational networks for accelerated MRI reconstruction," in *Lecture Notes in Computer Science (including subseries Lecture Notes in Artificial Intelligence and Lecture Notes in Bioinformatics)*, vol. 12262. Springer, 2020, pp. 64–73.

[18] F. Knoll, J. Zbontar, A. Sriram, M. J. Muckley, M. Bruno, A. Defazio, *et al.*, "FastMRI: A publicly available raw k-space and DICOM dataset of knee Images for accelerated MR image reconstruction using machine learning," *Radiology: Artificial Intelligence*, vol. 2, no. 1, p. e190 007, 2020.

[19] M. Uecker, P. Lai, M. J. Murphy, P. Virtue, M. Elad, J. M. Pauly, S. S. Vasanawala, and M. Lustig, "ESPIRiT – An eigenvalue approach to autocalibrating parallel MRI: Where SENSE meets GRAPPA," *Magnetic Resonance in Medicine*, vol. 71, no. 3, pp. 990–1001, 2014.

[20] S. S. Vasanawala, M. J. Murphy, M. T. Alley, P. Lai, K. Keutzer, J. M. Pauly, and M. Lustig, "Practical parallel imaging compressed sensing MRI: Summary of two years of experience

in accelerating body MRI of pediatric patients," in *Proc. International Symposium on Biomedical Imaging*, 2011.

[21] K. P. Pruessmann, M. Weiger, P. Boernert, and P. Boesiger, "Advances in sensitivity encoding with arbitrary k-space trajectories." *Magnetic Resonance in Medicine*, vol. 46, no. 4, pp. 638–651, 2001.

[22] A. Chambolle and T. Pock, "A first-order primal–dual algorithm for convex problems with applications to imaging," *Journal of Mathematical Imaging and Vision*, vol. 40, no. 1, pp. 120–145, 2011.

[23] F. Knoll, K. Bredies, T. Pock, and R. Stollberger, "Second order total generalized variation (TGV) for MRI," *Magnetic Resonance in Medicine*, vol. 65, no. 2, pp. 480–491, 2011.

[24] F. Chen, V. Taviani, I. Malkiel, J. Y. Cheng, J. I. Tamir, J. Shaikh, S. T. Chang, C. J. Hardy, J. M. Pauly, and S. S. Vasanawala, "Variable-density single-shot fast spin-echo MRI with deep learning reconstruction by using variational networks," *Radiology*, p. 180 445, 2018.

[25] V. Antun, F. Renna, C. Poon, B. Adcock, and A. C. Hansen, "On instabilities of deep learning in image reconstruction and the potential costs of AI," *Proceedings of the National Academy of Sciences*, p. 2019 07 377, 2020.

[26] F. Knoll, T. Murrell, A. Sriram, N. Yakubova, J. Zbontar, M. Rabbat, A. Defazio, M. J. Muckley, D. K. Sodickson, C. L. Zitnick, and M. P. Recht, "Advancing machine learning for MR image reconstruction with an open competition: Overview of the 2019 fastMRI challenge," *Magnetic Resonance in Medicine*, p. mrm.28 338, 2020.

8 Model-Based Deep-Learning Algorithms for Inverse Problems

Mathews Jacob, Hemant K. Aggarwal, and Qing Zou

8.1 Introduction

The model-based recovery of images from noisy and sparse multi-channel measurements is now a mature area with success in several application areas such as MRI [1], CT [2], PET [3], and microscopy [4]. These schemes rely on a numerical model of the measurement system, often termed as the forward model, that represents the physics of an acquisition. Image recovery is then posed as an optimization problem, where the objective is to improve the consistency between the ground-truth data and measurements obtained from the image using the forward model. Since the recovery from just a few measurements is an ill-posed problem, the general approach is to modify the objective function using priors that penalize solutions that fall outside the class of natural images. Carefully engineered priors, including the total variation [5, 6], patch-based nonlocal methods [7, 8], low-rank penalties [9–12], and priors learned from exemplary data [13] or from the measurements themselves [14] are widely used.

Several researchers have recently proposed exploiting the power of deep convolutional neural networks (CNNs) in image recovery. Some of these schemes customize existing CNN architectures (e.g., UNET [15] & ResNet [16]) to image recovery tasks [17–19]. The early methods trained a deep network to learn the inverse of the mapping and to exploit the extensive redundancy in the images. While these approaches showed the power of deep-learning algorithms, a challenge with these schemes is their high demand for training data. In addition, it is also challenging to incorporate the known physics of an acquisition. This chapter reviews model-based deep-learning strategies that combine the imaging physics with deep-learned modules to overcome the above challenges.

An alternative to the above direct inversion approaches is iterative algorithms that alternate between data-consistency enforcement and pre-trained CNN denoisers [20–24]. These schemes are motivated by plug-and-play methods [25, 26] that use denoiser priors. Unlike the use of off-the-shelf denoisers in classical plug-and-play methods in [25, 26], the above schemes learn the denoisers from Gaussian noise-corrupted example images. Similar approaches using generative-adversarial-network- (GAN-) based models [27] and projection-based networks were introduced [28]. A benefit of these schemes is that they are model-agnostic: the same approach will work for any measurement scheme. Figure 8.1 summarizes the MR image reconstruction process using direct inversion (Fig. 8.1(a)), a deep image prior [29] (Fig. 8.1(b)), and plug-and-play approaches (Fig. 8.1(c)).

Recently, end-to-end training schemes [30–35] have been introduced to offer improved performance over the above-mentioned plug-and-play models. These schemes unroll the networks [30] and train the learnable components of the network parameters in an end-to-end fashion, thus customizing the network to the specific forward model. Since the networks are fine-tuned to the specific forward model, the schemes significantly reduce the amount of training data. Most of the above models use different CNN networks at different iterations to increase the network capacity. By contrast, the sharing of the network weights across iterations enables a reduction in training data demand for comparable image quality [32, 36].

8.2 Model-Based Approaches that Rely on Shallow Learning

8.2.1 Image Formation and Forward Model

The acquisition process in most imaging schemes can be modeled by an operator $\mathcal{A}$ applied to the continuous domain image γ, denoted by $\mathbf{y} = \mathcal{A}(\gamma) + \mathbf{n}$. The general practice is to discretize the problem, thus considering the recovery of an image vector γ from its linear measurements, modeled by a linear matrix $\mathbf{A}$:

$$\mathbf{y} = \mathbf{A}\gamma + \mathbf{n} \tag{8.1}$$

The above equation is a numerical model for the imaging device and is often referred to as the forward model. In many imaging methods, the forward model is known accurately. However, there are applications where the forward model is unknown or only partially known. Examples include imaging in the presence of motion during the acquisition, trajectory errors and field inhomogeneity effects in MRI acquisitions, and unknown blur kernel in deconvolution problems.

8.2.2 Model-Based Algorithms

Model-based algorithms use information from the forward model (8.1) to recover the image γ from the measured data. The recovery of γ from the noise-corrupted measurements $\mathbf{y}$ is ill-posed when $\mathbf{A}$ is a rectangular matrix. The general practice in model-based imaging is to use prior information to constrain the reconstructions. The recovery is posed as a regularized optimization scheme:

$$\gamma = \arg\min_{\gamma} \underbrace{\|\mathbf{A}\gamma - \mathbf{y}\|_2^2}_{\text{data consistency}} + \lambda \underbrace{\mathcal{R}(\gamma)}_{\text{regularization}} \tag{8.2}$$

The regularization prior $\mathcal{R} : \mathbb{C}^n \to \mathbb{R}_{>0}$ is often used to restrict the solutions to the space of desirable images. For instance, the wavelet-based sparsity prior is often used to encourage the recovery of images γ that have few nonzero wavelet coefficients.

The prior $\mathcal{R}(\gamma)$ is a large scalar when γ is an image with several nonzero wavelet coefficients.

The above formulation can also be viewed as an a posteriori estimate of the value of γ that maximizes the posterior distribution $p(\gamma|\mathbf{y}) = p(\mathbf{y}|\gamma)p(\gamma)/p(\mathbf{y})$. The estimate is obtained by minimizing the negative log posterior,

$$-\log p(\gamma|\mathbf{y}) = \underbrace{-\log p(\mathbf{y}|\gamma)}_{\text{data consistency}} \underbrace{-\log p(\gamma)}_{\text{prior}}. \quad (8.3)$$

The first term is the data consistency term, assuming the noise vector $\mathbf{n}$ in Eq. (8.1) to be Gaussian distributed. The second term captures the prior information one has about the images. Since an accurate prior can offer improved recovery, a focus has been the design of priors. Several researchers have come up with carefully engineered priors in the past decade to model the data distribution $\log p(\gamma)$. For instance, classical Tikhonov regularization relies on a Gaussian prior on γ, resulting in $\mathcal{R}(\gamma) = \|\gamma\|^2$. Similarly, compressed sensing methods choose $\mathcal{R}(\gamma) = \|\mathbf{W}\gamma\|_{\ell_1}$ to promote sparse images in an appropriate transform domain [37] (e.g., wavelet domain, when $\mathbf{W}$ is the wavelet transform). Other choices include include the total variation [38] and related priors.

Regularization penalties are also often learned from example data. For instance, transform learning [39] methods learn transformations or dictionaries from pre-acquired images, which are very efficient in sparsely representing the images. These schemes are observed to yield improved reconstructions compared with analytically derived transforms. Similarly, blind dictionary/transform learning approaches estimate the dictionary from the measured data themselves, offering further improved performance. Low-rank approaches exploit the low-dimensional structure of images in a dynamic dataset, or patches in a static image or in the Fourier domain, to recover images from very few measurements [40–43].

8.2.3 Challenges with Traditional Model-Based Algorithms

The solution to Eq. (8.1) may be computed using a variety of optimization methods, including the alternating direction method of multipliers (ADMM), the fast iterative shrinkage thresholding algorithm (FISTA) [52], two-step iterative shrinkage thresholding (TwIST), or alternating minimization. For instance, the ADMM scheme considers the equivalent problem

$$\gamma = \arg\min_{\gamma, \mathbf{v}} \|\mathbf{A}\gamma - \mathbf{y}\|_2^2 + \lambda \, \mathcal{R}(\mathbf{v}) \quad \text{s.t} \quad \mathbf{v} = \gamma. \quad (8.4)$$

The above constrained optimization problem has an associated augmented Lagrangian given by

$$L_\beta(\gamma, \mathbf{v}, \mathbf{u}) = \|\mathbf{A}\gamma - \mathbf{y}\|_2^2 + \lambda R(\mathbf{v}) + \beta \|\gamma - \mathbf{v} - \mathbf{u}\|^2 - \beta \|\mathbf{u}\|^2. \quad (8.5)$$

The above problem is solved by alternating between the following steps:

$$\widehat{\boldsymbol{\gamma}} \leftarrow \arg\min_{\boldsymbol{\gamma}} L_\beta(\boldsymbol{\gamma}, \widehat{\mathbf{v}}, \mathbf{u}) = \arg\min_{\boldsymbol{\gamma}} \|\mathbf{A}\boldsymbol{\gamma} - \mathbf{y}\|_2^2 + \beta\|\boldsymbol{\gamma} - (\mathbf{v} - \mathbf{u})\|^2, \quad (8.6)$$

$$\widehat{\mathbf{v}} \leftarrow \arg\min_{\mathbf{v}} L_\beta(\widehat{\boldsymbol{\gamma}}, \mathbf{v}, \mathbf{u}) = \beta\|\mathbf{v} - \underbrace{(\boldsymbol{\gamma} - \mathbf{u})}_{\overline{\boldsymbol{\gamma}}}\|^2 + \lambda R(\mathbf{v}), \quad (8.7)$$

$$\mathbf{u} \leftarrow \mathbf{u} + (\widehat{\boldsymbol{\gamma}} - \widehat{\mathbf{v}}). \quad (8.8)$$

The second step of the above optimization scheme,

$$\widehat{\mathbf{v}} \leftarrow \arg\min_{\mathbf{v}} \beta\|\mathbf{v} - \underbrace{(\boldsymbol{\gamma} - \mathbf{u})}_{\overline{\boldsymbol{\gamma}}}\|^2 + \lambda R(\mathbf{v}) = \mathcal{D}_\beta(\overline{\boldsymbol{\gamma}}), \quad (8.9)$$

often (e.g., for the ℓ_1-norm) has an analytical solution $\mathcal{D}_\beta(\overline{\boldsymbol{\gamma}})$ that is called the proximal mapping. It can be viewed as a *denoising* step to reduce noise in the current solution $\overline{\boldsymbol{\gamma}}$, thus yielding $\widehat{\mathbf{v}}$. The factor β in Eq. (8.5) is a continuation parameter that is often interpreted as $1/\sigma_\beta^2$; σ_β is the variance of the *noise* in $\overline{\boldsymbol{\gamma}}$ in that specific iteration, and it decreases with iterations. The first step (8.6) involves an inversion step to reduce a cost function that is the linear combination of the data consistency error and the deviation from the *denoised* image $\mathbf{v}$.

The above steps can be viewed as an iterative denoising scheme, which is capitalized in the plug-and-play algorithms to be discussed in Section 8.4. The main challenge with the above convex optimization schemes is the high computational complexity resulting from the numerous iterations needed for the above algorithms to converge. In particular, the data consistency step (8.6) involves the evaluation of the forward model and its adjoint, which is often computationally expensive. Despite several advances in novel optimization algorithms, the high computational complexity of the above algorithms is a central bottleneck in applying them to practical problems.

Most of the current priors are designed with computational efficiency and mathematical tractability in mind. While these handcrafted and learned transform/dictionary priors have had great success in past years, they are not very effective in capturing the extensive nonlinear redundancies often present in natural datasets. In practice, real-world datasets may be modeled as points on low-dimensional manifolds in high-dimensional spaces. More flexible signal representations would be able to model the probability density more accurately, translating to improved performance.

8.3 Direct Inversion-Based Deep-Learning Algorithms

Deep-learning methods that rely on convolutional neural networks are emerging as powerful alternatives for image representation and recovery. The early approaches relied on direct inversion strategies [17–19, 28] which learn to recover the images from an approximate reconstruction (e.g., an inverse Fourier transform with zero filling):

$$\boldsymbol{\gamma} = \mathcal{T}_{\mathbf{w}}\left(\mathbf{A}^H \mathbf{y}\right), \quad (8.10)$$

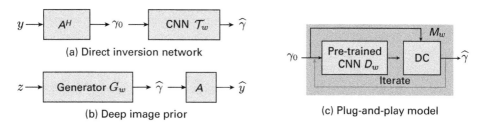

Figure 8.1 (a) A schematic diagram representing a direct inversion architecture. Here the neural network $\mathcal{T}_w$ learns to reconstruct a fully sampled image $\hat{\gamma}$ from the zero-filled reconstruction γ_0. (b) The deep-image-prior (DIP) approach, where the generator G_w learns to map a fixed random vector z to a measured k-space value y. In DIP, the loss between the predicted k-space value $\hat{y}$ and the observed k-space value y is minimized. (c) The general scheme followed in plug-and-play-prior approaches.

where $\mathcal{T}_\mathbf{w}$ is a learned CNN [44]. Figure 8.1(a) shows a visual representation of Eq. (8.3). The operator $\mathbf{A}^H(\cdot)$ transforms the measurement data to the image domain, since CNNs are designed to work in the image domain. The network parameters $\mathbf{w}$ are obtained during the supervised training procedure:

$$\mathbf{w}^* = \arg\min_{\mathbf{w}} \sum_{i=1}^{N_t} \|\boldsymbol{\gamma}_i - \mathcal{T}_\mathbf{w}\left(\mathbf{A}^H \mathbf{y}_i\right)\|^2. \tag{8.11}$$

Here $\boldsymbol{\gamma}_i$ and $\mathbf{y}_i$ are the fully sampled images and the corresponding undersampled measurements of the ith image in the training dataset, respectively. The minimization process will yield a CNN network learned solely from the data to invert the $\mathbf{A}$ operator for signals $\boldsymbol{\gamma}_i$ living on the image manifold/space. This is an entirely data-driven approach, which we will refer to as the direct inversion-based image recovery method. One of the main benefits of these schemes over conventional iterative strategies is the significant computational-complexity reduction. The image quality of these methods is also often better than that of conventional sparsity/low-rank priors. Such algorithms have been greatly successful in image denoising, where $\mathbf{A} = \mathbf{I}$. The CNN-based image denoisers are highly effective in denoising images compared with state-of-the-art methods such as block-matching and three-dimensional (3D) filtering (BM3D).

For many measurement operators (e.g., Fourier sampling, blurring, projection imaging), $\mathbf{A}^H\mathbf{A}$ is a translation-invariant operator; the convolutional structure makes it possible for CNNs to solve such problems [28]. However, the receptive field of the CNN has to be comparable with the support of the point spread function corresponding to $\mathbf{A}^H\mathbf{A}$. In applications involving Fourier sampling or projection imaging, the receptive field of the CNNs must be the same as that of the image; large networks such as UNET with several layers are required to obtain such a large receptive field. A challenge presented by such a large network with so many free parameters is the need for extensive training data to reliably train the parameters. Another challenge is that the CNN structure may not be well suited for problems such as parallel MRI, where $\mathbf{A}^H\mathbf{A}$ is not translation-invariant. Another challenge with this scheme is the lack of interpretability.

8.4 Model-Based Deep-Learning Image Recovery Using Plug-and-Play Methods

A highly successful framework for model-based image recovery is the plug-and-play formulation [26], which uses off-the-shelf image denoisers to replace the proximal operator D_β in (8.9). The plug-and-play formulation was later generalized using the concept of consensus equilibrium (CE) [45], which generalized the regularized inversion in Eq. (8.4) to include a wider variety of forward models and prior terms; the CE formulation does not require the terms to have an explicit cost function. Early methods relied on off-the-shelf image denoisers such as BM3D, assuming an input noise variance of σ_β. The great success of deep-learning denoisers has prompted many researchers to use pre-trained CNN blocks. The alternation between data-consistency and denoising steps can also be seen as a solution to the problem

$$\pmb{\gamma}^* = \arg\min_{\pmb{\gamma}} \|\mathbf{A}\pmb{\gamma} - \mathbf{y}\|^2 \quad \text{s.t.} \quad \pmb{\gamma} \in \mathcal{M}, \tag{8.12}$$

where $\mathcal{M}$ is the image set. The solution to this problem can be obtained by an iterative algorithm similar to the ADMM scheme described above, where the denoising–proximal mapping is replaced by a projection onto $\mathcal{M}$. We will now describe the popular CNN-based plug-and-play schemes that have been introduced recently.

8.4.1 Denoising Networks

The most popular approach is to pre-train CNNs (e.g., DCNN [46]) with additive Gaussian noise levels σ_β different from those of the training images; once trained, these algorithms are used within the model-based ADMM described above or similar proximal mapping algorithms. Figure 8.1(c) shows an example of utilizing a CNN network for image reconstruction. The main benefit of this scheme is the simplicity of the implementation. Such approaches have been widely used in applications such as image deblurring, CT reconstruction, image super-resolution, and MR image recovery. The CNN $\mathcal{D}_{\sigma_\beta}$ is trained to recover noise-free signals from their noise-perturbed versions. The learned mapping can be viewed as a contraction for points on the manifold $\mathcal{M}$.

8.4.2 Autoencoders

Autoencoders have been used to learn the operator $\mathcal{P}_\beta$ that projects onto the signal manifold. These methods are useful in multidimensional applications such as diffusion MRI [47] and MR spectroscopic imaging [48]. Here, one is interested in recovering a series of images γ_i; the intensity profile of each pixel is highly constrained and may be modeled by quantum mechanical simulations or diffusion models with few parameters. While the projection onto the signal manifold may be computed using iterative strategies, these approaches can be associated with high computational complexity. The above works approximate the projection step using autoencoders, which offer very fast inference. The autoencoder framework can be encouraged to form a contraction close to the manifold by (a) using bottleneck layers, (b) learning a denoising autoencoder, or (c) using regularization penalties. A benefit of the use of autoencoders or variational autoencoders would be the interpretability of the intermediate results because of the ability to view the latent mappings and their progression.

8.4.3 Generative Adversarial Networks (GANs)

Adversarial training is often used to learn a signal manifold [49, 50]; the training in GANs is formulated as a min–max optimization problem, where a CNN-based discriminator is used to detect deviations of the generated signals from the image class or manifold. After training, the generator $\mathcal{G}$ can synthesize an image $\mathbf{y} = \mathcal{G}(\mathbf{z})$ from latent variables $\mathbf{z}$ which follow a simple probability distribution (e.g., zero-mean unit Gaussian). The learned image manifold or class can be denoted as

$$\mathcal{M} = \{\mathbf{y} = \mathcal{G}(\mathbf{z}) | \mathbf{z} \sim p(\mathbf{z})\}. \tag{8.13}$$

Once learned, the generator can be applied to image recovery problems such as Eq. (8.12) [27]. Early approaches applied this generative model to inverse problems by solving

$$\mathbf{y} = \mathcal{G}(\mathbf{z}^*) \text{ where } \mathbf{z}^* = \arg\min_{\mathbf{z}} \|\mathbf{A}\mathcal{G}(\mathbf{z}) - \mathbf{y}\|^2. \tag{8.14}$$

The optimization problem relies on gradient descent to determine $\mathbf{z}^*$. A challenge with this direct approach is the slow nature of the convergence, resulting in high computational complexity during image reconstruction. Later methods [51] instead proposed the learning of a CNN-based projector $\widehat{\mathcal{P}}$ such that

$$\widehat{\mathcal{P}} = \arg\min_{\mathcal{P}} \sum_{i=1}^{N_t} \|\mathcal{G}\big(\mathcal{P}(\mathbf{y}_i)\big) - \mathbf{y}_i\|^2. \tag{8.15}$$

Once learned, the composition of the operators $\mathcal{D} = \mathcal{G} \circ \mathcal{P}$ can be assumed to be a projection onto the image manifold; it can be used as a *denoising* projection operator in plug-and-play algorithms described by Eqs. (8.6)–(8.8).

8.4.4 Benefits and Challenges of Plug-and-Play Methods

One of the main benefits of the plug-and-play formulations is that the CNN models are independent of the specific measurement operator $\mathbf{A}$; the algorithm can be adapted to arbitrary problems without retraining it to each setting. In addition, convergence results are available when the denoiser satisfies conditions including bounded Lipshitz constants, or when it is a projection [27, 51]. Elegant results on guaranteed signal recovery similar to those for compressed sensing are also available when the measurement operator satisfies restricted isometry-like conditions.

We note that plug-and-play methods are similar in nature to conventional sparse optimization algorithms; a large number of iterations are often needed to guarantee convergence. Thus, these methods do not enjoy the computational benefits offered by direct inversion schemes, especially when the forward operator and its adjoint are computationally expensive to compute.

Plug-and-play methods learn the prior distribution of the signals γ, denoted by $p(\gamma)$. The performance of the algorithm is dependent on the accuracy of the learned prior distribution to be learned. To obtain an accurate model for the signal distribution, large datasets and large networks are often needed. By contrast, the unrolled optimization schemes to be described in Section 8.5 consider the learning of priors $p(\gamma|y)$ [52], assuming the measurement model in Eq. (8.1).

The projection formulation in Eqs. (8.12) and (8.14) recovers an image in $\mathcal{M}$. A challenge with this scheme is the out-of-class generalization, when the signal $\gamma \notin \mathcal{M}$. This is an important problem, especially in applications such as medical imaging where the goal is to detect anomalies in the data. In particular, projection-based algorithms will provide visually pleasing images from the image manifold even when $\gamma \notin \mathcal{M}$. We note that in similar settings, classical compressed sensing methods often fail catastrophically, resulting in a non-sparse image, which might be a more desirable solution than a visually pleasing solution. Recent approaches that extend the range of the generator [53] offer some promise in this scenario.

8.5 Model-Based Deep-Learning Algorithms with Unrolling and End-to-End Optimization

The seminal work [30] as well as several follow-up works [31–35] consider the unrolling of several iterations of the optimization algorithm specified by Eqs. (8.6)–(8.8) or similar iterative proximal mapping algorithms. Figure 8.2 shows an example of an unrolled model-based deep-learning architecture. The unrolled structure can be viewed as a neural network architecture that combines known physics-based forward models, their adjoints, and learnable CNN modules. The parameters of the optimization algorithm, including step sizes and regularization parameters, can be set as trainable variables and can be trained along with the CNN modules using the training data, as

$$w^* = \arg\min_{w} \sum_{i=1}^{N_t} \|\gamma_i - \mathcal{M}_w\left(\mathbf{A}^H \mathbf{y}_i\right)\|^2, \qquad (8.16)$$

where $\mathcal{M}_w$ is the unrolled network. The supervised training strategy is very similar to direct inversion schemes. However, the unrolled architecture is more appropriate than the generic CNN (e.g., UNET) considered in Section 8.3 for solving a specific problem. In particular, physics-based forward models and optimization algorithms can

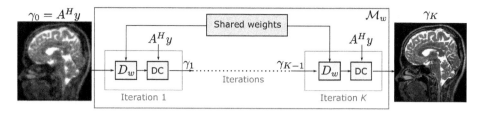

Figure 8.2 Unrolled model-based deep-learning architecture $\mathcal{M}_w$ in Eq. (8.16). The input is a zero-filled reconstruction γ_0. The network is unrolled for K iterations where each iteration contains a CNN D_w and a data-consistency (DC) step. The block D_w can be any CNN architecture such as ResNet or UNET. In this figure, all K blocks D_w share the network weights. However, it is also possible to arrange that they do not share the network weights and thus to increase the network capacity. The DC layer can be implemented as a gradient descent step or a more sophisticated conjugate gradient step. Depending upon the forward model, the DC step can also have an analytical solution.

ensure the consistency of the solution at each iteration with the measurements. The CNN can learn the complementary aspects from exemplar data.

In many cases, such as deconvolution and single-channel Fourier sampling, the data-consistency steps specified by Eq. (8.6) can be analytically solved. When (8.6) cannot be solved analytically (e.g., in multi-channel parallel MRI [32]), one can use an optimization algorithm (e.g., a conjugate gradient (CG) algorithm) to solve it). These optimization algorithms can be embedded as layers within the network [32, 54]. One can rely on auto differentiation to backpropagate the gradients through these layers or compute them analytically. A conjugate gradient algorithm can enforce data consistency more effectively than the steepest descent updates that are widely used in unrolling algorithms; unrolling schemes with CG can offer improved reconstruction in the unrolling setting, which typically restrict the number of unrolling steps for memory and computational efficiency. Recent approaches have considered the unrolling of complex algorithms, including structured low-rank algorithms [35, 54], low-rank minimization [55], and alternating algorithms, to solve bilinear optimization algorithms [56]

Many of the current unrolled schemes use different CNN blocks at each iteration of the network to improve capacity [31, 34]. By contrast, some researchers have argued reuse of the same CNN block [32, 35, 54] or, equivalently, sharing the parameters of the CNN networks at each iteration during training. The latter approach is similar to conventional or plug-and-play models and hence is more interpretable. It is observed that the image quality often improves monotonically as one proceeds through the iterations. By contrast, when different CNN blocks are used at different iterations, the output at each layer may not be very interpretable. Another benefit of sharing the CNN parameters with consequent reduced data demand is that unrolled schemes with shared parameters often require far fewer training data to offer a performance comparable with approaches that use different networks at each iteration.

Figure 8.3 shows a visual comparison of the reconstruction quality of compressed sensing and deep-learning methods at 4× acceleration using a two-dimensional (2D)

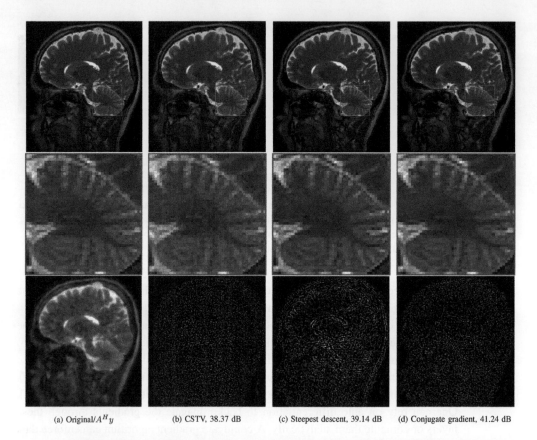

(a) Original/$A^H y$ (b) CSTV, 38.37 dB   (c) Steepest descent, 39.14 dB   (d) Conjugate gradient, 41.24 dB

Figure 8.3 For comparison, reconstruction quality using compressed sensing and model-based deep-learning algorithms at 4× acceleration using a 2D variable-density random sampling mask. (a) The ground-truth image, a zoomed cerebellum region, and zero-filled reconstruction. (b) Compressed sensing reconstruction using CSTV algorithm. (c) and (d) Reconstructions using model-based deep-learning architecture with steepest descent and CG used inside the DC layer, respectively. The numbers in the sub-captions show the PSNR values in dB. © 2019 IEEE. Reprinted, with permission, from [57].

variable-density random sampling mask. We observe that both the deep-learning methods in Figs. 8.3(c) and 8.3(d) give improved visual quality as well as higher peak-signal-to-noise ratio (PSNR) values as compared with the compressed sensing method CSTV [6]. We can also observe from the error maps that conjugate-gradient-based data consistency improves the reconstruction quality as compared with a steepest-descent-based data consistency step.

8.5.1 Benefits and Challenges

Several authors have observed empirically that the number of iterations required for unrolled optimization schemes to yield good performance is often orders of magnitude smaller than classical model-based and plug-and-play methods. For instance, many of the current unrolled optimization schemes rely on 10–20 iterations and have reported superior performance. The improved performance is often attributed to the CNN network being adapted to the specific forward model as well as to the optimization scheme. The computational efficiency of these algorithms during inference is reported to be comparable with that of direct inversion schemes.

Another benefit of unrolled optimization is the reduced data demand for training, as highlighted by the recent review [52]. For instance, training a generic denoiser to be used in plug-and-play methods amounts to learning $p(\gamma)$; training a GAN or auto encoder for this setting is hence associated with high sample complexity. By contrast, the training of CNN blocks using unrolled optimization amounts to learning conditional priors $p(\gamma|\mathbf{y})$ and can be achieved using only a few training examples.

The benefits of unrolled schemes over plug-and-play methods often come at the expense of generalization to other measurement operators. It is reported that networks trained with a specific $\mathbf{A}$ matrix perform poorly when used with other forward models (e.g., with different sampling patterns, scanners, or measurement conditions), which is a challenge associated with direct inversion schemes as well [58]. However, we note that many of the early unrolled approaches were trained with a variety of sampling operators drawn randomly from a set [31, 32, 34, 35, 54]; experiments show that this training strategy reduces the above sensitivity, the same network being sufficient to recover images at a range of acceleration factors [32, 54]. Another approach to reducing this sensitivity is model adaptation, where the models are fine-tuned to the new setting by re-training them with few images in the new setting [58, 59].

8.6 Model-Based Deep-Learning Reconstruction Without Pre-Learning

All the above discussed methods discussed above rely on training CNNs using training data. However, an elegant observation by Ulyanov et al. [29] shows that untrained neural networks can also be used to recover images in ill-posed inverse problems; this approach is termed as deep image prior (DIP). The ability of an untrained network to recover images is often attributed to the inherent bias of CNNs towards natural images. Since the original work, several authors have extended the DIP approach in various directions.

8.6.1 Single-Image Recovery using DIP

Ulyanov et al. [29] showed that the recovery of an image can be formulated as

$$\gamma^* = \arg\min_\theta \|\mathbf{A}\gamma - \mathbf{y}\|^2 \quad \text{s.t.} \quad \gamma = \mathcal{G}_\theta[\mathbf{z}]. \tag{8.17}$$

Here $\mathcal{G}_\theta$ is a CNN generator and $\mathbf{z}$ is a noise vector drawn from some noise space. The optimization is performed using stochastic gradient descent (SGD) or adaptive moment estimation (ADAM), starting with random initialization of the network weights. It is observed that the image quality improves with iterations at the beginning, while it degrades after a certain number of iterations because the network begins to fit the measurement noise in **y**. The common explanation for this behavior is that CNN networks can rapidly learn spatially smooth images, while it is more challenging for them to learn noise. To obtain the best image quality, early stopping is suggested.

A key benefit of this approach is that it does not need extensive training data. However, this benefit comes at the expense of performance, which leads to the performance of DIP not being comparable with the performance obtained from the previously discussed pre-trained networks, which require training data. In addition, DIP often results in longer run times compared with the unrolled and direct inversion approaches because of the need for ADAM or SGD optimization during reconstruction.

As discussed above, a challenge with the DIP framework is the need for manual stopping to obtain the best signal-to-noise ratio. Recently, it was shown that the DIP is asymptotically equivalent to a Gaussian-process prior as the number of channels goes to infinity [60]. On the basis of this insight, the authors of this chapter have introduced a posterior inference using stochastic gradient Langevin dynamics, where noise is added to the weights during training. The results show that this approach eliminates the need for early stopping, in addition to offering improved performance and providing uncertainty measures.

8.6.2 Inverse Problems Involving Multiple Images in a Manifold

Recently, the DIP framework was extended to dynamic imaging applications [61–63] where the images in a time series are modeled as the output of a generator,

$$\gamma_t(\mathbf{r}) = \mathcal{G}_\theta [\mathbf{z}_t]. \tag{8.18}$$

The variables $\mathbf{z}_t$ are assumed to be Gaussian noise variables in [61], as in [29], while they are optimized along with the weights of the deep CNN generator $\mathcal{G}_\theta$. The model in Eq. (8.18) is interpreted as a nonlinear mapping or lifting from a low-dimensional subspace $\mathbf{Z}$ to the image space [62, 63]. In free-breathing cardiac MRI, the images in the time series can be viewed as nonlinear functions of the cardiac and respiratory phases. The low-dimensional nature of the latent vectors enables exploitation of the nonlocal redundancies between images at different time points, thus facilitating the fusion of information between them as in earlier kernel low-rank methods [64, 65].

The network parameters θ and the latent variables $\mathbf{z}$ are jointly solved by minimizing the cost function

$$\mathcal{C}(\mathbf{z}, \theta) = \sum_{t=1}^{N} \|\mathcal{A}_t\left(\mathcal{G}_\theta[\mathbf{z}_t]\right) - \mathbf{y}_t\|^2 + \lambda_1 \underbrace{\|\nabla_\mathbf{z}\mathcal{G}_\theta\|^2}_{\text{network regularization}} + \lambda_2 \underbrace{\mathcal{R}(\mathbf{z})}_{\text{latent regularization}}. \tag{8.19}$$

The network regularization is an ℓ_2 penalty on the weights, which is observed to minimize the need for early stopping and offer improved performance. The latent vector regularization term involves a smoothness regularization to capitalize on the temporal smoothness of the images in the time series.

The above approach can also be extended to 3D applications, where the joint alignment and recovery of data from different slices is obtained using different acquisitions [66]. In this case, a Kullback–Leibler divergence term is used to encourage the latent vectors of all the slices to follow a zero-mean Gaussian distribution, facilitating the alignment of data from different slices.

8.7 Model-Based Deep-Learning Image Reconstruction: General Challenges, Current Solutions, and Opportunities

We now discuss the challenges associated with the current model-based deep-learning-based algorithms and discuss the current solutions. These challenges also present opportunities for future work which could further improve the performance and utility of the current methods for challenging applications.

8.7.1 Lack of Distortion-Free Training Data

While deep learned algorithms offer improved performance and reduced computational complexity during inference, compared with traditional algorithms, the main challenge in their widespread utility is the lack of fully sampled and noise-free training data in many situations. While more and more datasets are becoming public, the training data is often corrupted by noise. In addition, it is often challenging to acquire artifact-free or fully sampled images in many situations. For instance, when one is acquiring ultrahigh-resolution MRI data of the brain or the heart, it is often difficult to acquire fully sampled images because of the long scan time as well as subject motion. Similarly, it may be challenging to acquire blur-free images in deblurring applications. We will now briefly review the approaches to training deep networks in these contexts.

Learning of Deep Denoisers from Noisy Data
Several interesting approaches have been introduced for the training of deep denoisers. For instance, the Noise2Noise approach [67] relies on a pair of noisy images to train a denoiser without the need for clean images, which are not available in many applications, including microscopy. To avoid the need for paired measurements, the Noise2Void approach [68] was introduced. This is a blind spot method that excludes the central pixel from the network's receptive field, which minimizes the overfitting of the network to noise.

Another approach is to use Stein's unbiased risk estimate (SURE) [69], which is an unbiased estimate of the mean square error (MSE) that depends only on the recovered image and the noisy measurements. Assuming the additive noise n to be Gaussian distributed, the SURE [69] approach uses the loss function

$$\text{SURE}(\widehat{\gamma}, u) = \|\gamma - u\|_2^2 + 2\sigma^2 \nabla_u \cdot f_\Phi(u) - N\sigma^2, \tag{8.20}$$

which is an unbiased estimate of the true MSE, denoted by

$$\text{MSE} = \mathbb{E}_u \, \|\widehat{\gamma} - \gamma\|^2. \qquad (8.21)$$

Note that the expression (8.20) does not depend on the noise-free images γ; it depends only on the noisy images u and the estimates $\widehat{\gamma}$. In Eq. (8.20), $\nabla_u \cdot f_\Phi(u)$ represents the network divergence, which is often estimated using Monte-Carlo simulations [70]. Several researchers have adapted SURE as a loss function for the unsupervised training of deep image denoisers [71, 72], with performance approaching that of supervised methods.

Learning of Image Reconstruction Algorithms from Noisy and Ill-Posed Measurements

The SURE approach was extended to inverse problems with a rank-deficient measurement operator; this extension is termed generalized SURE (GSURE)) [73]. The GSURE approach provides an unbiased estimate of the projected MSE, which is the expected error of the projections in the range space of the measurement operator. This approach was recently used for inverse problems in [71]. The experiments in [71] show that the GSURE-based projected MSE is a poor approximation to the actual MSE in a highly undersampled setting. To improve performance, the authors trained the denoisers at each iteration in a message-passing algorithm in a layer-by-layer fashion, using classical SURE; this approach is termed LDAMP-SURE [71]. This approach approximates the residual aliasing errors at each iteration to be Gaussian random noise; since this assumption is often violated in many inverse problems, the performance of this layer-by-layer training approach is not as good as that of supervised methods.

Recently, the self-supervised learning using data undersampling (SSDU) approach [74], which has conceptual similarities to the blind-spot-based method [68], was introduced for the end-to-end training of unrolled algorithms. This approach partitions the measured k-space into two disjoint sets. The first set is used in the data consistency step, and the remaining set is used for measuring the error. Figure 8.4(a) summarizes the steps of the SSDU algorithm. This algorithm also uses different sampling operators for each image to improve the diversity of the observed k-space samples. This approach is more effective than using all the k-space measurements for data consistency and evaluating the error.

The ENSURE framework [75, 76] extends the SURE setting to learning with different sampling operators. Similarly to the classical SURE metrics [73, 77], the proposed ENSURE loss metric has a data consistency term and a divergence term. The data consistency term in ENSURE is the sum of the weighted projected losses [73]; the weighting depends on the class of sampling operators. The ENSURE metric is an unbiased estimate for the true image MSE and hence is a superior loss function, when compared with projected SURE [73], for training deep image reconstruction algorithms. Comparison of the above methods shows that the ENSURE approach can offer a performance that is comparable ith the supervised setting. Fig. 8.4(b) summarizes the implementation of the ENSURE loss function.

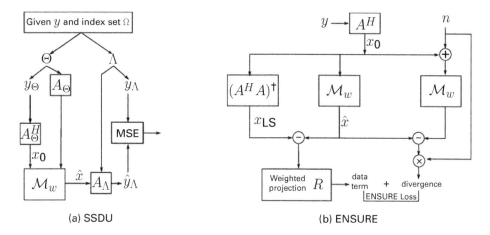

Figure 8.4 This figure summarizes two unsupervised model-based deep-learning approaches, SSDU and ENSURE. (a) Initially, SSDU partitions the available samples into two disjoint groups Θ and Λ. The first group Θ is used during training using the unrolled network $\mathcal{M}_w$. The second group Λ is used during loss function estimation using the MSE. (b) ENSURE consists of two terms. First the data term is calculated using a weighted projection R of the error between the network prediction $\hat{x}$ and a least squares solution x_{LS}; R is implemented using weighted least squares where the weights are the sampling density values. The divergence term is calculated using Monte-Carlo SURE by perturbing the zero-filled reconstruction x_0 using a standard Gaussian noise n and then passing through the network.

Learning of Generative Models from Noisy and Ill-Posed Measurements

Most GAN frameworks require fully sampled data. Once the training is complete, one can use the GAN framework in a plug-and-play mode as described in Section 8.4.3. Recent works show that GANs can also be trained from undersampled data. The ambient-GAN framework assumes that different images are acquired using different measurement operators $\mathbf{A}_i$, drawn randomly from a set. It uses a discriminator that differentiates the true measurements $\mathbf{y}_i = \mathbf{A}_i \mathbf{x}_i$ from the fake ones $\mathbf{A}_j \mathcal{G}(\mathbf{z}_j)$. In particular, the discriminator differentiates between $p(\mathbf{y})$ and $p(\mathbf{A}\mathcal{G}(\mathbf{z}))$. Once ambient-GAN is trained, it can be used for inference as in Section 8.4.3.

8.7.2 Vulnerability to Input Perturbations and Model Misfit

A general concern with deep-learning algorithms is their robustness to perturbations. Several examples have shown the fragility of deep-learning algorithms to subtle adversarial perturbations in the context of classification and object detection. Recent studies show that image reconstruction algorithms are also sensitive to adversarial perturbations, resulting in false positives (introduction of a fake feature in the image) and false negatives (missing or washout of features in the true image). In the context of model-based algorithms, such perturbations can result from either input perturbations or model misfit.

Adversarial Perturbations

In the recent works [78, 79], the authors argue that if a deep network provides good recovery of images close to the null space of $\mathbf{A}$, it will be sensitive to adversarial perturbation. The authors considered the setting when the network provides good recovery of two images $\mathbf{x}$ and $\mathbf{x}'$ close to the null space (i.e., $d(\mathbf{Ax}, \mathbf{Ax}') < \eta$) such that $d(\mathcal{M}(\mathbf{x}), \mathbf{x}) < \eta$ and $d(\mathcal{M}(\mathbf{x}'), \mathbf{x}') < \eta$). The authors argue that, in the above case, the Lipshitz constants will be high and hence the network will be vulnerable to false negatives (missing genuine features) and false positives (introducing artificial features). These results show that there is a trade-off between performance and stability. By contrast, conventional approaches such as sparse recovery are robust to such perturbations.

We note that plug-and-play methods, which rely on iterative projections onto the range space of the generator/denoiser, may be less sensitive to such perturbations. Using theoretical tools similar to restricted isometry conditions, it was shown that the recovery is robust to input noise if the measurement matrix $\mathbf{A}$ satisfies the set-restricted isometry (SREC) condition [27, 51]. In particular, this condition guarantees that the measurements of any two images $\mathbf{x}_1 = \mathcal{G}(\mathbf{z}_1)$ and $\mathbf{x}_2 = \mathcal{G}(\mathbf{z}_2)$ in the range space are distinct (i.e., $\|\mathbf{Ax}_i - \mathbf{Ax}_j\| > \delta > 0$). Since the difference between the images does not lie close to the null space of $\mathbf{A}$, robust recovery of the images are possible if SREC conditions are satisfied. Algorithms for designing the matrices that satisfy this condition were introduced in [51].

While direct inversion and unrolling-based methods that learn priors $p(\mathbf{x}|\mathbf{y})$ can offer improved performance and significantly reduced computational complexity over plug-and-play methods that use measurement-matrix-independent priors, the price to pay is the potential reduction in robustness. One possible countermeasure is to add regularization penalties in the reconstruction networks to keep the Lipshitz constants bounded, similarly to the adversarial training strategies used in other deep-learning areas. These schemes use adversarial training using a min–max optimization loss. In particular, the inner optimization tries to fabricate adversarial perturbations in order to maximize the error, while the outer loop aims to minimize the loss on the adversarially perturbed examples. These approaches aim to keep the Lipshitz constants low, which should minimize the vulnerability to adversarial perturbations. We note that such regularization approaches will result in reduced performance, especially for signals close to the null space of $\mathbf{A}$.

Vulnerability to Model Misfit

Another challenge associated with the direct inversion and unrolled optimization approaches is their vulnerability to model misfit. Specifically, the forward model may be different from the one used to train the networks. In the context of MRI, the data acquired from different MRI scanners or coils, or even from different measurement matrices, may result in potential degradation in performance.

One approach to minimizing the potential impact of misfitting might be to train with different forward models [32]. Large networks trained using extensive data from multiple settings (e.g., fastMRI data) show the ability to generalize across different acquisition conditions. Nevertheless, this approach may translate to lower overall performance when compared with schemes that have been trained with the specific sampling pattern, especially when training takes place with insufficient data.

Another approach is to adapt the pre-trained networks to the new setting using the measured data. Note that fully sampled images are usually not available to do the model adaptation. The unsupervised approaches discussed in Section 8.7.1 can be used to adapt the networks. For example, the SSDU approach is used in [80], while the SURE approach is used in [59]. Unlike [75], where a network is trained from scratch, the adaptation to a pre-trained network using only undersampled measurement is observed to be fast and to offer a performance that is close to the corresponding supervised strategy using extensive training data. While an adaptation using a single image is considered in [59], a generalization using ENSURE with multiple images can offer improved performance.

8.7.3 Joint Design of System Matrix and Image Recovery

We note that there is a close interplay between the learned image priors and the acquisition scheme as discussed in Section 8.7.2. In particular, recovery is sensitive to adversarial perturbations if the measurement operator cannot distinguish two images that live in the range space of the generator in projection-based approaches. Likewise, the learned priors could be adapted to the specific measurement operator to improve performance and robustness.

In many areas, there is extensive opportunity for designing the measurement system. For instance, the sampling pattern can be optimized to improve performance in MRI. Likewise the parameters of the optical system may be tuned to improve performance in computational photography systems. Since classical algorithms such as compressed sensing are computationally expensive, the general practice is to design a bound on the performance (e.g., a restricted isometry property). In many cases, these bounds that are independent of the specific algorithms may not be a good indicator of the true performance. The optimization of the system for a specific reconstruction algorithm such as compressed sensing is computationally expensive. In particular, the design of the sampling pattern involves a nested optimization strategy; the optimization of the sampling patterns is performed in an outer loop, while image recovery is performed in the inner loop to evaluate the cost associated with the sampling pattern. The use of deep-learning methods for image reconstruction offers an opportunity to speed up the computational design. Specifically, deep-learning inference schemes enable fast evaluation of the loss associated with each sampling pattern.

Several authors have added the forward model as a layer in the neural network and optimized for the system parameters using the auto-differentiation property. For instance, the recent LOUPE algorithm [81] jointly optimizes the sampling density in k-space and the reconstruction algorithm, while PILOT [82] and J-MODL [83] jointly optimize the sampling locations. The use of a model-based approach is empirically observed to improve the optimization landscape, facilitating convergence in [83]. Similar approaches have been used in optical system design [84, 85].

8.8 Summary

This chapter has provided a summary of some popular model-based deep-learning methods and their extensions. Section 8.1 briefly described classical model-based methods and their benefits as well as their limitations. Section 8.2 described how deep learning can help in overcoming some limitations of classical model-based methods. Section 8.3 discussed how to incorporate a pre-trained deep network as a regularizer using the plug-and-play approach. Section 8.4 described end-to-end training using a model-based deep-learning framework. That section also discussed some benefits and limitations of end-to-end training. Sections 8.5 and 8.6 discussed unsupervised model-based deep-learning approaches when a clean training dataset not available. Section 8.6 also discussed model mismatch issues as well as the joint design of acquisition and reconstruction framework.

References

[1] J. A. Fessler, "Model-based image reconstruction for MRI," *IEEE Signal Processing Magazine*, vol. 27, no. 4, pp. 81–89, 2010.

[2] I. A. Elbakri and J. A. Fessler, "Statistical image reconstruction for polyenergetic X-ray computed tomography," *IEEE Transactions on Medical Imaging*, vol. 21, no. 2, pp. 89–99, 2002.

[3] J. Verhaeghe, D. Van De Ville, I. Khalidov, Y. D'Asseler, I. Lemahieu, and M. Unser, "Dynamic PET reconstruction using wavelet regularization with adapted basis functions," *IEEE Transactions on Medical Imaging*, vol. 27, no. 7, pp. 943–959, 2008.

[4] F. Aguet, D. Van De Ville, and M. Unser, "Model-based 2.5-D deconvolution for extended depth of field in brightfield microscopy," *IEEE Transactions on Image Processing*, vol. 17, no. 7, pp. 1144–1153, 2008.

[5] T. Chan, A. Marqina, and P. Mulet, "Higher-order total variation-based image restoration," *SIAM Journal of Scientific Computing*, vol. 22, no. 2, pp. 503–516, 2000.

[6] S. Ma, W. Yin, Y. Zhang, and A. Chakraborty, "An efficient algorithm for compressed MR imaging using total variation and wavelets," in *Proc. Conference on Computer Vision and Pattern Recognition*, 2008, pp. 1–8.

[7] A. Buades, B. Coll, and J. M. Morel, "A non-local algorithm for image denoising," in *Proc. 2005 IEEE Computer Society Conference on Computer Vision and Pattern Recognition*, vol. II. IEEE Computer Society, 2005, pp. 60–65.

[8] Z. Yang and M. Jacob, "Nonlocal regularization of inverse problems: A unified variational framework," *IEEE Transactions on Image Processing*, vol. 22, no. 8, pp. 3192–3203, 2012.

[9] Z. P. Liang, "Spatiotemporal imaging with partially separable functions," in *Proc. 4th IEEE International Symposium on Biomedical Imaging: From Nano to Macro*, 2007, pp. 988–991.

[10] B. Zhao, J. P. Haldar, C. Brinegar, and Z. P. Liang, "Low rank matrix recovery for real-time cardiac MRI," in *Proc. 7th IEEE International Symposium on Biomedical Imaging: From Nano to Macro*, 2010.

[11] S. G. Lingala, Y. Hu, E. DiBella, and M. Jacob, "Accelerated dynamic MRI exploiting sparsity and low-rank structure: kt SLR," *IEEE Transactions on Medical Imaging*, vol. 30, no. 5, pp. 1042–1054, 2011.

[12] S. G. Lingala and M. Jacob, "A blind compressive sensing frame work for accelerated dynamic MRI," in *Proc. IEEE International Symposium on Biological Imaging* IEEE, 2012, pp. 1060–1063.

[13] S. Ravishankar and Y. Bresler, "L0 sparsifying transform learning with efficient optimal updates and convergence Guarantees," *IEEE Transactions on Signal Processing*, vol. 63, no. 9, pp. 2389–2404, 2015.

[14] S. G. Lingala and M. Jacob, "Blind compressive sensing dynamic MRI," *IEEE Transactions on Medical Imaging*, vol. 32, no. 6, pp. 1132–1145, 2013.

[15] O. Ronneberger, P. Fischer, and T. Brox, "U-net: Convolutional networks for biomedical image segmentation," in *Proc. International Conference on Medical Image Computing and Computer-Assisted Intervention*. Springer, 2015, pp. 234–241.

[16] K. He, X. Zhang, S. Ren, and J. Sun, "Deep residual learning for image recognition," in *Proc. IEEE Conference on Computer Vision and Pattern Recognition*, 2016, pp. 770–778.

[17] H. Chen, Y. Zhang, M. K. Kalra, F. Lin, Y. Chen, P. Liao, J. Zhou, and G. Wang, "Low-dose CT with a residual encoder–decoder convolutional neural network," *IEEE Transactions in Medical Imaging*, vol. 36, no. 12, pp. 2524–2535, 2017.

[18] Y. Han and J. C. Ye, "Framing U-Net via deep convolutional framelets: application to sparse-view CT," *IEEE Transactions on Medical Imaging*, vol. 37, no. 6, pp. 1418–1429, 2018.

[19] K. Zhang, W. Zuo, S. Member, Y. Chen, D. Meng, and L. Zhang, "Beyond a Gaussian denoiser: Residual learning of deep CNN for image denoising," *IEEE Transactions on Image Processing*, vol. 26, no. 7, pp. 3142–3155, 2017.

[20] L. Zhang and W. Zuo, "Image restoration: From sparse and low-rank priors to deep priors," *IEEE Signal Processing Magazine*, vol. 34, no. 5, pp. 172–179, 2017.

[21] G. Wang, J. C. Ye, K. Mueller, and J. A. Fessler, "Image reconstruction is a new frontier of machine learning," *IEEE Transactions on Medical Imaging*, vol. 37, no. 6, pp. 1289–1296, 2018.

[22] Y. Han and J. C. Ye, "Framing U-Net via deep convolutional framelets: Application to sparse-view CT," *IEEE Transactions on Medical Imaging*, vol. 37, no. 6, pp. 1418–1429, 2018.

[23] K. Zhang, W. Zuo, S. Gu, and L. Zhang, "Learning deep CNN denoiser prior for image restoration," in *Proc. IEEE Conference on Computer Vision and Pattern Recognition*, 2017, pp. 2808–2817.

[24] J. H. R. Chang, C.-L. Li, B. Poczos, B. V. K. V. Kumar, and A. C. Sankaranarayanan, "One network to solve them all – Solving linear inverse problems using deep projection models," in *IProc. EEE International Conference on Computer Vision*, 2017, pp. 1–12.

[25] S. Sreehari, S. V. Venkatakrishnan, B. Wohlberg, G. T. Buzzard, L. F. Drummy, J. P. Simmons, and C. A. Bouman, "Plug-and-play priors for bright field electron tomography and sparse interpolation," *IEEE Transactions on Computational Imaging*, vol. 2, no. 4, pp. 408–423, 2016.

[26] S. V. Venkatakrishnan, C. A. Bouman, and B. Wohlberg, "Plug-and-play priors for model based reconstruction," in *Proc. IEEE Global Conference on Signal and Information Processing*. IEEE, 2013, pp. 945–948.

[27] A. Bora, A. Jalal, E. Price, and A. G. Dimakis, "Compressed sensing using generative models," in *Proc. International Conference on Machine Learning*. PMLR, 2017, pp. 537–546.

[28] K. H. Jin, M. T. McCann, E. Froustey, and M. Unser, "Deep convolutional neural network for inverse problems in imaging," *IEEE Transactions on Image Processing*, vol. 29, pp. 4509–4522, 2017.

[29] D. Ulyanov, A. Vedaldi, and V. Lempitsky, "Deep image prior," in *Proc. IEEE Conference on Computer Vision and Pattern Recognition*, 2018, pp. 9446–9454.

[30] K. Gregor and Y. LeCun, "Learning fast approximations of sparse coding," in *Proc. International Conference on Machine Learning*. Omnipress, 2010, pp. 399–406.

[31] K. Hammernik, T. Klatzer, E. Kobler, M. P. Recht, D. K. Sodickson, T. Pock, and F. Knoll, "Learning a variational network for reconstruction of accelerated MRI data," *Magnetic Resonance in Medicine*, vol. 79, no. 6, pp. 3055–3071, 2017.

[32] H. K. Aggarwal, M. P. Mani, and M. Jacob, "MoDL: Model based deep learning architecture for inverse problems," *IEEE Transactions on Medical Imaging*, vol. 38, no. 2, pp. 394–405, 2019.

[33] P. Putzky and M. Willing, "Recurrent inference machines for solving inverse problems," in *arXiv*, 2017, pp. 1–12.

[34] J. Schlemper, J. Caballero, J. V. Hajnal, A. Price, and D. Rueckert, "A deep cascade of convolutional neural networks for MR image reconstruction," in *Proc. Conference on Information Processing in Medical Imaging*, 2017, pp. 647–658.

[35] A. Pramanik, H. Aggarwal, and M. Jacob, "Deep generalization of structured low-rank algorithms (Deep-SLR)," *IEEE Transactions on Medical Imaging*, vol. 39, no. 12, pp. 4186–4197, 2020.

[36] M. Mardani, H. Monajemi, V. Papyan, S. Vasanawala, D. Donoho, and J. Pauly, "Recurrent generative adversarial networks for proximal learning and automated compressive image recovery," in *Proc. IEEE Conference on Computer Vision and Pattern Recognition*, 2018.

[37] M. A. T. Figueiredo, R. D. Nowak, S. Member, and R. D. Nowak, "An EM algorithm for wavelet-based image restoration," *IEEE Transactions on Image Processing*, vol. 12, no. 8, pp. 906–916, 2003.

[38] Y. Hu and M. Jacob, "Higher degree total variation (HDTV) regularization for image recovery," *IEEE Transactions on Image Processing*, vol. 21, no. 5, pp. 2259–2271, 2012.

[39] S. Ravishankar and Y. Bresler, "Learning sparsifying transforms," *IEEE Transactions on Signal Processing*, vol. 61, no. 5, pp. 1072–1086, 2012.

[40] G. Ongie, S. Biswas, and M. Jacob, "Convex recovery of continuous domain piecewise constant images from non-uniform Fourier samples," *IEEE Transactions on Signal Processing*, vol. 66, no. 1, pp. 236–250, 2017.

[41] G. Ongie and M. Jacob, "Off-the-grid recovery of piecewise constant images from few Fourier samples," *SIAM Journal on Imaging Sciences*, vol. 9, no. 3, pp. 1004–1041, 2016.

[42] D. Lee, K. H. Jin, E. Y. Kim, S.-H. Park, and J. C. Ye, "Acceleration of MR parameter mapping using annihilating filter-based low rank Hankel matrix (ALOHA)," *Magnetic Resonance in Medicine*, vol. 76, no. 6, pp. 1848–1864, 2016.

[43] J. P. Haldar, "Low-rank modeling of local k-space neighborhoods (LORAKS) for constrained MRI," *IEEE Transactions on Medical Imaging*, vol. 33, no. 3, pp. 668–681, 2014.

[44] A. Mousavi and R. G. Baraniuk, "Learning to invert: Signal recovery via deep convolutional networks," in *Proc. IEEE International Conference on Acoustics, Speech, and Signal Processing*, 2017, pp. 2272–2276.

[45] G. T. Buzzard, S. H. Chan, S. Sreehari, and C. A. Bouman, "Plug-and-play unplugged: Optimization-free reconstruction using consensus equilibrium," *SIAM Journal on Imaging Sciences*, vol. 11, no. 3, pp. 2001–2020, 2018.

[46] K. Zhang, W. Zuo, S. Member, Y. Chen, D. Meng, and L. Zhang, "Beyond a Gaussian denoiser: Residual learning of deep CNN for image denoising," *IEEE Transactions on Image Processing*, vol. 26, no. 7, pp. 3142–3155, 2017.

[47] M. Mani, V. A. Magnotta, and M. Jacob, "qmodel: A plug-and-play model-based reconstruction for highly accelerated multi-shot diffusion MRI using learned priors," *Magnetic Resonance in Medicine*, vol. 86, no. 2, pp. 835–851, 2021.

[48] F. Lam, Y. Li, and X. Peng, "Constrained magnetic resonance spectroscopic imaging by learning nonlinear low-dimensional models," *IEEE Transactions on Medical Imaging*, vol. 39, no. 3, pp. 545–555, 2019.

[49] I. Goodfellow, J. Pouget-Abadie, M. Mirza, B. Xu, D. Warde-Farley, S. Ozair, A. Courville, and Y. Bengio, "Generative adversarial nets," in *Advances in Neural Information Processing Systems*, 2014, pp. 2672–2680.

[50] A. Creswell, T. White, V. Dumoulin, K. Arulkumaran, B. Sengupta, and A. A. Bharath, "Generative adversarial networks: An overview," *IEEE Signal Processing Magazine*, vol. 35, no. 1, pp. 53–65, 2018.

[51] A. Raj, Y. Li, and Y. Bresler, "Gan-based projector for faster recovery with convergence guarantees in linear inverse problems," in *Proc. IEEE/CVF International Conference on Computer Vision*, 2019, pp. 5602–5611.

[52] G. Ongie, A. Jalal, C. A. Metzler, R. G. Baraniuk, A. G. Dimakis, and R. Willett, "Deep learning techniques for inverse problems in imaging," *IEEE Journal on Selected Areas in Information Theory*, vol. 1, no. 1, pp. 39–56, 2020.

[53] G. Daras, J. Dean, A. Jalal, and A. G. Dimakis, "Intermediate layer optimization for inverse problems using deep generative models," *arXiv:2102.07364*, 2021.

[54] H. K. Aggarwal, M. P. Mani, and M. Jacob, "MoDL-MUSSELS: Model-based deep learning for multishot sensitivity-encoded diffusion MRI," *IEEE Transactions on Medical Imaging*, 2019.

[55] R. Cohen, Y. Zhang, O. Solomon, D. Toberman, L. Taieb, R. J. van Sloun, and Y. C. Eldar, "Deep convolutional robust pca with application to ultrasound imaging," in *Proc. IEEE International Conference on Acoustics, Speech and Signal Processing*. IEEE, 2019, pp. 3212–3216.

[56] M. Arvinte, S. Vishwanath, A. H. Tewfik, and J. I. Tamir, "Deep j-sense: Accelerated MRI reconstruction via unrolled alternating optimization," *arXiv:2103.02087*, 2021.

[57] H. K. Aggarwal, M. P. Mani, and M. Jacob, "MoDL: Model-based deep learning architecture for inverse problems," *IEEE Transactions on Medical Imaging*, vol. 38, no. 2, pp. 394–405, 2019.

[58] D. Gilton, G. Ongie, and R. Willett, "Model adaptation for inverse problems in imaging," *arXiv:2012.00139*, 2020.

[59] H. K. Aggarwal and M. Jacob, "Model adaptation for image reconstruction using generalized stein's unbiased risk estimator," *arXiv:2102.00047*, 2021.

[60] Z. Cheng, M. Gadelha, S. Maji, and D. Sheldon, "A bayesian perspective on the deep image prior," in *Proc. IEEE/CVF Conference on Computer Vision and Pattern Recognition*, 2019, pp. 5443–5451.

[61] K. H. Jin, H. Gupta, J. Yerly, M. Stuber, and M. Unser, "Time-dependent deep image prior for dynamic MRI," *arXiv:1910.01684*, 2019.

[62] Q. Zou, A. H. Ahmed, P. Nagal, S. Kruger, and M. Jacob, "Deep generative SToRM model for dynamic imaging," in *Proc. IEEE International Symposium on Biomedical Imaging*. IEEE, 2021, *arxiv.org/abs/2101.12366*.

[63] ——, "Alignment & joint recovery of multi-slice dynamic MRI using deep generative manifold model," *arXiv.org/abs/2101.08196*, 2021.

[64] A. H. Ahmed, R. Zhou, Y. Yang, P. Nagpal, M. Salerno, and M. Jacob, "Free-breathing and ungated dynamic MRI using navigator-less spiral storm," *IEEE Transactions on Medical Imaging*, vol. 39, no. 12, pp. 3933–3943, 2020.

[65] S. Poddar, Y. Q. Mohsin, D. Ansah, B. Thattaliyath, R. Ashwath, and M. Jacob, "Manifold recovery using kernel low-rank regularization: Application to dynamic imaging," *IEEE Transactions on Computational Imaging*, vol. 5, no. 3, pp. 478–491, 2019.

[66] Q. Zou, A. H. Ahmed, P. Nagpal, S. Priya, R. Schulte, and M. Jacob, "Generative storm: A novel approach for joint alignment and recovery of multi-slice dynamic MRI," *arXiv:2101.08196*, 2021.

[67] J. Lehtinen, J. Munkberg, J. Hasselgren, S. Laine, T. Karras, M. Aittala, and T. Aila, "Noise2Noise: Learning image restoration without clean data," in *Proc. International Conference on Machine Learning*, 2018, pp. 2965–2974.

[68] A. Krull, T.-O. Buchholz, and F. Jug, "Noise2Void-learning denoising from single noisy images," in *Proc. IEEE Conference on Computer Vision and Pattern Recognition*, 2019, pp. 2129–2137.

[69] C. M. Stein, "Estimation of the mean of a multivariate normal distribution," *Annals of Statistics*, vol. 9, no. 6, pp. 1135–1151, 1981.

[70] S. Ramani, T. Blu, and M. Unser, "Monte-Carlo SURE: A black-box optimization of regularization parameters for general denoising algorithms," *IEEE Transactions on Medical Imaging*, vol. 17, no. 9, pp. 1540–1554, 2008.

[71] C. A. Metzler, A. Mousavi, R. Heckel, and R. G. Baraniuk, "Unsupervised learning with Stein's unbiased risk estimator," *arXiv:1805.10531*, 2018.

[72] M. Zhussip, S. Soltanayev, and S. Y. Chun, "Training deep learning based image denoisers from undersampled measurements without ground truth and without image prior," in *Proc. IEEE/CVF Conference on Computer Vision and Pattern Recognition*, 2019, pp. 10 255–10 264.

[73] Y. C. Eldar, "Generalized SURE for exponential families: Applications to regularization," *IEEE Transactions on Signal Processing*, vol. 57, no. 2, pp. 471–481, 2008.

[74] B. Yaman, S. A. H. Hosseini, S. Moeller, J. Ellermann, K. Uğurbil, and M. Akçakaya, "Self-supervised learning of physics-guided reconstruction neural networks without fully sampled reference data," *Magnetic Resonance in Medicine*, vol. 84, no. 6, pp. 3172–3191, 2020.

[75] H. K. Aggarwal, A. Pramanik, and M. Jacob, "Ensure: A general approach for unsupervised training of deep image reconstruction algorithms," *arXiv:2010.10631*, 2020.

[76] ——, "ENSURE: Ensemble Stein's unbiased risk estimator for unsupervised learning," in *Proc. IEEE International Conference on Acoustics, Speech, and Signal Processing*, 2021, *https://arxiv.org/abs/2010.10631*.

[77] D. L. Donoho and I. M. Johnstone, "Adapting to unknown smoothness via wavelet shrinkage," *Journal of the American Statistical Association*, vol. 90, no. 432, pp. 1200–1224, 1995.

[78] N. M. Gottschling, V. Antun, B. Adcock, and A. C. Hansen, "The troublesome kernel: Why deep learning for inverse problems is typically unstable," *arXiv:2001.01258*, 2020.

[79] V. Antun, F. Renna, C. Poon, B. Adcock, and A. C. Hansen, "On instabilities of deep learning in image reconstruction and the potential costs of AI," *Proceedings of the National Academy of Science*, vol. 117, no. 48, pp. 30 088–30 095, 2020.

[80] B. Yaman, S. A. H. Hosseini, S. Moeller, J. Ellermann, K. Uğurbil, and M. Akçakaya, "Self-supervised learning of physics-guided reconstruction neural networks without fully sampled reference data," *Magnetic Resonance in Medicine*, vol. 84, no. 6, pp. 3172–3191, 2020.

[81] C. D. Bahadir, A. V. Dalca, and M. R. Sabuncu, "Learning-based optimization of the under-sampling pattern in MRI," in *Proc. International Conference on Information Processing in Medical Imaging*. Springer, 2019, pp. 780–792.

[82] T. Weiss, O. Senouf, S. Vedula, O. Michailovich, M. Zibulevsky, and A. Bronstein, "Pilot: Physics-informed learned optimized trajectories for accelerated MRI," *arXiv:1909.05773*, 2019.

[83] H. K. Aggarwal and M. Jacob, "J-modl: Joint model-based deep learning for optimized sampling and reconstruction," *IEEE Journal of Selected Topics in Signal Processing*, vol. 14, no. 6, pp. 1151–1162, 2020.

[84] V. Sitzmann, S. Diamond, Y. Peng, X. Dun, S. Boyd, W. Heidrich, F. Heide, and G. Wetzstein, "End-to-end optimization of optics and image processing for achromatic extended depth of field and super-resolution imaging," *ACM Transactions on Graphics*, vol. 37, no. 4, pp. 1–13, 2018.

[85] J. Chang, V. Sitzmann, X. Dun, W. Heidrich, and G. Wetzstein, "Hybrid optical–electronic convolutional neural networks with optimized diffractive optics for image classification," *Scientific Reports*, vol. 8, no. 1, pp. 1–10, 2018.

9 *k*-Space Deep Learning for MR Reconstruction and Artifact Removal

Mehmet Akcakaya, Gyutaek Oh, and Jong Chul Ye

Inspired by the success of deep learning in computer vision tasks, deep-learning approaches for various magnetic resonance imaging (MRI) problems have been extensively studied in recent years. Early deep-learning studies for MRI reconstruction and enhancement were mostly based on image-domain learning. However, because an MR signal is acquired in the *k*-space domain, researchers have demonstrated that deep neural networks can be directly designed in *k*-space to utilize the physics of MR acquisition. In Section 9.1, the recent trend of *k*-space deep learning for MRI reconstruction and artifact removal will be reviewed. First, scan-specific *k*-space learning, which is inspired by parallel MRI, will be covered. Then, we will provide an overview of data-driven *k*-space learning. Subsequently, unsupervised learning for MRI reconstruction and motion-artifact removal will be discussed.

9.1 Introduction

For the last decade, compressed sensing (CS) [1] has been a main research thrust for several problems in MRI, such as accelerated MRI reconstruction, artifact removal, etc. Compressed-sensing-based algorithms exploit the sparsity of signals in specific transform domains (e.g., wavelet transforms) and reconstruct high-quality MR images from lower-quality MR data. Although CS methods show high performance, high computational complexity and the difficulty of hyperparameter tuning are the drawbacks of CS methods.

Recently, as deep-learning approaches have shown outstanding performance in computer vision tasks [2, 3], there have been many attempts to study deep-learning approaches for MRI reconstruction and enhancement problems. These attempts have shown significant performance gain as well as reduced runtime complexity.

In particular, accelerated MRI reconstruction using deep learning is a major research topic. Fig. 9.1 shows the types of existing deep-learning approaches for accelerated MRI reconstruction. Early deep-learning approaches for MRI reconstruction were based on image-domain enhancement, as depicted in Fig. 9.1(a) [4]. Image-domain learning methods have shown improved performance and reduced runtime compared with CS methods. However, because image-domain learning only serves as post-processing and does not treat raw measurement data (which exists in

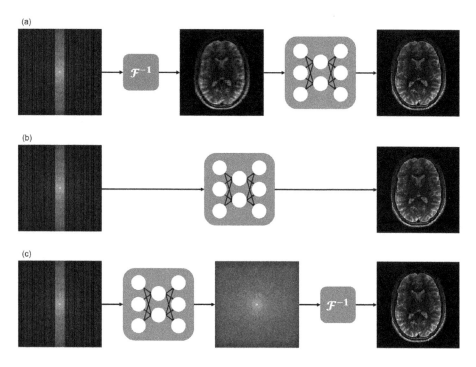

Figure 9.1 Data-driven approaches for accelerated MRI reconstruction: (a) image-domain learning, (b) direct mapping from k-space to image, and (c) k-space learning.

k-space), there can be a loss of information in reconstructed images, especially in high-frequency regions.

There was an attempt to estimate Fourier transforms through fully connected layers [5], as shown in Fig. 9.1(b). Although this work outperformed conventional methods and reduced the runtime complexity, it could be applied only to relatively small images because fully connected layers require a large amount of memory.

However, some researchers have studied the reconstruction of the original k-space from downsampled k-space directly [6, 7]. Such k-space deep-learning methods interpolate k-space samples that have not been acquired, and the reconstructed image can be obtained by inverse Fourier transformation of the interpolated k-space. Because k-space deep learning treats k-space data directly, it is possible to reconstruct high-frequency details more accurately than image-domain learning. Furthermore, k-space learning is based on convolutional neural networks (CNNs), which do not contain fully connected layers, and thus it does not require large amounts of memory. Inspired by the success of k-space deep learning for accelerated MRI, k-space learning has been employed for other MRI problems, such as echo planar imaging (EPI) ghost-artifact correction. Moreover, other deep-learning approaches which deal with the properties of k-space have been studied [8, 9].

We can divide k-space deep-learning strategies into two approaches: scan-specific learning and data-driven learning. Building on k-space-based parallel imaging methods [10], scan-specific methods perform k-space interpolation using neural networks that have been trained on the corresponding calibration scan data. Such data-driven learning requires training data for neural network training, but, once the network is trained, reconstruction can be done instantaneously at runtime.

This chapter is organized as follows. First, scan-specific k-space deep learning will be introduced in Section 9.2. Subsequently, in Section 9.3, the description of data-driven approaches for accelerated MRI reconstruction and motion-artifact removal will be presented. Last, a summary of the chapter will be provided in Section 9.4.

9.2 Scan-Specific k-Space Learning

Early work in k-space deep learning focused on scan-specific training strategies [6, 11, 12]. Methodologically, these build on conventional linear k-space parallel imaging reconstruction [10]. Logistically, the scan-specific approach was motivated by concerns about the generalizability of database-trained models, which remains an ongoing research challenge [13, 14], as well as by the lack of large publicly available datasets of raw multi-coil k-space data at the time. Note that this second challenge has since been largely resolved both by the availability of large public databases of k-space data [15] and by the introduction of learning methods that can train larger models on databases of undersampled data [8, 16, 17].

In this section, we will cover scan-specific k-space deep-learning approaches for MRI reconstruction.

9.2.1 Scan-Specific Neural Networks for k-Space Interpolation

Lengthy scan times remain a challenge in MRI. There are several methods for reconstructing accelerated MRI scans that have been undersampled beyond the Nyquist rate, using additional information. Parallel imaging is the most commonly used accelerated MRI method [10, 18, 19]. These methods utilize variations in the sensitivity profiles of receiver coils in reconstruction [18]. These differences are estimated from scan-specific calibration data [10, 19], after which reconstruction is performed either as an inverse problem in the image domain [19] or a linear interpolation problem in k-space [10]. In this subsection, we will explain how scan-specific neural networks can be used to improve on the latter k-space interpolation approaches.

Multi-Coil MRI Acquisition Model

Modern MRI scanners are equipped with multiple receiver coils [20]. Each of these coils is sensitive to a different part of the field of view. The image in the ith coil, denoted by $\mathbf{m}_i$, is given as the desired image $\mathbf{m}$ modulated pixel wise with the diagonal entries of a coil sensitivity profile, $\mathbf{C}_i$ [19]. Thus, for the ith coil, the received k-space signal is given as

$$\mathbf{s}_{i,\Lambda} = \mathbf{F}_\Lambda \mathbf{C}_i \mathbf{m} + \mathbf{n}_i, \tag{9.1}$$

where $\mathbf{F}_\Lambda$ is a partial Fourier sampling operator that samples the locations specified by the index set Λ, and $\mathbf{n}_i$ is the measurement noise in the ith coil.

Parallel imaging methods use linear algorithms to reconstruct such undersampled datasets using differences in the coil sensitivity profiles. The two most commonly used parallel imaging approaches are sensitivity encoding (SENSE) [19] and generalized autocalibrating partially parallel acquisition (GRAPPA) [10]. Both rely on a uniform undersampling of k-space. For reconstruction, the former approach is formulated as a least squares problem for estimating $\mathbf{m}$ from $\{\mathbf{s}_{i,\Lambda}\}_{i=1}^{n_c}$, where n_c is the number of coils. This requires estimation of the coil sensitivities, $\mathbf{C}_i$, and is performed using low-resolution calibration images acquired and generated on a subject-specific basis [19]. The GRAPPA reconstruction formulates this problem as an interpolation problem in k-space using linear shift-invariant convolutional kernels. These kernels are used to estimate missing k-space data from data acquired across all coils, effectively capturing the coil geometry. As with SENSE, these convolutional kernels are estimated from a small amount of calibration data, referred to as the autocalibration signal (ACS) [10]. While the formulation of these parallel imaging methods may theoretically allow for acceleration rates up to the number of receiver coils, in practice the acceleration rates are limited due to dependencies between coil coverage and coil geometry dependent noise amplification [19].

k-Space Interpolation in Parallel Imaging

We will first look at the k-space interpolation strategy in GRAPPA [10]. Let R be the acceleration rate with uniform undersampling, and let $s(k_x, k_y, i)$ denote the k-space point at position (k_x, k_y) in the ith coil. For notational convenience, let

$$\mathcal{N}(k_x, k_y) = \{(k_x - b_x \Delta k_x, k_y - R b_y \Delta k_y, i)$$
$$: b_x \in \{-B_x, \ldots, B_x\}, b_y \in \{-B_y, \ldots B_y\}, i \in \{1, \ldots, n_c\}\} \quad (9.2)$$

denote a neighborhood of sampled points around the k-space location (k_x, k_y) across all coils, where B_x and B_y are pre-specified integer-valued kernel sizes. We define $\tilde{\mathbf{s}}_{\mathcal{N}(k_x, k_y)}$ to be the column vector whose entries are elements $s(a, b, c)$ with $(a, b, c) \in \mathcal{N}(k_x, k_y)$. With this notation, GRAPPA estimates the missing points in a uniformly undersampled k-space acquisition by interpolating the acquired points in its vicinity as

$$s(k_x, k_y - m \Delta k_y, i) = \mathbf{g}_{m,i} \tilde{\mathbf{s}}_{\mathcal{N}(k_x, k_y)}, \quad (9.3)$$

where $\mathbf{g}_{m,i}$ is a row vector that contains the corresponding linear convolutional weights for estimating the mth skipped line in coil i. The GRAPPA convolution kernel is obtained by solving a linear least squares problem, where the calibration points in the ACS region are used as regressors and regressand in Eq. (9.3).

Robust Artificial Neural Networks for *k*-Space Interpolation

Like other linear parallel imaging approches, GRAPPA suffers from noise amplification at high acceleration rates [19]. Robust artificial neural networks for k-space interpolation (RAKI) is a scan-specific machine learning approach for improved k-space reconstruction that tackles this challenge [6]. This approach trains CNNs on scan-specific calibration data, and uses these for interpolating missing k-space points from acquired ones. Thus, RAKI extends the linear convolutions in GRAPPA to using nonlinear CNNs for k-space interpolation [6]. The nonlinear interpolation strategy in k-space addresses the issue of noise, in both the target and source points, derived from the calibration data [21] and provides a better lower-dimensional approximation [6]; it was shown to improve upon the noise amplification associated with GRAPPA [6].

For processing in RAKI using CNNs, complex k-space data is mapped to the real field, leading to a total of $2n_c$ input channels from n_c coils. This slightly affects the way in which the neighborhood is defined in Eq. (9.2), as i now ranges from 1 to $2n_c$. To this end, let $\tilde{\mathcal{N}}$ denote this new definition adapted to the real field. We also define $\tilde{\mathbf{s}}_{\tilde{\mathcal{N}}(k_x,k_y)}$ analogously as a two-dimensional (2D) patch. Additionally, let

$$\mathcal{U}(k_x, k_y, j) = \{(k_x, k_y - m\Delta k_y, j) : m \in \{1, \ldots R-1\}\} \tag{9.4}$$

be the $R-1$ missing phase-encoding points adjacent to (k_x, k_y) in channel j, and let $\tilde{\mathbf{s}}_{\mathcal{U}(k_x,k_y,j)}$ be the patch whose entries are elements $s(a, b, c)$ with $(a, b, c) \in \mathcal{U}(k_x, k_y, j)$. Then RAKI estimates the missing k-space points using acquired data via CNNs as

$$\tilde{\mathbf{s}}_{\mathcal{U}(k_x,k_y,j)} = f_j\big(\tilde{\mathbf{s}}_{\tilde{\mathcal{N}}(k_x,k_y)}\big) \tag{9.5}$$

where $f_j(\cdot)$ denotes a CNN that uses acquired data to estimate the missing points in channel j. A representative CNN architecture, used in [6], is depicted in Fig. 9.2, though deeper architectures are also possible depending on the availability of calibration data [22].

The CNN for RAKI was trained using the mean square error (MSE) loss over the calibration data. Let $\mathbf{y}_j$ denote the target points in ACS region, and $\mathbf{y}_{\text{source}}$ denote the source points in the ACS region [6, 10]. The training loss for channel j network is given by

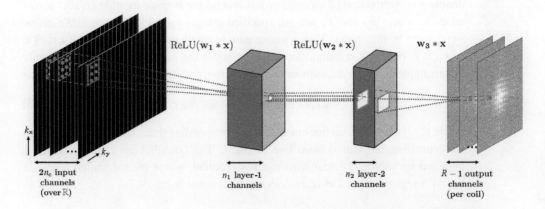

Figure 9.2 A representative CNN architecture that implements the k-space interpolation in RAKI. The input to the CNN is the undersampled k-space over all the n_c coils, which is embedded over $\mathbb{R}$, resulting in $2n_c$ input channels for the network. The weights of the neural network layers are $\mathbf{w}_1$, $\mathbf{w}_2$, and $\mathbf{w}_3$; $\mathbf{x}$ is the input tensor to each layer. Nonlinear activations are used in all but the last layer.

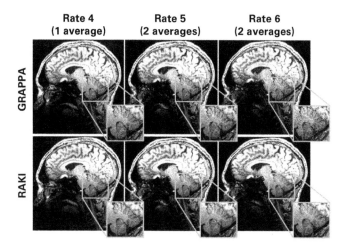

Figure 9.3 A representative slice from a high-resolution (0.6 mm isotropic) 7T MPRAGE acquisition, where all acquisitions were performed with prospective acceleration. The CNN-based RAKI method visibly reduces noise amplification compared with the linear GRAPPA reconstruction.

$$\min_{\boldsymbol{\theta}_j} \ \|\mathbf{y}_j - f_j(\mathbf{y}_{\text{source}}; \boldsymbol{\theta}_j) - G_j(\mathbf{y}_{\text{source}}; \boldsymbol{\theta}_j)\|_2^2. \tag{9.6}$$

Note that even though fully sampled datasets are not needed for learning the RAKI network, this approach still amounts to supervised training over the fully sampled calibration range, which is similar to its linear parallel imaging counterpart.

Figure 9.3 shows representative RAKI and GRAPPA reconstructions for high-resolution brain imaging datasets acquired with prospective undersampling. Acceleration rates of 5 and 6 were acquired with two averages to improve the SNR for clearer visualization of any residual artifacts. This ensures that the reconstructed structures are not buried under noise at these higher acceleration rates, and thus the recovery of fine structures can be clearly assessed. All other imaging parameters are reported in [6]. For these data, RAKI visibly reduces the noise amplification compared with GRAPPA. Note that quantitative metrics are unavailable in this case, since fully sampled reference data cannot be acquired in a reasonable scan time at this resolution.

As with the relationship between RAKI and GRAPPA, self-consistent RAKI (sRAKI) was introduced to address the noise amplification associated with linear parallel imaging reconstruction using SPIRiT [23, 24]. Thus, the self-consistency term was modified to incorporate a nonlinear CNN $G(\cdot)$ instead of the linear CNN G_s. This method was shown to reduce noise amplification and blurring artifacts in high-resolution coronary MRI compared with nonregularized and regularized SPIRiT reconstructions, respectively [24].

Later, the sRAKI method was formulated as an unrolled recurrent neural network to incorporate additional regularization [25]. It also used a densely connected CNN for the self-consistency operation. This method was trained using a variant [26] of the self-supervised learning framework introduced in [16, 17]. The results showed improved reconstruction compared with both nonregularized and regularized SPIRiT reconstructions, with further noise reduction compared with nonregularized sRAKI.

Simultaneous Multi-Slice Imaging

Simultaneous multi-slice (SMS) or multiband (MB) imaging is an image acceleration technique where multiple slices are excited and acquired at the same time [27]. An advantage of SMS/MB imaging is that there is no inherent signal-to-noise ratio (SNR) loss because the acceleration is gained by exciting multiple slices simultaneously [28, 29]. The information from these multiple slices is then resolved in a manner similar to parallel imaging, using the redundancies among the multiple sensors in the receiver coil arrays used in MRI. Several reconstruction strategies have been proposed [27], but k-space methods [29–31] remain the more commonly used approaches, having found a use in large-scale National Institutes of Health (NIH) projects such as the Human Connectome Project [32] for fast acquisition in functional MRI (fMRI) and diffusion MRI.

Several strategies have been proposed for using scan-specific learning for SMS imaging [33–35]. These methods utilize the RAKI framework with several modifications to the calibration and network architectures. In [33], a readout concatenation approach was used [29, 36]. This effectively encodes the slice acceleration as undersampling in an extended readout concatenated domain, allowing the use of k-space interpolation. Readout concetanation also has inherent interslice leakage blocking properties [36, 37]. An alternative line of work [35] considered the use of a split-slice GRAPPA type of k-space projection [31]. The split-slice GRAPPA method explicitly enforces a leakage-blocking constraint during the calibration stage. Further improvements to the network architecture, such as the addition of batch normalization and dropout layers, were also investigated [35].

Residual Connection for Improved Interpretability

Though the improvements from RAKI and its variants over their conventional linear parallel-imaging counterparts are visually clear, the source of the improvement is hard to characterize. To this end, a more interpretable methodology was proposed in [34]. This method, called residual RAKI, combines the advantages of the linear GRAPPA and nonlinear RAKI reconstructions while maintaining the scan-specificity of these components. A residual CNN that implements a skip connection incorporating a linear convolution in parallel with a multi-layer CNN is used for reconstruction. In this setting, the linear skip connection implements linear k-space interpolation, as in GRAPPA, while the multi-layer CNN estimates and compensates for the imperfections, such as noise amplification, that arise from the linear component. All components in this residual CNN are trained jointly on scan-specific calibration data.

While residual RAKI is generally applicable to all acquisitions where RAKI can be used, Fig. 9.4 depicts a representative example of residual RAKI applied to an SMS fMRI dataset. This dataset simulates an SMS = 16 acquisition by combining two slice groups prospectively acquired with SMS = 8 [34]. Since no fully sampled reference exists for these acquisitions, the high-quality SMS = 8 reconstruction

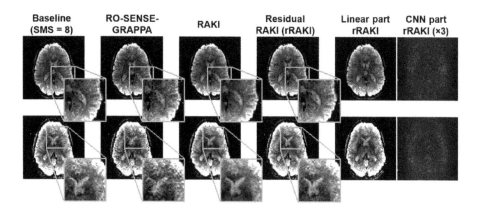

Figure 9.4 Two representative slices from an SMS = 16 3 tesla fMRI acquisition simulated from two prospective SMS = 8 acquisitions. The original SMS = 8 reconstruction is shown as a baseline. Residual RAKI improves upon both linear parallel-imaging (RO-SENSE-GRAPPA) and the conventional RAKI in this application, in terms of noise amplification and blurring artifacts, respectively. Furthermore, its explicit separation of linear and nonlinear components facilitates an interpretable reconstruction.

results obtained with standard methods are provided for baseline comparison [32]. At SMS = 16, the linear RO-SENSE-GRAPPA reconstruction suffers from visible artifacts and reconstruction noise, which are reduced in RAKI and residual RAKI. Furthermore, residual RAKI leads to sharper images than RAKI, while reducing artifacts and noise amplification in an interpretable manner.

k-Space Completion Methods

Though k-space interpolation methods remain the prevalent approach for k-space-based reconstruction in parallel imaging, there have been several works that recast this type of reconstruction into a low-rank matrix completion problem. Simultaneous autocalibrating and k-space estimation (SAKE) was an early work in this line, where local k-space neighborhoods across coils were reshaped into a block Hankel form; the low-rank properties were used for subsequent reconstruction subject to data consistency [38]. The low-rank matrix modeling of local k-space neighborhoods (LORAKS) is another approach that builds on these ideas, utilizing well-defined finite image support and image phase constraints instead of correlations across multiple coils [39]. A further generalization of LORAKS is the annihilating filter-based low-rank Hankel matrix approach (ALOHA), which extends the finite-support constraint to transform domains [40], and is detailed in the following section.

Among these methods, the LORAKS method was also combined with the RAKI framework in a strategy called LORAKI [41]. This method unrolls the LORAKS iterations and replaces the subject-specific linear convolutions with a CNN in accordance with RAKI [6]. The results show improvement over both conventional LORAKS and k-space interpolation strategies.

9.3 k-Space Deep Learning using Training Data

In data-driven approaches for various MRI problems (e.g., accelerated MRI reconstruction), a neural network is first trained with a large amount of data during the training phase, after which the trained neural network is used for unseen (test) data. Because the models used in data-driven methods do not have to be optimized for each set of data, they show significantly reduced inference time compared with scan-specific methods.

9.3.1 Data-Driven k-Space Deep Learning for k-Space Interpolation

As discussed in Section 9.2.1, the long acquisition time due to the MR acquisition physics is a major challenge in MRI. To reduce the scan time of MRI, subsampling in k-space is usually employed during acquisition. However, aliasing artifacts in the spatial domain are unavoidable if there is k-space undersampling.

As explained in Section 9.1, it is possible to reconstruct accelerated MRI via k-space deep learning. In this section, we will introduce the interpolation of the missing k-space samples using deep learning. First, ALOHA [40], which is a CS algorithm that provides a theoretical background for data-driven k-space learning, will be introduced. Next, the theoretical background of data-driven k-space deep learning will be described. Last, we will explain the details and results of data-driven k-space learning.

Forward Model for Accelerated MRI

The spatial Fourier transform of a function $m : \mathbb{R}^2 \to \mathbb{R}$ is defined by

$$\widehat{s}(k) = \mathbf{F}[\mathbf{m}](k) := \int_{\mathbb{R}^2} e^{-\iota k \cdot r} \mathbf{m}(r) dr,$$

with spatial frequency $k \in \mathbb{R}$ and $\iota = \sqrt{-1}$. Let $\{k_n\}_{n=1}^N$ for some integer $N \in \mathbb{N}$ be a collection of a finite number of sampling points of the k-space conforming to the Nyquist sampling rate. Then, $\widehat{\mathbf{m}} \in \mathbb{C}^N$ can be defined as

$$\widehat{\mathbf{s}} = \begin{bmatrix} \widehat{s}(k_0) & \cdots & \widehat{s}(k_{N-1}) \end{bmatrix}. \tag{9.7}$$

For a given undersampling pattern Λ for accelerated MR acquisition, let the downsampling operator $\mathcal{P}_\Lambda : \mathbb{C}^N \to \mathbb{C}^N$ be defined as

$$[\mathcal{P}_\Lambda[\widehat{\mathbf{s}}]]_i = \begin{cases} \widehat{\mathbf{s}}[i], & i \in \Lambda \\ 0, & \text{otherwise} \end{cases}, \tag{9.8}$$

so that the undersampled k-space data can be expressed by

$$\mathbf{s} := \mathcal{P}_\Lambda[\widehat{\mathbf{s}}] = \mathbf{F}_\Lambda[\mathbf{s}], \tag{9.9}$$

where $\mathbf{F}_\Lambda = \mathcal{P}_\Lambda \mathbf{F}$. Then, the goal of accelerated MRI is to recover the unknown signal $\mathbf{m}(r)$ from the subsampled Fourier measurement $\mathbf{s}$.

Annihilating Filter-Based Low-Rank Hankel Matrix Approach

One of the most important discoveries in the theory of ALOHA [40, 42, 43] is that if the underlying signal $\mathbf{m}(r)$ in the image domain is sparse and described as a signal with a finite rate of innovations (FRI) s [44], the associated Hankel matrix in the k-space is low-rank [40, 42, 43]. Therefore, if some k-space data are missing, it is possible to recover the missing elements using low-rank Hankel matrix completion approaches, so a suitable weighted Hankel matrix can be reconstructed with the missing elements.

For simplicity, here we consider one-dimensional (1D) signals, but the extension to two dimensions is straightforward [45]. In addition, to avoid a separate treatment of boundary conditions, we assume that they are periodic. Now, let $\mathbb{H}_d(\widehat{\mathbf{s}})$ denote the Hankel matrix constructed from the k-space measurement $\widehat{\mathbf{s}}$ in Eq. (9.7):

$$\mathbb{H}_d(\widehat{\mathbf{s}}) = \begin{bmatrix} \widehat{s}[0] & \widehat{s}[1] & \cdots & \widehat{s}[d-1] \\ \widehat{s}[1] & \widehat{s}[2] & \cdots & \widehat{s}[d] \\ \vdots & \vdots & \ddots & \vdots \\ \widehat{s}[N-1] & \widehat{s}[N] & \cdots & \widehat{s}[d-2] \end{bmatrix} \quad (9.10)$$

Under the assumption of sparsity in the image domain, the missing k-space interpolation problem can be addressed by solving the following optimization problem:

$$(P) \quad \min_{\widehat{z} \in \mathbb{C}^N} \quad \text{RANK } \mathbb{H}_d(\widehat{z}) \quad (9.11)$$

$$\text{s.t.} \quad \mathcal{P}_\Lambda[\widehat{\mathbf{s}}] = \mathcal{P}_\Lambda[\widehat{z}],$$

where

$$\widehat{z} = \begin{bmatrix} \widehat{z}(k_0) & \cdots & \widehat{z}(k_{N-1}) \end{bmatrix}^T. \quad (9.12)$$

From ALOHA to k-Space Deep Learning

Now we will demonstrate the relationship between ALOHA and the encoder–decoder architecture. Consider the following image-regression problem under the low-rank Hankel matrix constraint:

$$\min_{\widehat{z} \in \mathbb{C}^N} \quad \|\mathbf{m} - \mathbf{F}^{-1}[\widehat{z}]\|^2 \quad (9.13)$$

$$\text{s.t.} \quad \text{RANK } \mathbb{H}_d(\widehat{z}) = r, \quad \mathcal{P}_\Lambda[\widehat{\mathbf{s}}] = \mathcal{P}_\Lambda[\widehat{z}], \quad (9.14)$$

where r denotes the estimated rank. In the above formulation, the cost in Eq. (9.13) is defined in the image domain in order to minimize the errors in the image domain, while the low-rank Hankel matrix constraint in Eq. (9.14) is implied in the k-space after k-space weighting.

Note that the Hankel matrix $\mathbb{H}_d(\widehat{z})$ is complex-valued since it is composed of k-space samples. For simplicity, we will take it as real-valued for the rest of the chapter. For the detailed method of converting a complex-valued Hankel matrix to a real one, see the original paper [7].

The above low-rank constraint optimization problem was addressed using a learning-based signal representation in the theory of deep convolutional framelets [45]. More specifically, let the Hankel-structured matrix $\mathbb{H}_d(\widehat{z})$ for any $\widehat{z} \in \mathbb{C}^N$ have the singular-value decomposition $U\Sigma V^\top$, where $U = [u_1 \cdots u_Q] \in \mathbb{R}^{N \times Q}$ and $V = [v_1 \cdots v_Q] \in \mathbb{R}^{d \times Q}$ denote the left and the right singular vector basis matrices, respectively, and $\Sigma = (\sigma_{ij}) \in \mathbb{R}^{Q \times Q}$ is the diagonal matrix with singular values. Now, consider a matrix pair $\Psi, \tilde{\Psi} \in \mathbb{R}^{d \times Q}$ that satisfies the following low-rank projection constraint:

$$\Psi \tilde{\Psi}^\top = P_{R(V)}, \tag{9.15}$$

where $P_{R(V)}$ denotes the projection matrix onto the range space of V. Also, generalized pooling and unpooling matrices $\Phi, \tilde{\Phi} \in \mathbb{R}^{N \times M}$ that satisfy the following condition should be introduced:

$$\tilde{\Phi} \Phi^\top = I_N. \tag{9.16}$$

Using Eqs. (9.15) and (9.16), it is possible to obtain the following matrix equality:

$$\mathbb{H}_d(\hat{z}) = \tilde{\Phi} \Phi^\top \mathbb{H}_d(\hat{z}) \Psi \tilde{\Psi}^\top = \tilde{\Phi} C \tilde{\Psi}^\top \tag{9.17}$$

where

$$C := \Phi^\top \mathbb{H}_d(\hat{z}) \Psi \in \mathbb{R}^{N \times Q}. \tag{9.18}$$

If we take the generalized inverse of the Hankel matrix, Eq. (9.17) is transformed into the framelet basis representation [45]. Specifically, the resulting framelet basis representation can be equivalently represented by a single-layer encoder–decoder convolution architecture:

$$\hat{z} = (\tilde{\Phi} C) \circledast g(\tilde{\Psi}), \quad C = \Phi^\top (\hat{z} \circledast h(\Psi)), \tag{9.19}$$

where $\circledast$ denotes the multi-channel-input–multi-channel-output convolution. The second and first parts of Eq. (9.19) correspond to the encoder and decoder layers with convolution filters $h(\Psi) \in \mathbb{R}^{d \times Q}$ and $g(\tilde{\Psi}) \in \mathbb{R}^{dQ}$, which are obtained by reordering the matrices Ψ and $\tilde{\Psi}$ [7]. In [46], it was shown that recursive application of the encoder–decoder operations across the layers increases the net length of the convolutional filters.

Since Eq. (9.19) gives the general form of the signals that are associated with a Hankel-structured matrix, we are interested in using it to estimate bases for k-space interpolation. Specifically, we consider a complex-valued space $\mathcal{H}$ determined by the filters Ψ and $\tilde{\Psi}$:

$$\mathcal{H}(\Psi, \tilde{\Psi}) = \Big\{ z \in \mathbb{C}^N \mid z = (\tilde{\Phi} C) \circledast g(\tilde{\Psi}),$$

$$C = \Phi^\top (z \circledast h(\Psi)) \Big\}. \tag{9.20}$$

Then the ALOHA formulation P_A can be equivalently represented by

$$(P_A') \quad \min_{\hat{z} \in \mathcal{H}(\Psi, \tilde{\Psi})} \min_{\Psi, \tilde{\Psi}} \|\mathbf{m} - \mathbf{F}^{-1}[\hat{z}]\|^2$$

$$\text{s.t.} \quad \mathcal{P}_\Lambda[\hat{s}] = \mathcal{P}_\Lambda[\hat{z}].$$

In other words, ALOHA aims to find the optimal filter $\Psi, \tilde{\Psi}$ and the associated k-space signal $\hat{z} \in \mathcal{H}(\Psi, \tilde{\Psi})$ that satisfy the data-consistency conditions. In contrast with the standard CS approaches, in which the signal presentation in the image domain is applied separately from the data-consistency constraint in k-space, the success of ALOHA compared with CS can contribute to a more efficient signal representation in the k-space domain that simultaneously ensure data consistency in the k-space domain.

Now, the observation can be easily extended to parallel imaging. If $\{\hat{s}_i\}_{i=1}^{P}$ denotes the k-space measurements from P receiver coils, the following extended Hankel-structured matrix is low-ranked [40]:

$$\mathbb{H}_{d|P}(\hat{\mathbf{S}}) = [\mathbb{H}_d(\hat{s}_1) \quad \cdots \quad \mathbb{H}_d(\hat{s}_P)], \tag{9.21}$$

where

$$\hat{\mathbf{S}} = [\hat{s}_1 \quad \cdots \quad \hat{s}_P] \in \mathbb{C}^{N \times P}.$$

Then, similarly to the single-channel cases, the data-driven decomposition of the extended Hankel matrix in Eq. (9.21) can be represented by concatenating each set of k-space data along the channel direction and applying a deep neural network for the given multi-channel data. Therefore, except for the numbers of input and output channels, the network structure for parallel imaging data is the same as for single-channel k-space learning.

Generalization and Depth

The problem formulation in (P'_A) should be separated into two phases. The first phase is the learning phase, to estimate $\mathbf{\Psi}, \widetilde{\mathbf{\Psi}}$ from the training data. The second is the inference phase, to estimate the interpolated signal $\hat{z}$ for the given filter set $\mathbf{\Psi}, \widetilde{\mathbf{\Psi}}$. Although the relationship between ALOHA and the encoder–decoder architecture has been demonstrated, it is not clear how this relationship would translate when the training is performed using multiple training datasets. In view of the fact that the sparsity prior in dictionary learning enables the selection of suitable basis functions from the dictionary for any given input, one may conjecture that there should be similar mechanisms in deep neural networks that allow adaptation to the specific input signals.

In fact, it has been shown that ReLU offers combinatorial convolution-frame basis selection depending on each input image [7, 46]. More specifically, thanks to ReLU, a trained filter set creates a large number of partitions in the input space, as shown in Fig. 9.5, in which each region shares the same linear signal representation. Therefore, depending on each set of k-space input data, a specific region and the associated linear representation are selected. In addition, the number of input space partitions and the associated linear representation increases exponentially with depth, channel, and skip connection. Through the synergistic use of efficient signal representation in the k-space domain, this enormous expressivity from the same filter sets can make the k-space deep neural network more powerful than conventional image-domain learning.

Implementation and Results

The network backbone for k-space learning follows the U-Net [2] which consists of convolution, batch normalization, ReLU, and contracting path connection with

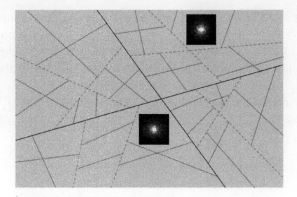

Figure 9.5 An example of $\mathbb{R}^2$ input space partitioning for the case of a two-channel three-layer ReLU neural network. Depending on the input k-space data, a partition and its associated linear representation are selected.

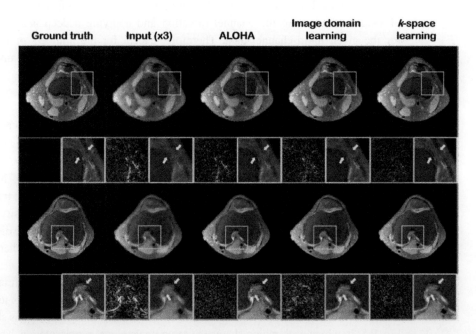

Figure 9.6 Reconstruction results from a Cartesian trajectory at $R = 3$ in multi-coils. The left-hand and right-hand boxes below each result (the right-hand boxes contain two arrows) illustrate the difference and enlarged views, respectively. The difference images are amplified by a factor 5.

concatenation. Here, the input and output are the complex-valued k-space data, while $\mathbb{R}[\cdot]$ and $\mathbb{R}^{-1}[\cdot]$ denote the operators that convert a complex-valued input into two-channel real-valued signals, and vice versa. For parallel imaging, multi-coil k-space data are concatenated along the channel direction.

Figure 9.6 shows accelerated MRI reconstruction results using CS, image-domain learning, and k-space learning. To downsample the k-space, a Cartesian trajectory with a Gaussian sampling pattern was used, and the net acceleration factor R was set

Table 9.1. Quantitative comparison from Cartesian trajectory at $R = 3$.

Metric	Input	ALOHA	Image domain learning	k-space learning
PSNR (dB)	34.20	36.10	35.85	**36.99**
NMSE ($\times 10^{-2}$)	2.34	1.58	1.65	**1.32**
SSIM	0.77	0.79	0.78	**0.81**
Time (seconds/slice)	—	16.56	0.12	0.14

to about 3. For quantitative evaluation, the normalized mean square error (NMSE), peak SNR (PSNR), and structural similarity index metric (SSIM) were used. Because ALOHA and k-space deep learning directly interpolate missing k-space data, these methods preserve textures and detail structures more clearly than the image-domain learning, as shown in Fig. 9.6. However, ALOHA is about 100 times slower than the k-space deep learning, as shown in Table 9.1. Furthermore, also as shown in Table 9.1, the k-space deep learning shows the best performance in terms of average PSNR, NMSE, and SSIM values.

9.3.2 MR Motion-Artifact Removal

As mentioned in previous sections, the scan time of MRI is relatively long; thus motion artifacts due to the patient's motion are often an issue. For example, transient severe motion (TSM) occurs because of the acute transient dyspnea in gadoxetic acid (Gd-EOB-DTPA)-enhanced MR. In fact, motion artifacts are considered a major problem in MRI acquisitions.

Recently, deep-learning approaches for MRI motion-artifact reduction (MAR) have been proposed [47, 48]. However, most deep-learning approaches for motion-artifact correction are based on simulated motion-artifact data because it is difficult to obtain matched clean and artifact images in many imported real-world applications. For example, in arterial image degradation due to TSM in Gd-EOB-DTBA-enhanced MR, motion-free paired images cannot be obtained. Therefore, it is difficult to apply them to real MR situations. To overcome the lack of paired data, some approaches exploited unpaired real motion-artifact data [49]. However, these algorithms interpret the motion-artifact problem as a style transfer problem, so there exists no explicit motion-artifact rejection mechanism. Therefore, its applications to a real dataset exhibit limited performance.

Below, we review a deep-learning method for MR motion-artifact correction that does not require matched motion-free and motion-artifact images, but still offers high-quality motion-artifact correction by employing an explicit motion-artifact rejection mechanism [9].

Motion Artifacts as Sparse k-space Outliers

In [50, 51] it was shown that various MRI artifacts, such as zipper artifacts or non-rigid cardiac motion, appear as sparse outliers in k-space. Here, we will provide a brief review of these important findings.

According to existing works related to motion artifacts, [50, 51], when there is transient motion, this incurs displacements in k-space along the phase encoding direction at a few specific time instances. This results in the following k-space data:

$$\widehat{s}_e(k_x,k_y) = \begin{cases} \widehat{s}(k_x,k_y)e^{-\iota\Phi(k_y)}, & k_y \in \mathbb{K} \\ \widehat{s}(k_x,k_y), & \text{otherwise} \end{cases}, \quad (9.22)$$

where $\iota = \sqrt{-1}$, $\widehat{s}_e(k_x,k_y)$ and $\widehat{s}(k_x,k_y)$ refer to the motion-corrupted and motion-free k-space data, respectively, with k_x and k_y the indices along the read-out and phase encoding directions, respectively. Furthermore, $\Phi(k_y)$ is the displacement (in radians) at the phase-encoding index k_y, and $\mathbb{K}$ denotes the phase encoding indices where the displacements occur. Equation (9.22) shows that motion artifacts cause phase variations, which appear as k-space sparse outliers along the phase encoding direction.

Figure 9.7 shows the sparse outliers in k-space with 2D or 3D acquisitions. If the image is acquired by 2D imaging, sparse outliers appear along the phase encoding direction, as shown in Fig. 9.7(a). Conversely, 3D imaging requires two-dimensional phase encoding steps, and sparse outliers appear in the 3D k-space volume, as shown in Fig. 9.7(b). To address these sparse outliers in 3D imaging, a neural network could be designed to handle the whole 3D volume. However, it is difficult to handle a 3D volume during training because it requires a large amount of graphics processing unit (GPU) memory usage. So, instead of using the 3D volume during training, it is possible to use cross sections which contain a frequency encoding axis and one of the phase encoding axes (e.g., the axial plane in Fig. 9.7(b)).

Specifically, if a 1D Fourier transform is applied to the 3D k-space volume along the phase encoding direction (see Fig. 9.7(b)), it is converted to the stack of 2D k-space

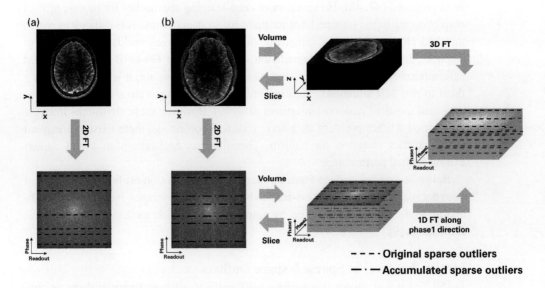

Figure 9.7 Sparse outliers in k-space along the phase encoding direction: (a) 2D imaging, and (b) 3D imaging.

data of all cross sections. Moreover, sparse outliers in 3D volume are transformed into the accumulated sparse outliers in the 2D k-space of cross sections which contain a frequency encoding axis and a phase encoding axis, as shown in Fig. 9.7(b). Therefore, the sparse outliers can be considered as 1D outliers in 2D k-space along the phase encoding direction. Because 3D motion artifacts can be considered as sparse outliers in 2D k-space, it is possible to correct the 3D motion by using 2D images after applying a 1D Fourier transform along the phase encoding direction.

Bootstrap Aggregation for Motion-Artifact Correction

Bootstrap aggregation is a classic machine learning technique, which uses the bootstrap sampling and aggregation of the results to improve the accuracy of the base learner [52]. The rationale for bootstrap aggregation is that it may be easier to train multiple simple weak learners and combine them into a more complex learner than to train one strong learner.

In the context of MR deep learning, a bootstrap aggregation can be represented as follows [53]:

$$\tilde{\mathbf{m}} = \sum_{n=1}^{N} w_n G_\Theta(\mathcal{P}_{\Lambda,n}[\hat{\mathbf{s}}]), \tag{9.23}$$

where $\hat{\mathbf{s}}$ denotes the k-space data, $\tilde{\mathbf{m}}$ denotes the reconstructed images, $\mathcal{P}_{\Lambda,n}$ refers to the nth k-space subsampling, G_Θ is a reconstruction network parameterized by Θ that reconstructs an image from subsampled k-space data, and w_n is the nth weighting factor. In other words, the final reconstructed image is acquired by aggregation of the individual reconstruction results from each subsampled k-space.

In [53], it was demonstrated that bootstrap aggregation can provide higher-quality image reconstruction in CS MRI than a single strong deep learner. In addition, the advantage of bootstrap aggregation in Eq. (9.23) is that it can play an important role in the context of motion-artifact removal, since the sparse outlier model in Eq. (9.22) leads to the following key observation:

$$\mathcal{P}_{\Lambda,n}[\hat{\mathbf{s}}] \simeq \mathcal{P}_{\Lambda,n}[\hat{\mathbf{s}}_e], \tag{9.24}$$

for some sampling instance $\mathcal{P}_{\Lambda,n}$. This is so because, when using 1D subsampling along the phase encoding direction, the subsampling operation $\mathcal{P}_{\Lambda,n}$ can remove many sparse outliers of $\hat{\mathbf{s}}_e$ in Eq. (9.22), so that it becomes similar to the subsampled k-space data from the clean image. Accordingly, it reduces the contribution of the motion artifacts, and in consequence the resulting bootstrap aggregation estimate in Eq. (9.23) may become much closer to the artifact-free image.

When the neural network is trained in a supervised manner, there exists the possibility that bias may be introduced because some artifact-corrupted k-space data remain after subsampling. To address this issue, an unpaired learning framework using optimal transport-driven cycleGAN [8, 54] can be employed for reconstruction from the subsampled k-space data.

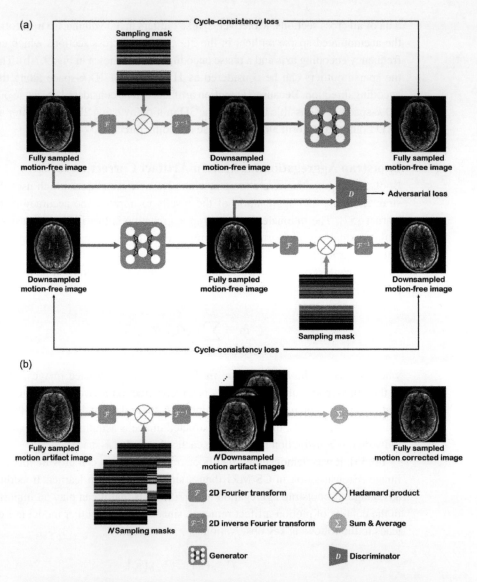

Figure 9.8 Overall flow of the MAR method using bootstrap subsampling and aggregation. (a) In the training phase, a clean image is downsampled in the k-space domain and the network is trained to convert the aliased image to the original fully sampled image. (b) In the testing phase, N aliased images are generated from one artifact image by random subsampling in the k-space domain. Then, each image is processed by the trained network. These reconstructed images are aggregated so that a motion corrected image can be obtained.

Figure 9.8 shows the flow of the MAR using bootstrap subsampling and aggregation. Here, N refers to the subsampling aggregation factor. During the training phase in Fig. 9.8(a), the neural network is trained for accelerated MRI reconstruction using only *artifact-free* data, so that it can reconstruct a high-quality image from any aliased image using randomly undersampled k-space data. In particular, in the upper branch of cycleGAN, the clean images are first converted into a Fourier spectrum and are then downsampled by k-space subsampling, so that the downsampled images have been generated by inverse Fourier transform. The aliased image from downsampling then becomes the input of the network. In the lower branch, first a downsampled artifact image, which is different from the image in the upper branch, passes through the network and then the output is downsampled in the Fourier domain. Furthermore, the reconstructed clean image and the real clean image become inputs of the discriminator, so that the discriminator distinguishes them as real or fake images. Therefore, the network learns how to convert downsampling artifact images to realistic fully sampled images by unpaired learning.

On the other hand, in the inference phase the images with motion artifacts are used as illustrated in Fig. 9.8(b). First, a motion-artifact image is converted to Fourier domain data, and then N random subsamplings are applied to obtain multiple aliased images. The subsampling of k-space can delete some k-space outliers with a phase error due to the motion. Therefore, when the downsampled artifact images are reconstructed by the trained network, it is possible to obtain images with reduced motion artifact since the network has been trained to reconstruct fully sampled artifact-free images. Then the neural network outputs from each aliased image are aggregated to obtain the final reconstruction.

Implementation and Results

To generate the simulated rigid-motion artifact, the phase error along the phase encoding direction can be formulated as Eq. (9.22) with

$$\Phi(k_y) = \begin{cases} k_y \Delta_k, & |k_y| > k_0 \\ 0, & \text{otherwise} \end{cases}, \qquad (9.25)$$

where Δ_k is the degree of motion at k-space line k, and k_0 is the delay time of the phase error due to the centric k-space filling [47].

Figure 9.9 shows the simulated 2D random-motion-artifact correction results for both quantitative and qualitative comparison. Here, MARC [47] is a method for the reduction of motion artifacts in liver MRI using a convolutional neural network. Next, Cycle-MedGAN V2.0 [49] is an unpaired learning method which is based on cycle-GAN [55] for motion-artifact correction and can therefore be trained with unpaired data. As shown in Fig. 9.9, simulated random motion artifacts appear in the input images. MARC reduces the motion artifacts in MR images, but some severe artifacts still remain in the output images of MARC. Next, the motion-artifact correction fails when Cycle-MedGAN V2.0 is applied, as shown in Fig. 9.9. However, the method using bootstrap subsampling and aggregation successfully corrects motion artifacts in MR images.

Table 9.2 shows the average quantitative metric values of motion-artifact correction methods. MARC shows the highest metric values because it is trained through supervised learning with paired data. The method using bootstrap aggregation shows quantitative results comparable with MARC.

Table 9.2. Quantitative comparison of various methods for simulated data. The values in the table are average values for the whole test data.

	PSNR (dB)	SSIM
Input	31.2510	0.7685
MARC	35.3905	0.9228
Cycle-MedGAN V2.0	26.6952	0.8279
Bootstrap subsampling and aggregation	35.0652	0.9360

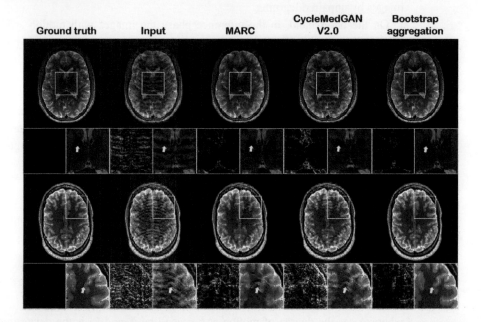

Figure 9.9 Motion-artifact correction results or brain data using various methods with simulated 2D random motion artifact. The left and right boxes below each result illustrate the difference images and enlarged views, respectively. The difference images are amplified by a factor 5.

9.4 Summary and Outlook

Recently, k-space deep learning has emerged as a promising strategy for accelerated MRI reconstruction. In this chapter, we covered these strategies on the basis of whether they are trained on a scan-specific manner or using large databases. The former group of methods is built on linear parallel-imaging methods that perform k-space reconstruction via interpolation. The latter group of techniques uses the ALOHA framework to learn an interpolation kernel over a larger database. Both approaches hold promise, and their utility depends on the target applications.

Scan-specific methods are useful when the acquisition of large datasets is not feasible, for instance when using experimental sequences on small sample sizes. It also addresses certain concerns about generalization that have been pointed out in the literature [13, 14], such as changes in undersampling patterns, acceleration rates, SNR, or anatomical features between the training and test data [56, 57], as well as the lack of examples with rare or subtle pathologies [14]. However, they also come with downsides such as the computational burden of training for each scan [22]. Furthermore, the CNN architectures are often more compact, and so avoid the acquisition of large amounts of calibration data. Database-trained methods, on the other hand, often can utilize deeper and more specialized network structures and provide improved performance [58]. There are ongoing efforts to reconcile these two approaches [26, 59], which may further synergistically improve the promise of k-space deep-learning methods.

References

[1] M. Lustig, D. L. Donoho, J. M. Santos, and J. M. Pauly, "Compressed sensing MRI," *IEEE Signal Processing Magazine*, vol. 25, no. 2, p. 72, 2008.

[2] O. Ronneberger, P. Fischer, and T. Brox, "U-net: Convolutional networks for biomedical image segmentation," in *Proc. International Conference on Medical Image Computing and Computer-Assisted Intervention*. Springer, 2015, pp. 234–241.

[3] B. Dong and Z. Shen, "Image restoration: A data-driven perspective," in *Proc. International Congress of Industrial and Applied Mathematics*, 2015, pp. 65–108.

[4] S. Wang, Z. Su, L. Ying, X. Peng, S. Zhu, F. Liang, D. Feng, and D. Liang, "Accelerating magnetic resonance imaging via deep learning," in *Proc. IEEE 13th International Symposium on Biomedical Imaging*. IEEE, 2016, pp. 514–517.

[5] B. Zhu, J. Z. Liu, S. F. Cauley, B. R. Rosen, and M. S. Rosen, "Image reconstruction by domain-transform manifold learning," *Nature*, vol. 555, no. 7697, p. 487, 2018.

[6] M. Akçakaya, S. Moeller, S. Weingärtner, and K. Uğurbil, "Scan-specific robust artificial-neural-networks for k-space interpolation (RAKI) reconstruction: Database-free deep learning for fast imaging," *Magnetic Resonance in Medicine*, vol. 81, no. 1, pp. 439–453, 2019.

[7] Y. Han, L. Sunwoo, and J. C. Ye, "k-Space deep learning for accelerated MRI," *IEEE Transactions on Medical Imaging*, vol. 39, no. 2, pp. 377–386, 2019.

[8] G. Oh, B. Sim, H. Chung, L. Sunwoo, and J. C. Ye, "Unpaired deep learning for accelerated mri using optimal transport driven cyclegan," *IEEE Transactions on Computational Imaging*, vol. 6, pp. 1285–1296, 2020.

[9] G. Oh, J. E. Lee, and J. C. Ye, "Unsupervised MR motion artifact deep learning using outlier-rejecting bootstrap aggregation," *arXiv:2011.06337*, 2020.

[10] M. A. Griswold, P. M. Jakob, R. M. Heidemann, M. Nittka, V. Jellus, J. Wang, B. Kiefer, and A. Haase, "Generalized autocalibrating partially parallel acquisitions (GRAPPA)," *Magnetic Resonance in Medicine*, vol. 47, no. 6, pp. 1202–1210, 2002.

[11] M. Akçakaya, S. Moeller, S. Weingärtner, and K. Uğurbil, "Scan-specific deep learning with robust artificial-neural-networks for k-space interpolation (RAKI) for improved parallel imaging," in *Proc. International Society for Magnetic Resonance in Medicine Workshop on Machine Learning*, 2018.

[12] ——, "Subject-specific convolutional neural networks for accelerated magnetic resonance imaging," in *Proc. 2018 IEEE International Joint Conference on Neural Networks*, 2018.

[13] Y. C. Eldar, A. O. Hero III, L. Deng, J. Fessler, J. Kovacevic, H. V. Poor, and S. Young, "Challenges and open problems in signal processing: Panel discussion summary from ICASSP 2017," *IEEE Signal Processing Magazine*, vol. 34, pp. 8–23, 2017.

[14] F. Knoll, T. Murrell, A. Sriram, N. Yakubova, J. Zbontar, M. Rabbat, A. Defazio, M. J. Muckley, D. K. Sodickson, C. L. Zitnick, and M. P. Recht, "Advancing machine learning for MR image reconstruction with an open competition: Overview of the 2019 fastMRI challenge," *Magnetic Resonance in Medicine*, 2020.

[15] J. Zbontar, F. Knoll, A. Sriram, T. Murrell, Z. Huang, M. J. Muckley, *et al.*, "fastMRI: An open dataset and benchmarks for accelerated MRI," *arXiv:1811.08839*, 2018.

[16] B. Yaman, S. Hosseini, S. Moeller, J. Ellermann, K. Uğurbil, and M. Akçakaya, "Self-supervised physics-based deep learning MRI reconstruction without fully-sampled data," in *Proc. International Symposium on Biomedical Imaging*. IEEE, 2020, pp. 921–925.

[17] B. Yaman, S. A. H. Hosseini, S. Moeller, J. Ellermann, K. Ugurbil, and M. Akcakaya, "Self-supervised learning of physics-guided reconstruction neural networks without fully-sampled reference data," *Magnetic Resonance in Medicine*, vol. 84, pp. 3172–3191, 2020.

[18] D. K. Sodickson and W. J. Manning, "Simultaneous acquisition of spatial harmonics SMASH: Fast imaging with radiofrequency coil arrays," *Magnetic Resonance in Medicine*, vol. 38, no. 4, pp. 591–603, 1997.

[19] K. P. Pruessmann, M. Weiger, M. B. Scheidegger, and P. Boesiger, "SENSE: Sensitivity encoding for fast MRI," *Magnetic Resonance in Medicine*, vol. 42, pp. 952–962, 1999.

[20] P. B. Roemer, W. A. Edelstein, C. E. Hayes, S. P. Souza, and O. M. Mueller, "The NMR phased array," *Magnetic Resonance in Medicine*, vol. 16, no. 2, pp. 192–225, 1990.

[21] Y. Chang, D. Liang, and L. Ying, "Nonlinear GRAPPA: A kernel approach to parallel MRI reconstruction," *Magnetic Resonance in Medicine*, vol. 68, no. 3, pp. 730–740, 2012.

[22] C. Zhang, S. A. H. Hosseini, S. Weingärtner, K. Uğurbil, S. Moeller, and M. Akçakaya, "Optimized fast GPU implementation of robust artificial-neural-networks for k-space interpolation (RAKI) reconstruction," *PLoS One*, vol. 14, no. 10, p. e0223 315, 2019.

[23] S. A. H. Hosseini, S. Moeller, S. Weingärtner, K. Uğurbil, and M. Akçakaya, "Accelerated coronary MRI using 3D SPIRiT-RAKI with sparsity regularization," in *Proc. IEEE International Symposium on Biomedical Imaging*. IEEE, 2019, pp. 1692–1695.

[24] S. A. H. Hosseini, C. Zhang, S. Weingärtner, S. Moeller, M. Stuber, K. Ugurbil, and M. Akçakaya, "Accelerated coronary MRI with sRAKI: A database-free self-consistent neural network k-space reconstruction for arbitrary undersampling," *PLoS One*, vol. 15, no. 2, p. e0229 418, 2020.

[25] S. H. Hosseini, B. Yaman, C. Zhang, K. Uğurbil, S. Moeller, and M. Akçakaya, "Scan-specific accelerated MRI reconstruction using recurrent neural networks in a regularized self-consistent framework," in *Proc. 2020 IEEE 17th International Symposium on Biomedical Imaging Workshops*, 2020, pp. 1–4.

[26] S. A. H. Hosseini, B. Yaman, S. Moeller, and M. Akçakaya, "High-fidelity accelerated MRI reconstruction by scan-specific fine-tuning of physics-based neural networks," in *Proc. IEEE Engineering in Medicine Biology Society*, 2020, pp. 1481–1484.

[27] M. Barth, F. Breuer, P. J. Koopmans, D. G. Norris, and B. A. Poser, "Simultaneous multislice (SMS) imaging techniques," *Magnetic Resonance in Medicine*, vol. 75, no. 1, pp. 63–81, 2016.

[28] F. A. Breuer, M. Blaimer, R. M. Heidemann, M. F. Mueller, M. A. Griswold, and P. M. Jakob, "Controlled aliasing in parallel imaging results in higher acceleration

(CAIPIRINHA) for multi-slice imaging," *Magnetic Resonance in Medicine*, vol. 53, no. 3, pp. 684–691, 2005.

[29] S. Moeller, E. Yacoub, C. A. Olman, E. Auerbach, J. Strupp, N. Harel, and K. Uğurbil, "Multiband multislice GE-EPI at 7 tesla, with 16-fold acceleration using partial parallel imaging with application to high spatial and temporal whole-brain fMRI," *Magnetic Resonance in Medicine*, vol. 63, no. 5, pp. 1144–1153, 2010.

[30] K. Setsompop, B. A. Gagoski, J. R. Polimeni, T. Witzel, V. J. Wedeen, and L. L. Wald, "Blipped-controlled aliasing in parallel imaging for simultaneous multislice echo planar imaging with reduced g-factor penalty," *Magnetic Resonance in Medicine*, vol. 67, no. 5, pp. 1210–1224, 2012.

[31] S. F. Cauley, J. R. Polimeni, H. Bhat, L. L. Wald, and K. Setsompop, "Interslice leakage artifact reduction technique for simultaneous multislice acquisitions," *Magnetic Resonance in Medicine*, vol. 72, no. 1, p. 93, 2014.

[32] D. C. Van Essen, S. M. Smith, D. M. Barch, T. E. Behrens, E. Yacoub, K. Ugurbil *et al.*, "The WU-Minn Human Connectome Project: An overview," *Neuroimage*, vol. 80, pp. 62–79, 2013.

[33] C. Zhang, S. Moeller, S. Weingärtner, K. Uğurbil, and M. Akçakaya, "Accelerated simultaneous multi-slice MRI using subject-specific convolutional neural networks," in *Proc. 2018 52nd Asilomar Conference on Signals, Systems, and Computers*, 2018, pp. 1636–1640.

[34] C. Zhang, S. A. Hossein Hosseini, S. Moeller, S. Weingärtner, K. Ugurbil, and M. Akcakaya, "Scan-specific residual convolutional neural networks for fast mri using residual raki," in *Proc. 2019 53rd Asilomar Conference on Signals, Systems, and Computers*, 2019, pp. 1476–1480.

[35] A. S. Nencka, V. E. Arpinar, S. Bhave, B. Yang, S. Banerjee, M. McCrea, N. J. Mickevicius, L. T. Muftuler, and K. M. Koch, "Split-slice training and hyperparameter tuning of RAKI networks for simultaneous multi-slice reconstruction," *Magnetic Resonance in Medicine*, vol. 85, no. 6, pp. 3272–3280, 2021.

[36] O. B. Demirel, S. Weingärtner, S. Moeller, and M. Akçakaya, "Improved simultaneous multislice cardiac MRI using readout concatenated k-space SPIRiT (ROCK-SPIRiT)," *Magnetic Resonance in Medicine*, vol. 85, no. 6, pp. 3036–3048, 2021.

[37] S. Moeller, P. Pisharady Kumar, J. Andersson, M. Akcakaya, N. Harel, R. E. Ma, X. Wu, E. Yacoub, C. Lenglet, and K. Ugurbil, "Diffusion imaging in the post HCP era," *Journal of Magnetic Resonance Imaging*, 2020.

[38] P. J. Shin, P. E. Larson, M. A. Ohliger, M. Elad, J. M. Pauly, D. B. Vigneron, and M. Lustig, "Calibrationless parallel imaging reconstruction based on structured low-rank matrix completion," *Magnetic Resonance in Medicine*, vol. 72, no. 4, pp. 959–970, 2014.

[39] J. P. Haldar, "Low-rank modeling of local k-space neighborhoods (LORAKS) for constrained MRI," *IEEE Transactions on Medical Imaging*, vol. 33, no. 3, pp. 668–681, 2014.

[40] K. H. Jin, D. Lee, and J. C. Ye, "A general framework for compressed sensing and parallel MRI using annihilating filter based low-rank hankel matrix," *IEEE Transactions on Computational Imaging*, vol. 2, no. 4, pp. 480–495, 2016.

[41] T. H. Kim, P. Garg, and J. P. Haldar, "Loraki: Autocalibrated recurrent neural networks for autoregressive MRI reconstruction in k-space," *arXiv*, 2019.

[42] G. Ongie and M. Jacob, "Off-the-grid recovery of piecewise constant images from few fourier samples," *SIAM Journal on Imaging Sciences*, vol. 9, no. 3, pp. 1004–1041, 2016.

[43] J. C. Ye, J. M. Kim, K. H. Jin, and K. Lee, "Compressive sampling using annihilating filter-based low-rank interpolation," *IEEE Transactions on Information Theory*, vol. 63, no. 2, pp. 777–801, 2016.

[44] M. Vetterli, P. Marziliano, and T. Blu, "Sampling signals with finite rate of innovation," *IEEE Transactions on Signal Processing*, vol. 50, no. 6, pp. 1417–1428, 2002.

[45] J. C. Ye, Y. Han, and E. Cha, "Deep convolutional framelets: A general deep learning framework for inverse problems," *SIAM J. Imaging Sci.*, vol. 11, no. 2, pp. 991–1048, 2018.

[46] J. C. Ye and W. K. Sung, "Understanding geometry of encoder-decoder CNNs," in *Proc. International Conference on Machine Learning*. PMLR, 2019, pp. 7064–7073.

[47] D. Tamada, M.-L. Kromrey, S. Ichikawa, H. Onishi, and U. Motosugi, "Motion artifact reduction using a convolutional neural network for dynamic contrast enhanced MR imaging of the liver," *Magnetic Resonance in Medical Sciences*, vol. 19, no. 1, p. 64, 2020.

[48] I. Oksuz, J. Clough, B. Ruijsink, E. Puyol-Antón, A. Bustin, G. Cruz, C. Prieto, D. Rueckert, A. P. King, and J. A. Schnabel, "Detection and correction of cardiac MRI motion artefacts during reconstruction from k-space," in *Proc. International Conference on Medical Image Computing and Computer-Assisted Intervention*. Springer, 2019, pp. 695–703.

[49] K. Armanious, A. Tanwar, S. Abdulatif, T. Küstner, S. Gatidis, and B. Yang, "Unsupervised adversarial correction of rigid MR motion artifacts," in *Proc. 2020 IEEE 17th International Symposium on Biomedical Imaging*. IEEE, 2020, pp. 1494–1498.

[50] K. H. Jin, J.-Y. Um, D. Lee, J. Lee, S.-H. Park, and J. C. Ye, "MRI artifact correction using sparse+ low-rank decomposition of annihilating filter-based Hankel matrix," *Magnetic Resonance in Medicine*, vol. 78, no. 1, pp. 327–340, 2017.

[51] M. A. Bernstein, K. F. King, and X. J. Zhou, *Handbook of MRI Pulse Sequences*. Elsevier, 2004.

[52] L. Breiman, "Bagging predictors," *Machine Learning*, vol. 24, no. 2, pp. 123–140, 1996.

[53] E. Cha, G. Oh, and J. C. Ye, "Geometric approaches to increase the expressivity of deep neural networks for MR reconstruction," *IEEE Journal of Selected Topics in Signal Processing*, vol. 14, no. 6, pp. 1292–1305, 2020.

[54] B. Sim, G. Oh, J. Kim, C. Jung, and J. C. Ye, "Optimal transport driven cycleGAN for unsupervised learning in inverse problems," *SIAM Journal on Imaging Sciences*, vol. 13, no. 4, pp. 2281–2306, 2020.

[55] J.-Y. Zhu, T. Park, P. Isola, and A. A. Efros, "Unpaired image-to-image translation using cycle-consistent adversarial networks," in *Proc. IEEE International Conference on Computer Vision*, 2017, pp. 2223–2232.

[56] F. Knoll, K. Hammernik, E. Kobler, T. Pock, M. P. Recht, and D. K. Sodickson, "Assessment of the generalization of learned image reconstruction and the potential for transfer learning," *Magnetic Resonance in Medicine*, vol. 81, no. 1, pp. 116–128, 2019.

[57] M. J. Muckley, B. Riemenschneider, A. Radmanesh, S. Kim, G. Jeong, J. Ko, Y. Jun, H. Shin, D. Hwang, M. Mostapha et al., "State-of-the-art machine learning MRI reconstruction in 2020: Results of the second fastMRI challenge," *arXiv preprint arXiv:2012.06318*, 2020.

[58] F. Knoll, K. Hammernik, C. Zhang, S. Moeller, T. Pock, D. K. Sodickson, and M. Akçakaya, "Deep-learning methods for parallel magnetic resonance imaging reconstruction," *IEEE Signal Processing Magazine*, vol. 37, no. 1, pp. 128–140, 2020.

[59] B. Yaman, S. A. H. Hosseini, and M. Akçakaya, "Zero-shot self-supervised learning for MRI reconstruction," *arXiv:2102.07737*, 2021.

10 Deep Learning for Ultrasound Beamforming

Ruud J. G. van Sloun, Jong Chul Ye, and Yonina C. Eldar

10.1 Introduction and Relevance

Diagnostic imaging plays a critical role in healthcare, serving as a fundamental asset for timely diagnosis, disease staging, and management as well as for treatment choice, planning, guidance, and follow-up. Among the diagnostic imaging options, ultrasound imaging [1] is uniquely positioned, being a highly cost-effective modality that offers the clinician an unmatched and invaluable level of interaction, enabled by its real-time nature. Its portability and cost-effectiveness permits point-of-care imaging at the bedside, in emergency settings, rural clinics, and developing countries. Ultrasonography is increasingly used across many medical specialties, spanning from obstetrics, cardiology, and oncology to acute and intensive care, with a market share that is globally growing.

On the technological side, ultrasound probes are becoming increasingly compact and portable, with an expanding market demand for low-cost "pocket-sized" devices (i.e., "the stethoscope model") [2]. Transducers are miniaturized, allowing, e.g., in-body imaging for interventional applications. At the same time, there is a strong trend towards three-dimensional (3D) imaging [3] and the use of high-frame-rate imaging schemes [4]; both are accompanied by dramatically increasing data rates, which impose a heavy burden on the probe-system communication and subsequent image reconstruction algorithms. Systems today offer a wealth of advanced applications and methods, including shear wave elasticity imaging [5], ultra-sensitive Doppler [6], and ultrasound localization microscopy for super-resolution microvascular imaging [7, 8].

With the demand for high-quality image reconstruction and signal extraction from fewer transmissions that facilitate fast imaging, and a push towards compact probes, modern ultrasound imaging leans heavily on innovations in powerful digital receive-channel processing. Beamforming, the process of mapping received ultrasound echoes to the spatial image domain, naturally lies at the heart of the ultrasound image formation chain. In this chapter, we discuss why and when deep-learning methods can play a compelling role in the digital beamforming pipeline, and then show how these data-driven systems can be leveraged for improved ultrasound image reconstruction [9].

This chapter is organized as follows. Section 10.2 briefly introduces various scanning modes in ultrasound. Then, in Section 10.3, we describe methods and the rationale for digital receive beamforming. In Section 10.4 we elaborate on the opportunities of deep learning for ultrasound beamforming, and in Section 10.5 we review various deep network architectures. We then turn to typical approaches for training in Section 10.6. Finally, in Section 10.7 we discuss several future directions.

10.2 Ultrasound Scanning in a Nutshell

Ultrasound imaging is based on the pulse-echo principle. First, a radio-frequency (RF) pressure wave is transmitted into the medium of interest through a multi-element ultrasound transducer. These transducers are typically based on piezoelectric (preferably single-crystal) mechanisms or capacitive micromachined ultrasonic transducer (CMUT) technology. After insonification, the acoustic wave backscatters owing to inhomogeneities in the medium properties, such as its density and speed of sound. The resulting reflections are recorded by the same transducer array and used to generate a so-called "brightness-mode" (B-mode) image through a signal processing step termed beamforming. We will elaborate on this step in Section 10.3. The achievable resolution, contrast, and overall fidelity of B-mode imaging depend on the array aperture and geometry, element sensitivity, and bandwidth. Transducer geometries include linear, curved, or phased arrays. The latter are mainly used for extended fields of view from limited acoustic windows (e.g., imaging of the heart between the ribs), enabling the use of angular beam steering due to a smaller pitch, namely, the distance between elements. The elements effectively sample the aperture: using a pitch of half the wavelength (i.e., spatial Nyquist-rate sampling) avoids the appearance of grating lobes (spatial aliasing) in the array response. Two-dimensional ultrasound imaging is based on 1D arrays, while 3D imaging makes use of 2D matrix designs.

Given a transducer's physical constraints, getting the most out of the system requires careful optimization across its entire imaging chain. At the front end, this starts with the design of appropriate transmit schemes for wave field generation. At this stage, crucial trade-offs are made in which the frame rate, imaging depth, and attainable axial and lateral resolution are weighted carefully against each other. Improved resolution can be achieved through the use of higher pulse modulation frequencies and bandwidths, yet these shorter wavelengths suffer from increased absorption and thus lead to reduced penetration depth. Likewise, high frame rates can be reached by exploiting parallel transmission schemes based on, e.g., planar or diverging waves. However, the use of such unfocused transmissions comes with the cost of loss in lateral resolution compared with line-based scanning with tightly focused beams. As such, optimal transmit schemes depend on the application. We will briefly elaborate on three common transmit schemes for ultrasound B-mode imaging, an illustration of which is given in Fig. 10.1.

10.2.1 Focused Transmits/Line Scanning

In line scanning, a series of E transmit events is used, with each transmit e, to produce a single depth wise line in the image by focusing the transmitted acoustic energy along that line. Such focused transmits are typically achieved using a sub-aperture of

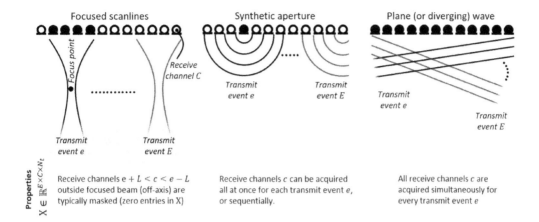

Figure 10.1 An illustration of three transmit types. The black transducers are actively emitting signals. The white transducers are inactive. Transmit events are denoted by e, receive channels by c and, for focused scanlines, $2L + 1$ is the size of the active aperture in terms of elements.

transducer elements $c \in \{e - L, \ldots, e + L\}$, excited with time-delayed RF pulses. By choosing these transmit delays per channel appropriately, the beam can be focused towards a given depth and (for phased arrays) angle. Focused line scanning is the most common transmit design in commercial ultrasound systems, enjoying improved lateral resolution and image contrast compared with unfocused transmits. Line-by-line acquisition is, however, time consuming (every lateral line requires a distinct transmit event), thus upper bounding the frame rate of this transmit mode by the number of lines, imaging depth, and speed of sound. This constraint can be relaxed via multi-line parallel transmit approaches, at the expense of reduced image quality.

10.2.2 Synthetic Aperture

Synthetic aperture transmit schemes allow for synthetic dynamic transmit focusing by acquiring echoes with the full array, following near-spherical wave excitation using individual transducer elements c. By performing E such transmits (typically $E = C$, the number of array elements), the image reconstruction algorithm (i.e., the beamformer) has full access to all transmit–receive pairs, enabling retrospective focusing in both the transmit and the receive. Sequential transmission with all elements is, however, time consuming, and acoustic energy delivered into the medium is limited (preventing, e.g., harmonic imaging applications). Synthetic aperture imaging also finds application in phased array intravascular ultrasound (IVUS) imaging, where one can only transmit and receive using one element or channel at a time owing to catheter-based constraints.

10.2.3 Plane Wave Ultrafast

Today, an increasing amount of ultrasound applications rely on high frame-rate (dubbed *ultrafast*) parallel imaging based on plane waves or diverging waves. Among these are ultrasound localization microscopy, highly sensitive Doppler, shear wave elastography, and (blood) speckle tracking. Where the former two mostly exploit the incredible vastness of data to obtain accurate signal statistics, the latter two leverage high-speed imaging to track ultrasound-induced shear waves or tissue motion to estimate elasticity, strain, or flow parameters. In plane wave imaging, planar waves are transmitted to insonify the full region of interest in a single transmit. Typically, several plane waves with different angles are compounded to improve image quality. For small-footprint phased arrays, diverging waves are used. These diverging waves are based on a set of virtual focus points that are placed behind the array, acting as virtual spherical sources. In that context, one can also interpret diverging-wave imaging as a synthetic aperture technique.

With the expanding use of ultrafast transmit sequences in modern ultrasound imaging, a strong burden is placed on the subsequent receive channel processing. High data rates not only raise substantial hardware complications related to data storage and data transfer; in addition, the corresponding unfocused transmissions require advanced receive beamforming to reach satisfactory image quality.

10.3 Digital Ultrasound Beamforming

We now describe how the received and digitized channel data is used to reconstruct an image in the digital domain, via a digital signal processing algorithm called beamforming.

10.3.1 Digital Beamforming Model and Framework

Consider an array of C channels, and E transmit events, resulting in $E \times C$ measured and digitized RF signal vectors containing N_t samples. Denote $X \in \mathbb{R}^{E \times C \times N_t}$ as the resulting received RF data cube. Individual transmit events e can be tilted planar waves, diverging waves, focused scanlines through transmit beamforming, or any other desired pressure distribution in transmission. The goal of beamforming is to map this time-array-domain "channel" data cube to the spatial domain, through a processor $f(\cdot)$:

$$Y = f(X), \tag{10.1}$$

where $Y \in \mathbb{R}^{N_x \times N_y}$ denotes the beamformed spatial data, with N_x and N_y the numbers of pixels in the axial and lateral directions, respectively. In principle, all beamforming architectures in ultrasound can be formulated according to Eq. (10.1) and the illustration in Fig. 10.2. The different approaches vary in their parameterization of $f(\cdot)$. Most are composed of a geometrical time-to-space migration of individual channels, and a subsequent combiner–processor. We will now describe some of the most common parameterizations for ultrasound beamforming. In Section 10.5, we will see that these conventional signal processing methods can also directly inspire parameterizations comprising deep neural networks (DNNs).

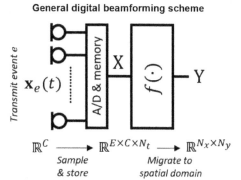

Figure 10.2 General beamforming scheme; $x_e(t)$ labels the data for transmit event e.

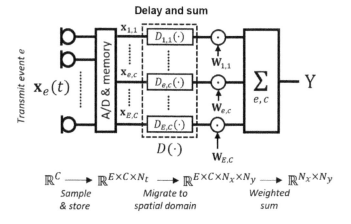

Figure 10.3 Delay-and-sum beamforming.

10.3.2 Delay-and-Sum

The industry-standard beamforming algorithm is delay-and-sum beamforming (DAS). This beamforming is commonplace owing to its low complexity, allowing for real-time image reconstruction at the expense of nonoptimal image quality. Its processing can in general be written as (see Fig. 10.3 for an illustration):

$$Y = \sum_{E,L} W \odot D(X), \quad (10.2)$$

where $W \in \mathbb{R}^{E \times C \times N_x \times N_y}$ is an apodization weight tensor and $\odot$ denotes an element-wise product. Note that apodization weights can be complex-valued when $D(X)$ is IQ-demodulated, allowing for phase shifting by W. In the remainder of this chapter, we will (without loss of generality) only consider real weights applied to RF data. Here, $D(\cdot)$ is a focusing function that migrates the RF time signal to space for each transmit event and channel, mapping X from $\mathbb{R}^{E \times C \times N_t}$ to $\mathbb{R}^{E \times C \times N_x \times N_y}$. This mapping is obtained by applying geometry-based time delays to the RF signals with the aim of time-aligning the received echoes from the set of focal points.

Delay-and-sum beamforming is typically employed in a so-called dynamic receive beamforming mode, in which the focal points change as a function of scan depth. In a specific variant of dynamic receive beamforming, *pixel-based beamforming*, each pixel is a focus point. Note that, unlike what its name suggests, dynamic does not mean that the beamformer is dynamically updating $D(\cdot)$ and W on the fly. The name stems from the varying time delays across fast-time[1] (and therefore depth) to dynamically move the focal point deeper.

As said above, channel delays are used to time-align the received echoes from a given position and are determined by ray-based wave propagation, which in turn is dictated by the array geometry, transmit design (plane wave, diverging wave, focused, synthetic-aperture), the position of interest, and an estimate of the speed of sound. For each focal point $\{r_x, r_y\}$, channel c, and transmit event e, we can write the time of flight as

$$\tau_{e,c,r_x,r_y} = \tau_{e,c,\mathbf{r}} = \frac{\|\mathbf{r}_e - \mathbf{r}\|_2 + \|\mathbf{r}_c - \mathbf{r}\|_2}{v}, \tag{10.3}$$

where $\tau_{e,c,\mathbf{r}}$ is the time of flight for an imaging point $\mathbf{r}$, $\mathbf{r}_c$ is the position vector of the receiving element in the array, and v is the speed of sound in the medium. The vector $\mathbf{r}_e$ depends on the transmit sequence: for focused transmits it is the position vector of the (sub)aperture center for transmit e; for synthetic aperture transmits it is the position vector of the eth transmitting element; for diverging waves it is the position vector of the eth virtual source located behind the array; for plane waves it is dictated by the transmit angle.

For any focus point $\{r_x, r_y\}$, the response at channel c for a given transmit event e is thus given by

$$\mathbf{z}_{e,c}[r_x, r_y] = D_{e,c}(\mathbf{x}_{e,c}; \tau_{e,c,r_x,r_y}), \tag{10.4}$$

where $\mathbf{x}_{e,c}$ denotes the received signal for the eth transmit event and cth channel, and $D_{e,c}(\mathbf{x}; \tau)$ migrates $\mathbf{x}_{e,c}$ from time to space on the basis of the geometry-derived delay τ. To achieve high-resolution delays in the discrete domain, Eq. (10.4) is typically implemented using interpolation or fractional delays with polyphase filters. Alternatively, delays can be implemented in the Fourier domain, which, as we shall discuss later, has practical advantages for, e.g., compressed sensing applications [10].

After migrating X to the spatial domain the apodization tensor W is applied, and the result is (coherently) summed across the e (transmit-event) and c (channel) dimensions. The design of the apodization tensor W inherently poses a compromise between the main lobe width and the side lobe intensity, or, equivalently, resolution and contrast and/or clutter. This can be intuitively understood from the far-field Fourier relationship between the beam pattern and array aperture: analogously to filtering in frequency, the

---

[1] In ultrasound imaging a distinction is made between slow-time and fast-time imaging: slow-time refers to a sequence of snapshots (i.e., across multiple transmit–receive events) at the pulse repetition rate, whereas fast-time refers to the time axis of the received RF signal for a given transmit event.

properties of a beamformer (a spatial filter) are dictated by the sampling (aperture pitch), the filter length (aperture size), and the coefficients (apodization weights). Typical choices include Hamming-style apodizations, suppressing side lobes at the expense of a wider main lobe. In commercial systems, the full (depth- and position-dependent) weight tensor is carefully engineered and fine-tuned on the basis of the transducer design (e.g., pitch, size, center frequency, near- and far-field zones) and the imaging application (e.g., cardiac, obstetrics, general imaging, or even intravascular imaging).

10.3.3 Adaptive Beamforming

Adaptive beamforming aims to overcome the inherent trade-off between sidelobe levels and the resolution of static DAS beamforming by making the apodization weight tensor W fully data-adaptive, i.e., $W \triangleq W(X)$. The adaptation of W is based on estimates of the array signal statistics, which are typically calculated either instantaneously on a spatiotemporal block of data or through recursive updates. Note that, in general, the performance of these methods is bounded by the bias and variance of the statistics estimators. The latter can be reduced by using either a larger block of samples (assuming some degree of spatiotemporal stationarity), which comes at the cost of reduced spatiotemporal adaptivity, or techniques such as subarray averaging.

We will now briefly review some typical adaptive beamforming structures (Fig. 10.4). In general, we distinguish between beamformers that act on individual channels and those that use the channel statistics to compute a single weighting factor across all channels, and so are called post-filters.

Minimum Variance (MV)

A popular adaptive beamforming method is the minimum-variance distortionless response (MVDR), or Capon, beamformer. The MVDR beamformer acts on individual channels, with optimal weights $W \in \mathbb{R}^{E \times C \times N_x \times N_y}$ defined as those that, for each transmit event e and location $[r_x, r_y]$, minimize the total signal variance or power while maintaining a distortionless response in the direction or focal point of interest. This amounts to solving

$$\hat{w}_{mv,e}[r_x,r_y] = \arg\min_{w} w^H R_{x_e[r_x,r_y]} w \qquad (10.5)$$
$$\text{s.t. } w^H \mathbf{1} = 1,$$

where $\hat{w}_{mv,e}[r_x,r_y] \in \mathbb{R}^C$ is the weight vector for a given transmit event and location, and $R_{x_e[r_x,r_y]}$ denotes the estimated channel covariance matrix for transmit event e and location $[r_x, r_y]$. Solving Eq. (10.5) requires inversion of the covariance matrix, the number of operations for which grows cubically with the number of array channels. This makes MV beamforming computationally much more demanding than delay-and-sum beamforming, in particular for large arrays. In practice this results in a significantly longer reconstruction time and thereby deprives ultrasound of the interactability that makes it so appealing compared with, e.g., MRI and CT. To boost image quality, eigenspace-based MV beamforming [11] performs an eigendecomposition of the covariance matrix and also subsequent signal subspace selection before inversion. This further increases the computational complexity. While significant progress has been made to decrease the computational time of MV beamforming algorithms [12, 13], real-time implementation remains a major challenge. In addition, MV beamforming relies on accurate estimates of the signal statistics, which (as mentioned above) requires some form of spatiotemporal averaging.

Coherence Factor (CF)

Coherence factor (CF) weighting [14] also applies content-adaptive apodization weights W, however, with a specific structure: the weights across different channels are identical and tied. Coherence factor weighting thus in practice acts as a post-filter after DAS beamforming. The pixel-wise weighting in this post-filter is based on a "coherence factor": the ratio between the coherent and incoherent energies across the array channels. This type of weighting does however suffer from artifacts when the signal-to-noise ratio (SNR) is low and estimation of the coherent energy is challenging [15]. This is particularly problematic for unfocused techniques such as pulsed-wave Doppler or synthetic-aperture imaging.

Wiener

The Wiener beamformer produces a minimum mean squared error (MMSE) estimate of the signal amplitude A stemming from a particular direction or location of interest:

$$\hat{\mathbf{w}}_e[r_x,r_y] = \arg\min_{\mathbf{w}} E(|A - \mathbf{w}^H \mathbf{x}_e[r_x,r_y]|^2). \tag{10.6}$$

The solution is [16]:

$$\hat{\mathbf{w}}_e[r_x,r_y] = \frac{|A|^2}{|A|^2 + (\mathbf{w}_{mv,e}[r_x,r_y])^H R_{n_e[r_x,r_y]} \mathbf{w}_{mv,e}[r_x,r_y]} \mathbf{w}_{mv,e}[r_x,r_y], \tag{10.7}$$

where $R_{n_e[r_x,r_y]}$ is the noise covariance matrix. Wiener beamforming is thus equivalent to a MVDR beamformer followed by a (CF-like) post-filter that scales the output as a function of the remaining noise power after MVDR beamforming (the second term in the denominator). Note that the Wiener beamformer requires estimates of both the signal power and noise covariance matrix. The latter can be estimated, e.g., by assuming i.i.d. white noise, i.e., $R_{n_e[r_x,r_y]} = \sigma_n^2 I$, and calculating σ_n from the mean squared difference between the MVDR beamformed output and the channel signals.

Iterative Maximum-a-Posteriori (iMAP)

Chernyakova et al. proposed an iterative maximum-a-posteriori (iMAP) estimator [17], which formalizes post-filter-based methods as a MAP problem by incorporating a statistical prior for the signal of interest. Assuming a zero-mean Gaussian random variable of interest with variance σ_a^2 that is uncorrelated to the noise (also Gaussian, with variance σ_n^2), the beamformer can be derived as

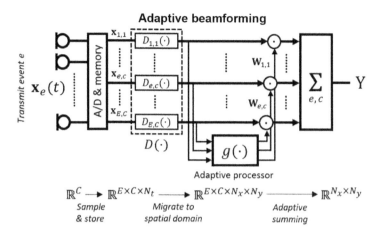

Figure 10.4 Adaptive beamforming.

$$\hat{w}_e[r_x, r_y] = \frac{\sigma_a[r_x, r_y]^2}{M\sigma_a[r_x, r_y]^2 + M\sigma_n[r_x, r_y]^2} \mathbf{1}. \qquad (10.8)$$

The signal and noise variances are estimated in an iterative fashion: first a beamformed output is produced according to Eq. (10.8), and then the noise variance is estimated as the mean squared difference from the individual channels. This process is repeated until a stopping criterion is met. Note that while iMAP performs iterative estimation of the statistics, the beamformer itself has strong similarity to Wiener post-filtering and CF beamforming.

10.4 Deep-Learning Opportunities

In this section we will elaborate on some of the fundamental challenges of ultrasound beamforming and the role that deep-learning solutions can play in overcoming these challenges [9].

10.4.1 Opportunity 1: Improving Image Quality

As we saw in the previous section, classic adaptive beamformers are derived on the basis of specific modeling assumptions and knowledge of the signal statistics. This limits the performance of model-based adaptive beamformers, which are bounded by:

1. The accuracy and precision of the estimated signal statistics obtained using data sampled from a strongly non-stationary ultrasound RF process. In practice, only limited samples are available, making accurate estimation challenging.
2. The adequacy of the simple linear acquisition model (including the use of a homogeneous speed of sound) and the assumptions on the statistical structure of the desired signals and noise (uncorrelated Gaussian). In practice the speed of

sound is heterogeneous and the noise statistics are highly complex and correlated with the signal, e.g., via multiple scattering, which leads to reverberation and haze in the image.

In addition, specifically for MVDR beamforming, the complexity of the required matrix inversions hinders real-time implementation. Deep learning can play an important role in addressing these issues, enabling:

1. Drastic acceleration of the slow model-based approaches such as MVDR, using accelerated neural network implementations as function approximators.
2. High-performance beamforming outputs without explicit estimation of signal statistics, obtained by exploiting useful priors learned from previous examples (the training data).
3. Data-driven nonlinear beamforming architectures that are not bounded by modeling assumptions on linearity and noise statistics. While analytically formalizing the underlying statistical models of ultrasound imaging is highly challenging, and its optimization probably intractable, deep learning circumvents this by learning powerful models directly from the data.

10.4.2 Opportunity 2: Enabling Fast and Robust Compressed Sensing

Beamforming in the digital domain requires sampling the signals received at the transducer elements and transmitting the samples to a back-end processing unit. To achieve sufficient delay resolution for focusing, hundreds of channel signals are typically sampled at 4–10 times their bandwidth, i.e., the sampling rate may severely exceed the Nyquist rate. This problem becomes even more pressing for 3D ultrasound imaging based on matrix transducer technology, where the direct streaming of the channel data from thousands of sensors would lead to data rates of thousands of gigabits per second. Today's technology thus relies on microbeamforming or time-multiplexing to keep data rates manageable. The former compresses data from multiple (adjacent) transducer elements into a single line, thereby in effect reducing the number of receive channels and limiting the attainable resolution and image quality. The latter communicates only a subset of the channel signals to the back-end of the system for every transmit event, yielding reduced frame rates.

To overcome these data-rate challenges without compromising image quality and frame rates, significant research effort has been focused on compressed sensing for ultrasound imaging. Compressed sensing permits low-data-rate sensing (below the Nyquist rate) with strong signal recovery guarantees under specific conditions [18, 19]. In general, one can perform compressed sensing along three axes in ultrasound: (1) fast-time, (2) slow-time, and (3) channels/array elements. We denote an undersampled measured RF data cube as $X_u \in \mathbb{R}^{E_u \times C_u \times N_u}$, with $E_u C_u N_u < ECN$.

Sub-Nyquist Fast-Time Sampling

To perform sampling-rate reduction across fast-time, one can consider the received signals within the framework of the finite rate of innovation (FRI) [19, 20].

Tur *et al.* [21] modeled the received signal at each element as a finite sum of replicas of the transmitted pulse backscattered from reflectors. The replicas are fully described by their unknown amplitudes and delays, which can be recovered from the signals' Fourier series coefficients. The latter can be computed from low-rate samples of the signal using compressed sensing (CS) techniques [18, 19]. In [22, 23], the authors extended this approach and introduced compressed ultrasound beamforming. It was shown that the beamformed signal follows an FRI model and thus can be reconstructed from a linear combination of the Fourier coefficients of the received signals. Moreover, these coefficients can be obtained from low-rate samples of the received signals taken according to the Xampling framework [24–26]. Chernyakova and Eldar showed that this Fourier domain relationship between the beam and the received signals holds irrespective of the FRI model. This leads to a general concept of frequency-domain beamforming (FDBF) [10], which is equivalent to beamforming in time. This technique allows one to sample the received signals at their effective Nyquist rate without assuming a structured model; thus, it avoids the oversampling dictated by the digital implementation of beamforming in time. Assuming that the beam follows an FRI model, the received signals can be sampled at sub-Nyquist rates, leading to an up to 28-fold reduction in sampling rate, i.e., $N_t = 28 N_u$ [27–29].

Channel and Transmit-Event Compression

Significant research effort has also been invested in the exploration of sparse array designs ($C_u < C$) [30, 31] and the efficient sparse sampling of transmit events ($E_u < E$). It has been shown that with proper sparse array selection and a process called convolutional beamforming, the beam pattern can be preserved using far fewer elements than the standard uniform linear array [32]. Typical designs include sparse periodic arrays [33] or fractal arrays [34]. In [35] a randomly subsampled set of receive transducer elements was used, and the authors of [36] propose learned dictionaries for improved CS-based reconstruction from subsampled RF lines. In [37], the authors used deep learning to optimize channel selection and slow-time sampling for B-mode and downstream color-Doppler processing, respectively. Reduction in both time sampling and spatial sampling can be achieved by compressed Fourier-domain convolutional beamforming, leading to reductions in the data of two orders of magnitude [38].

Beamforming and Image Recovery after Compression

After compressive acquisition, dedicated signal recovery algorithms are used to perform image reconstruction from the undersampled dataset. Before deep learning became popular, these algorithms relied on priors or regularizers (e.g., on sparsity in some domain) to solve the typically ill-posed optimization problem in a model-based (iterative) fashion. They assume knowledge about the measurement point spread function (PSF) and other system parameters. However, as mentioned before, the performance of model-based algorithms is bounded by the accuracy of the modeling assumptions, including the acquisition model and statistical priors. In addition, iterative solvers are time-consuming, hampering real-time implementation. Today, deep learning is increasingly used to overcome these challenges [39, 40]: (1) deep learning can be used to learn complex statistical models (explicitly or implicitly) directly from training data, (2) neural networks can serve as powerful function approximators that accelerate iterative model-based implementations.

10.4.3 Opportunity 3: Beyond MMSE with Task-Adaptive Beamforming

Classically, beamforming is posed as a signal recovery problem under spatially white Gaussian noise. In that context, optimal beamforming is defined as beamforming that best estimates the signal of interest in a MMSE sense. However, beamforming is rarely the last step in the processing chain of ultrasound systems. It is typically followed by demodulation (envelope detection), further image enhancement, spatiotemporal processing (e.g., motion estimation), and image analysis. In that regard it may be more meaningful to define optimality of the beamformer with respect to its downstream task. We refer to this as task-adaptive beamforming. We can define several such tasks in ultrasound imaging. First, we have tasks that focus on further enhancement of the images after beamforming. This includes downstream processing for, e.g., de-speckling, de-convolution, or super-resolution, which all have different needs and requirements from the beamformer output. For instance, the performance of deconvolution or super-resolution algorithms is for a large part determined by the invertibility of the point-spread-function model, shaped by the beamformer. Beyond image enhancement, one can think of motion estimation tasks (requiring temporal consistency of the speckle patterns with a clear spatial signature that can be tracked) or, even further downstream, applications such as segmentation or computer-aided diagnosis (CAD).

One may wonder how to optimize beamforming for such tasks in practice. Fortunately, today many of these downstream processing tasks are performed using convolutional neural networks or derivatives therefrom. If the downstream processor is indeed a neural architecture or another algorithm through which one can easily backpropagate gradients, one can directly optimize a neural-network beamformer with respect to its downstream objective or loss through deep-learning methods. In this case, *deep* not only refers to the layers in individual networks, but also to the stack of beamforming and downstream neural networks. If backpropagation is nontrivial, one could resort to Monte-Carlo gradient estimators based on, e.g., the REINFORCE estimator [41] or more generally through reinforcement learning algorithms [42].

10.4.4 A Brief Overview of the State-of-the-Art

The above opportunities have spurred an ever-growing collection of papers from the research community. In Table 10.1, we provide an overview of a selection of these in the context of these opportunities and the challenges they aim to address. Many of these papers simultaneously address more than one challenge or opportunity.

Table 10.1. Overview of some current literature and its main focus in terms of the opportunities defined in Section 10.4.

References	Real-time high image quality (main goals)	Compressed sensing (subsampling axis)	Task-adaptive (considered tasks)
	Opportunity and focus		
[43]	Reduce off-axis scattering	—	—
[44]	—	Transmit and channels	—
[45]	—	Channels	—
[46]	Boost resolution or suppress speckle	—	Deconvolution and speckle reduction
[47]	Boost contrast and resolution	Transmit and channels	—
[48]	—	—	Segmentation
[49]	Accelerate coherence imaging	—	—
[50] and [51]	—	Channels and fast-time Fourier coefficients	—
[37]	—	Channels and slow-time	Doppler and B-mode
[52]	Suppress speckle	—	Speckle reduction

10.4.5 Public Datasets and Open Source Code

To support the development of deep-learning-based solutions in the context of these opportunities, a challenge on ultrasound beamforming by deep learning (CUBDL) was organized. For public raw ultrasound channel datasets as well as open source code we refer the reader to the challenge website [53] and a paper describing the datasets, methods and tools [54].

10.5 Deep-Learning Architectures for Ultrasound Beamforming

10.5.1 Overview and Common Architectural Choices

Having set the scope and defined opportunities, we now turn to some of the most common implementations of deep learning in ultrasound beamforming architectures. As in computer vision, most neural architectures for ultrasound beamforming are based on 2D convolutional building blocks. With that, they thus rely on translational equivariance or spatial symmetry of the input data. It is worth noting that this symmetry only holds to a certain extent for ultrasound imaging, as its point-spread function in fact changes as a function of location. When operating on raw channel data before time–space migration (the time-of-flight correction), these effects become even more pronounced. Most architectures also restrict the receptive field of the neural network, i.e., the beamforming outputs for a given spatial location (line or pixel) are computed using on a selected subset of X. This is either implicit, through the depth and size of the selected convolutional kernels, or explicit, by for example providing only a selected number of depth slices [45] as an input to the network.

In the following we will discuss a number of architectures. We explicitly specify their input, architectural design choices, and output to clarify on which part of the beamforming chain they are acting on and how they are acting. We point out that the discussion below is not exhaustive but examples have been selected to illustrate and cover the spectrum of approaches and design choices from an educational perspective.

10.5.2 DNN Directly on Channel Data

In [48], deep learning was used directly on the raw channel data. The DNN thus has to learn both the geometry-based time–space migration (the TOF correction) as well as the subsequent beam-summing of channels to yield a beamformed image. In particular, the former consideration makes this task particularly challenging - the processor is not a typical image-to-image mapping but rather a time–space migration. In addition to yielding a beamformed output, the network in parallel also provides a segmentation mask, which is subsequently used to enhance the final image by masking regions in the image that are classified as anechoic:

DNN directly on channel data, by Nair *et al.* [48]

Acquisition type: Single plane wave imaging.
Input: Complex IQ demodulated data cube $X_{in} \in \mathbb{C}^{1 \times C \times N_t}$, reformatted to real inputs $X_{in} \in \mathbb{R}^{C \times N_t \times 2}$ before being fed to the network.
Architecture: U-net variant consisting of a single VGG encoder and two parallel decoders with skip connections that map to B-mode and segmentation outputs, respectively. The encoder has 10 convolutional layers (kernel size = 3×3) with batch normalization and downsamples the spatial domain via 2×2 max-pooling layers while simultaneously increasing the number of feature channels. The two decoders both comprise nine convolutional layers and perform spatial upsampling to map the feature space back to the desired spatial domain.
Output: RF B-mode data and pixel-wise class probabilities (segmentation), for the full image $Y_{bf} \in \mathbb{R}^{N_x \times N_y}$ and for $Y_{seg} \in \mathbb{R}^{N_x \times N_y \times 1}$, respectively.

10.5.3 DNN for Beam-Summing

Hybrid Architectures: Geometry and Learning

While the work in [48] replaces the entire beamformer by a DNN, most of today's embodiments of ultrasound beamforming with deep learning use post-delayed channel

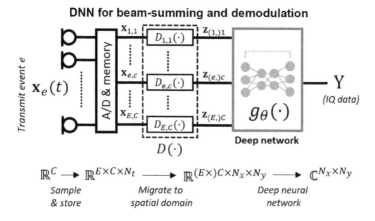

Figure 10.5 DNN replacing the beam-summing processor.

data. That is, the time–space migration is a deterministic pre-processing step that relies on geometry-based TOF correction. This holds for all the specific architectures that we will cover in what follows. In that sense, they are all hybrid model-based and data-driven beamforming architectures.

Learning Improved Beam-Summing

We will now discuss designs that replace only the beam-summing stage by a deep network, i.e., after TOF correction, as illustrated in Fig. 10.5. In [44–46, 55, 56], the authors apply deep convolutional neural networks to perform improved channel aggregation and beam-summing after TOF correction:

> **DNN beam-summing, by Khan et al. [45]**
>
> **Acquisition type**: Line scanning, so the first dimension (transmit events or lines E) has been mapped to the lateral dimension N_y during the time–space migration and TOF correction.
> **Input**: For each depth or axial location, a TOF-corrected RF data cube $Z_{in} \in \mathbb{R}^{C \times 3 \times N_y}$. The input data cube comprises a stack of three axial slices centered around the axial location of interest.
> **Architecture**: Thirty-seven convolutional layers (kernel size = 3 × 3), of which all but the last have batch normalization and ReLU activations.
> **Output**: IQ data for each axial location $\mathbf{y}_{n_x} \in \mathbb{R}^{2 \times 1 \times N_y}$, where the first dimension contains the in-phase (I) and quadrature (Q) components, so that we can also define $\mathbf{y}_{n_x} \in \mathbb{C}^{N_x \times 1}$.

Kessler and Eldar [50] used a similar strategy, except that pre-processing TOF correction was performed in the Fourier domain. This allows efficient processing when sensing only a small set of the Fourier coefficients of the receive channel data. The authors then used a deep convolutional network to perform beam-summing, mapping the TOF-corrected channel data into a single beamformed RF image without aliasing artifacts:

> **DNN beam-summing by Kessler and Eldar [50]**
>
> **Acquisition type**: Phased array line scanning, so the first dimension (transmit events or angular lines E) was mapped to the lateral dimension N_y during the time–space migration/TOF correction. Reconstruction is in the polar domain, i.e., N_x refers to radial position and N_y to angular position.
> **Input**: TOF-corrected RF data cube $Z_{in} \in \mathbb{R}^{C \times N_x \times 3}$. The input data cube comprises data corresponding to three angles centered around the angular position of interest.
> **Architecture**: U-net variant with three contracting blocks and three expanding blocks of convolutional layers with parametric ReLU (PReLU) activations.
> **Output**: RF data for each angle of interest $\mathbf{y}_{n_y} \in \mathbb{R}^{N_x \times 1}$.

10.5.4 DNN as an Adaptive Processor

The architecture that we will discuss now was inspired by the MV beamforming architecture. Instead of replacing the beamforming process entirely, the authors in [47, 57] proposed using a deep network as an artificial agent that calculates the optimal apodization weights W on the fly, given the received pre-delayed channel signals at the array $D(X)$. See Fig. 10.6 for an illustration. By only replacing this bottleneck component in the MVDR beamformer, and constraining the problem further by promoting close-to-distortionless response during training (i.e., $\Sigma_c w_c \approx 1$), this solution is highly data-efficient, interpretable, and has the ability to learn powerful models from only a few images [47]:

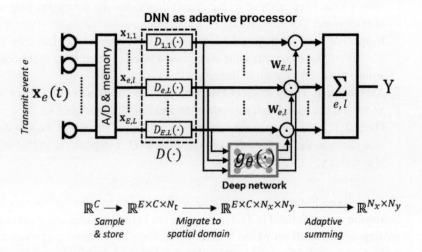

Figure 10.6 DNN replacing the adaptive processor of classical adaptive beamforming methods.

> **DNN as adaptive processor, by Luijten *et al.* [47, 57]**
>
> **Acquisition types**: Single plane wave imaging and synthetic aperture (intravascular ultrasound). For the latter, a virtual aperture is constructed by combining the received signals of multiple transmits and receives.
> **Input**: TOF-corrected RF data cube $Z \in \mathbb{R}^{1 \times C \times N_x \times N_y}$.
> **Architecture**: Four convolutional layers comprising 128 nodes for the input and output layers and 32 nodes for the hidden layers. The kernel size of the filters is 1×1, making the receptive field of the network a single pixel. In practice, this is thus a per-pixel fully connected layer across the array channels. The activation functions are antirectifiers [58], which, unlike ReLUs, preserve both the positive and negative signal components at the expense of a dimensionality increase.
> **Output**: An array apodization tensor $\mathbf{W} \in \mathbb{R}^{1 \times C \times N_x \times N_y}$, which is subsequently multiplied (element wise) with the network inputs Z to yield a beamformed output Y.

Complexity, Inference Speed, and Stability

Since pixels are processed independently by the network, a large amount of training data is available per acquisition. Inference is fast and real-time rates are achievable on a GPU-accelerated system. For an array of 128 elements, the adaptive calculation of a set of apodization weights through MV beamforming requires $> N^3 (= 2\,097\,152)$ floating point operations (FLOPS), while the deep-learning architecture requires only 74 656 FLOPS [47], in practice leading to a more than 400 times speed-up in reconstruction time. Compared with MV beamforming, the deep network is qualitatively more robust, with less observed artifactual reconstructions that stem from, e.g., unstable computations of the inverse autocorrelation estimates in MV.

10.5.5 DNN for Fourier-Domain Beam-Summing

Several model-based beamforming methods process ultrasound channel data in the frequency domain [59, 60]. In this spirit, the authors of [43, 61, 62] used deep networks to perform wideband beam-summing by processing individual discrete Fourier transform (DFT) bins of the TOF-corrected and axially windowed RF signals. Each DFT bin is processed using a distinct neural network. After processing in the Fourier domain, the channel signals are summed for each window and a beamformed RF scanline is reconstructed using an inverse short-time Fourier transform (see Fig. 10.7):

> **DNN for Fourier-domain beam-summing by Luchies and Byram [43, 61, 62]**
>
> **Acquisition type**: Line scanning, where each transmit event produces one lateral scanline. The first dimension of $\mathbf{X} \in \mathbb{R}^{E \times C \times N_t}$ is thus directly mapped to the lateral dimension N_y in the TOF-correction step: $Z = D(X) \in \mathbb{R}^{C \times N_x \times N_y}$.
> **Input**: Fourier transform of an axially windowed (window length S) and TOF-corrected RF data cube for a single scanline, i.e., $\tilde{Z}_{n_x,n_y} = \mathcal{F}(Z_{n_x,n_y}) \in \mathbb{C}^{C \times S \times 1}$,

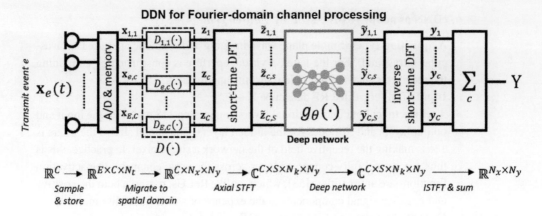

Figure 10.7 DNN for Fourier-domain beam-summing.

with $Z_{n_x,n_y} \in \mathbb{R}^{C \times L_x \times 1}$. Before being fed to the network, $\tilde{Z}_{n_x,n_y}$ is converted to real values by stacking the real and imaginary components, yielding $\tilde{Z}_{n_x,n_y} \in \mathbb{R}^{2C \times S \times 1}$.

Architecture: S identical fully connected neural networks, one for each DFT bin. Each neural network of this stack thus takes the $2C$ channel values corresponding to that bin as its input. The S networks all have five fully-connected layers; the hidden layers having 170 neurons and ReLU activations. Each network then returns $2C$ channel values, the real and imaginary components of that frequency bin after processing.

Output: Processed Fourier components of the axially windowed and TOF-corrected RF data cube for a single scanline: $\tilde{Y}_{n_x,n_y} \in \mathbb{C}^{C \times S \times 1}$. To obtain a beamformed image, the C channels are summed and an inverse short-time Fourier transform is used to compound the responses of all axial windows.

10.5.6 Post-Filtering after Beam-Summing

We will now discuss some post-filter approaches, i.e., methods applied after channel beam-summing but before envelope detection and brightness compression. Several post-filtering methods have been proposed for compounding beamsummed RF outputs from multiple transmit events e with some spatial overlap (e.g., multiple plane or diverging waves). Traditionally, multiple transmit events are compounded by coherent summing (i.e., after transmit delay compensation). Today, deep learning is increasingly used to replace the coherent summing step. In [63], the authors performed neural network compounding from a small number of transmits, and trained towards an image obtained by coherently summing a much larger number of transmit events. In [64], the authors posited compounding as an inverse problem, which they subsequently solved using a model-based deep network inspired by proximal gradient methods. Post-filtering has also been used to, e.g., remove aliasing artifacts due to sub-Nyquist sampling on beamformed 1D RF lines [51]. The most common method for performing sparse recovery and solving the L1 minimization problem is by the use of compressed sensing algorithms such as ISTA and NESTA [19]. However, these algorithms typically suffer from high computational load and do not always ensure high-quality recovery. Here we discuss the process of unfolding an iterative algorithm as the layers of a deep network for sparse recovery. The authors of [51] built an efficient network by unfolding the ISTA algorithm for sparse recovery, on the basis of the previously suggested LISTA [65]. Using their technique they recovered both spatially and temporally sub-Nyquist sampled ultrasound data, after delaying it in the frequency domain, using a simple, computationally efficient, and interpretable deep network:

DNN as a recovery method by Mamistvalov and Eldar [51]

Acquisition type: Line scanning, so the first dimension (transmit events or lines E) is mapped to the lateral dimension N_y during the time–space migration/TOF correction.

Input: Frequency-domain delayed and summed data (or frequency-domain convolutionally beamformed data), after appropriate inverse Fourier transform and appropriate zero padding to maintain the desired temporal resolution. The input data is a vector, $Z_{in} \in \mathbb{R}^{N_{st} \times 1}$, where N_{st} is the traditionally used number of samples for beamforming. The recovery is done for each image line separately.

Architecture: Simple architecture of an unfolded ISTA algorithm, consisting of 30 layers, each including two convolutional layers that mimic the matrix multiplications of ISTA and one soft-thresholding layer. One last convolutional layer is added to recover the actual beamformed signal from the recovered sparse code.

Output: Beamformed signal for each image line without artifacts caused by sub-Nyquist sampling, $Z_{out} \in \mathbb{R}^{N_{st} \times 1}$.

10.6 Training Strategies and Data

10.6.1 Training Data

The model parameters of the above beamforming networks are optimized using training data that consist of simulations, or *in vitro/in vivo* data, or a combination thereof. We will now discuss some strategies for selecting training data, and in particular generating useful training targets.

Simulations

Training ultrasound beamformers using simulated data is appealing, since various ultrasound simulation toolboxes, such as Field II [66], *k*-wave [67], and the Matlab

Ultrasound Toolbox (MUST),[2] allow for the flexible generation of input-target training data. Simulations can be used to generate pairs of RF data, each pair comprising a realistic imaging mode based on the actual hardware and probe (the input), and a second mode of the same scene with various more desirable properties (the target). One can get creative with the latter, and we here list a number of popular approaches found in the literature:

1. Target imaging mode with unrealistic, yet desired, array and hardware configuration:
    a. higher frequencies/shorter wavelengths (without increased absorption) to improve target resolution [64];
    b. larger array aperture to improve target resolution [56].
2. Removal of undesired imaging effects such as off-axis scattering from target RF data [61].
3. Targets constructed directly from a simulation scene or object:
    a. point targets on a high-resolution simulation grid [68];
    b. masks of medium properties, e.g., anechoic region segmentations [48].

When relying solely on simulations, one has to be careful to avoid a catastrophic domain shift when deploying the neural models on real data. Increasing the realism of the simulations, mixing simulations with real data, using domain-adaptation methods, or limiting the neural networks' receptive field can help combat domain-shift issues.

Real Data

Training targets from real acquisitions are typically based on high-quality (yet computationally complex) model-based solutions such as MV beamforming or extended or full acquisitions in a compressed sensing setup. In the former, training targets are generated offline by running powerful but time-consuming model-based beamformers on the training dataset of RF inputs. The goal of deep-learning-based beamforming is then to achieve the same performance as these model-based solutions, but at much faster inference rates. In the compressed sensing setup, training targets are generated by (DAS/MV) beamforming of the full (not compressed) set of RF measurements X. In this case, the objective of a deep-learning beamformer is to reproduce these beamformed outputs on the basis of compressed or undersampled measurements X_u. As discussed in Section 10.4, compression can entail fast-time sub-Nyquist sampling, imaging with sparse arrays, or limiting the number of transmit events in, e.g., plane-wave compounding. Real data is available on the aforementioned CUBDL Challenge website [53].

10.6.2 Loss Functions and Optimization

In this subsection, we will discuss typical loss functions used to train deep networks for ultrasound beamforming. Most networks are trained by directly optimizing the

---

[2] https://www.biomecardio.com/MUST/, by Damien Garcia.

loss on the beamformer output in the RF or IQ domain. Others indirectly optimize the beamformer output in a task-adaptive fashion by optimizing some downstream loss after additional processing. For training, some variant of stochastic gradient descent (SGD), often adaptive moment estimation (ADAM), is used, and some form of learning-rate decay is also common. Stochastic gradient descent operates on mini batches, which comprise either full input–output image pairs or some collection of patches or slices or cubes extracted from full images. In the following, we will (without loss of generality) use $Y^{(i)}$ and $Y_t^{(i)}$ to refer to respectively the network outputs and the targets for a sample i.

Loss Functions for Beamformed Outputs

Considering image reconstruction as a pixel wise regression problem under a Gaussian likelihood model, perhaps the most commonly used loss function is the MSE (or ℓ_2 norm) with respect to the target pixel values:

$$\mathcal{L}_{MSE} = \frac{1}{I} \sum_{i=0}^{I-1} \left\| Y^{(i)} - Y_t^{(i)} \right\|_2^2. \tag{10.9}$$

If one would like to penalize strong deviations less stringently, e.g., to be less sensitive to outliers, one can consider a likelihood model that decays less strongly for large deviations such as the Laplace distribution. Under that model, the negative log-likelihood loss function is the mean absolute error (or ℓ_1 norm):

$$\mathcal{L}_{l1} = \frac{1}{I} \sum_{i=0}^{I-1} \left\| Y^{(i)} - Y_t^{(i)} \right\|_1. \tag{10.10}$$

A commonly adopted variant of the MSE loss is the signed-mean-squared logarithmic error (SMSLE), proposed by Luijten et al. [57]. This metric compresses the large dynamic range of backscattered ultrasound RF signals to promote accurate reconstructions across the entire dynamic range:

$$\mathcal{L}_{SMSLE} = \frac{1}{2I} \sum_{i=0}^{I-1} \left\| \log_{10} \left(Y^{(i)} \right)^+ - \log_{10} \left(Y_t^{(i)} \right)^+ \right\|_2^2 \tag{10.11}$$

$$+ \left\| \log_{10} \left(Y^{(i)} \right)^- - \log_{10} \left(Y_t^{(i)} \right)^- \right\|_2^2, \tag{10.12}$$

where $(\cdot)^+$ and $(\cdot)^-$ yield the magnitudes of the positive and negative parts, respectively. Thus far, we have only covered pixel wise losses that consider every pixel as an independent sample. These losses have no notion of spatial context and do not measure structural deviations. The structural similarity index (SSIM) [69] aims to quantify the perceived changes in structural information, luminance, and contrast. Similarly to the SMSLE approach, Kessler and Eldar [50] proposed an SSIM loss for ultrasound beamforming that acts on the log-compressed positive and negative parts of the beamformed RF signals:

$$\mathcal{L}_{SSIM} = \frac{1}{2I} \sum_{i=0}^{I-1} \left(1 - SSIM\left(\log_{10}\left(Y^{(i)}\right)^{+}, \log_{10}\left(Y_t^{(i)}\right)^{+} \right) \right) \quad (10.13)$$

$$+ \left(1 - SSIM\left(\log_{10}\left(Y^{(i)}\right)^{-}, \log_{10}\left(Y_t^{(i)}\right)^{-} \right) \right), \quad (10.14)$$

where, when luminance, contrast, and structure are weighted equally, $SSIM$ is defined as:

$$SSIM(a,b) = \frac{(2\mu_a \mu_b + \epsilon_1)(2\sigma_{ab} + \epsilon_2)}{(\mu_a^2 + \mu_b^2 + \epsilon_1)(\sigma_a^2 + \sigma_b^2 + \epsilon_2)}, \quad (10.15)$$

with μ_a, μ_b, σ_a, σ_b, and σ_{ab} the means, standard deviations, and cross-correlation of a and b; ϵ_1, ϵ_2 are small constants needed to stabilize the division.

Beyond distance measurements between pixel values, some authors make use of specific adversarial optimization schemes that aim to match the distributions of the targets and generated outputs [64]. These approaches make use of a second neural network, the adversary or discriminator, that is trained to discriminate between images that are drawn from the distribution of targets and those that are generated by the beamforming neural network. This is achieved by minimizing the binary cross-entropy classification loss between the predictions of the neural network and the labels (target or generated), evaluated on batches that contain both target images and generated images. At the same time, the beamforming network is trained to maximize this loss, thereby attempting to fool the discriminator. The rationale here is that the probability distributions of target images and beamformed network outputs match (or strongly overlap) whenever this neural discriminator can no longer distinguish images from either distribution. The beamforming network and discriminator thus play a min–max game, expressed by the following optimization problem for the neural network parameters estimated across the training data distribution $P_\mathcal{D}$:

$$\hat{\theta}, \hat{\Psi} = \operatorname*{argmin}_{\psi} \operatorname*{argmax}_{\theta} \left\{ -\mathbb{E}_{(X, Y_t) \sim P_\mathcal{D}} \left[\log(D_\psi(Y_t)) + \log(1 - D_\psi(f_\theta(X))) \right] \right\},$$
(10.16)

where θ represents the parameters of the beamforming network $f_\theta(\cdot)$ and ψ represents the parameters of the discriminator D_ψ. It is important to realize that merely matching distributions does not guarantee accurate image reconstructions. That is why adversarial losses are often applied on input–output and input–target pairs (matching, e.g., their joint distributions) or used in combination with additional distance metrics such as those discussed earlier in this section. The relative contributions or weighting of these individual loss terms is typically selected empirically.

Task-Adaptive Optimization

As discussed in Section 10.4, one can also optimize the parameters of beamforming architectures using a downstream task-based loss, i.e.,

$$\hat{\theta}, \hat{\phi} = \operatorname*{argmin}_{\theta} \left\{ \mathbb{E}_{(X, s_{task}) \sim P_\mathcal{D}} \left[\mathcal{L}_{task} \left\{ g_\phi \left(f_\theta(X) \right), s_{task} \right\} \right] \right\}, \quad (10.17)$$

where s_t denotes some target task, $\mathcal{L}_{task}\{a,b\}$ is a task-specific loss function between outputs a and targets b, and ϕ are the parameters of the task function g_ψ, which can be a neural network. Examples of such tasks include segmentation [48], for which $\mathcal{L}_{task}$ could be a dice loss, or motion estimation (Doppler), for which $\mathcal{L}_{task}$ could be an MSE penalty [37].

10.7 New Research Opportunities

10.7.1 Multi-Functional Deep Beamformer

Although deep beamformers provide impressive performance and ultra-fast reconstruction, one of the downsides of the deep beamformers is that a distinct model is needed for each type of desired output. For instance, to obtain DAS outputs, a model is needed which mimics DAS; similarly, for minimum-variance-based beamforming (MVBF) a separate model is needed. Although the architecture of the model could be the same, separate weights need to be stored for each output type. Given that hundreds/thousands of B-mode optimizations/settings are used in the current high-end commercial systems, one may wonder whether we need to store thousands of deep models in the scanner to deal with various B-mode settings.

To address this issue, the authors of [46] recently proposed a *switchable* deep beamformer architecture using adaptive instance normalization (AdaIN) layers, as shown in Fig. 10.8. Specifically, AdaIN was originally proposed as an image style-transfer method, in which the mean and variance of the feature vectors are replaced by those of the style reference image [70]. Suppose that a multi-channel feature tensor at a specific layer is represented by

$$X = \begin{bmatrix} x_1 & \cdots & x_C \end{bmatrix} \in \mathbb{R}^{HW \times C}, \tag{10.18}$$

where C is the number of channels in the feature tensor X and $x_i \in \mathbb{R}^{HW \times 1}$ refers to the ith column vector of X, which represents the vectorized feature map of size of $H \times W$ at the ith channel. Then, AdaIN [70] converts the feature data at each channel using the following transform:

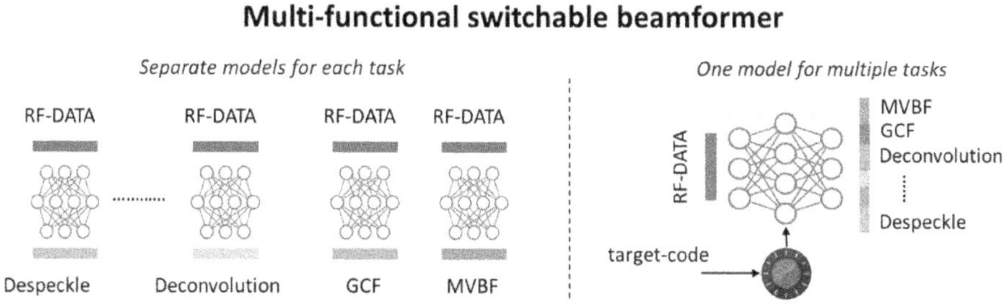

Figure 10.8 A switchable deep beamformer that uses an AdaIN layer.

$$z_i = \mathcal{T}(\boldsymbol{x}_i, \boldsymbol{y}_i), \quad i = 1, \ldots, C \tag{10.19}$$

where

$$\mathcal{T}(\boldsymbol{x}, \boldsymbol{y}) := \frac{\sigma(\boldsymbol{y})}{\sigma(\boldsymbol{x})} (\boldsymbol{x} - m(\boldsymbol{x})\mathbf{1}) + m(\boldsymbol{y})\mathbf{1}, \tag{10.20}$$

where $\mathbf{1} \in \mathbb{R}^{HW}$ is the HW-dimensional vector composed of ones, and $m(\boldsymbol{x})$ and $\sigma(\boldsymbol{x})$ are the mean and standard deviation of $\boldsymbol{x} \in \mathbb{R}^{HW}$; $m(\boldsymbol{y})$ and $\sigma(\boldsymbol{y})$ refer to the target-style-domain mean and standard deviation, respectively. Equation (10.20) implies that the mean and variance of the feature in the input image are normalized in such a way that they match the mean and variance of the style image feature. Although Eq. (10.20) looks heuristic, it was shown that this transform is closely related to the optimal transport between two Gaussian probability distributions [71, 72].

Inspired by this, in [46] it was demonstrated that a *single* deep beamformer with AdaIN layers can learn target images from various styles. Here, a "style" refers to a specific output processing, such as DAS, MVBF, deconvolution image, despeckled images, etc. Once the network is trained, the deep beamformer can then generate various style outputs by simply changing the AdaIN code. Furthermore, the AdaIN code generation is easily performed with a very light AdaIN code generator, so the additional memory overhead at the training step is minimal. Once the neural network is trained, we need only the AdaIN codes, without the generator, which makes the system even simpler.

10.7.2 Unsupervised Learning

As discussed before, most existing deep-learning strategies for ultrasound beamforming are based on supervised learning, thus relying predominantly on paired input–target datasets. However, in many real-world imaging situations, access to paired images (input channel data and a corresponding desired output image) is not possible. For example, to improve the visual quality of ultrasound images acquired using a low-cost imaging system we need to scan exactly the same field of view using a high-end machine, which is nontrivial. For denoising or artifact removal, the actual ground truth is not known *in vivo*, so supervised learning approaches are typically left with simulation datasets for training. This challenge has spurred a growing interest in developing an unsupervised learning strategy where channel data from low-end system or artifact corruption can be used as inputs, using surrogate performance metrics (based on high-quality images from different machines and imaging conditions or statistical properties) to train networks.

One possible approach for addressing this problem is to adopt unpaired style-transfer strategies based on, e.g., cycleGANs – a technique that has shown itself successful for many image-domain quality improvements, and also in ultrasound. For example, the authors in [73, 74] employed such a cycleGAN to improve the image quality from that of a portable ultrasound images using high-end unmatched image data. In general, approaches that drive training by matching distribution properties (e.g., through discriminator networks as in cycleGAN) rather than strict input–output pairs hold promise for such applications.

References

[1] T. L. Szabo, *Diagnostic Ultrasound Imaging: Inside Out*. Academic Press, 2004.

[2] J. M. Baran and J. G. Webster, "Design of low-cost portable ultrasound systems," in *Proc. Annual International Conference of the IEEE Engineering in Medicine and Biology Society*. IEEE, 2009, pp. 792–795.

[3] J. Provost, C. Papadacci, J. E. Arango, M. Imbault, M. Fink, J.-L. Gennisson, M. Tanter, and M. Pernot, "3D ultrafast ultrasound imaging in vivo," *Physics in Medicine and Biology*, vol. 59, no. 19, p. L1, 2014.

[4] M. Tanter and M. Fink, "Ultrafast imaging in biomedical ultrasound," *IEEE Transactions on Ultrasonics, Ferroelectrics, and Frequency Control*, vol. 61, no. 1, pp. 102–119, 2014.

[5] J. Bercoff, M. Tanter, and M. Fink, "Supersonic shear imaging: A new technique for soft tissue elasticity mapping," *IEEE Transactions on Ultrasonics, Ferroelectrics, and Frequency Control*, vol. 51, no. 4, pp. 396–409, 2004.

[6] C. Demené, T. Deffieux, M. Pernot, B.-F. Osmanski, V. Biran, J.-L. Gennisson, L.-A. Sieu, A. Bergel, S. Franqui, J.-M. Correas et al., "Spatiotemporal clutter filtering of ultrafast ultrasound data highly increases Doppler and ultrasound sensitivity," *IEEE Transactions on Medical imaging*, vol. 34, no. 11, pp. 2271–2285, 2015.

[7] C. Errico, J. Pierre, S. Pezet, Y. Desailly, Z. Lenkei, O. Couture, and M. Tanter, "Ultrafast ultrasound localization microscopy for deep super-resolution vascular imaging," *Nature*, vol. 527, no. 7579, p. 499, 2015.

[8] K. Christensen-Jeffries, O. Couture, P. A. Dayton, Y. C. Eldar, K. Hynynen, F. Kiessling, M. O'Reilly, G. F. Pinton, G. Schmitz, M.-X. Tang et al., "Super-resolution ultrasound imaging," *Ultrasound in Medicine and Biology*, vol. 46, no. 4, pp. 865–891, 2020.

[9] R. J. Van Sloun, R. Cohen, and Y. C. Eldar, "Deep learning in ultrasound imaging," *Proceedings of the IEEE*, vol. 108, no. 1, pp. 11–29, 2019.

[10] T. Chernyakova and Y. Eldar, "Fourier-domain beamforming: The path to compressed ultrasound imaging," *IEEE Transactions on Ultrasonics, Ferroelectrics, and Frequency Control*, vol. 61, no. 8, pp. 1252–1267, 2014.

[11] B. M. Asl and A. Mahloojifar, "Eigenspace-based minimum variance beamforming applied to medical ultrasound imaging," *IEEE Transactions on Ultrasonics, Ferroelectrics, and Frequency Control*, vol. 57, no. 11, pp. 2381–2390, 2010.

[12] K. Kim, S. Park, J. Kim, S. Park, and M. Bae, "A fast minimum variance beamforming method using principal component analysis," *IEEE Transactions on Ultrasonics, Ferroelectrics, and Frequency Control*, vol. 61, no. 6, pp. 930–945, June 2014.

[13] M. Bae, S. B. Park, and S. J. Kwon, "Fast minimum variance beamforming based on legendre polynomials," *IEEE Transactions on Ultrasonics, Ferroelectrics, and Frequency Control*, vol. 63, no. 9, pp. 1422–1431, 2016.

[14] R. Mallart and M. Fink, "Adaptive focusing in scattering media through sound-speed inhomogeneities: The van Cittert Zernike approach and focusing criterion," *Journal of the Acoustical Society of America*, vol. 96, no. 6, pp. 3721–3732, 1994.

[15] C. C. Nilsen and S. Holm, "Wiener beamforming and the coherence factor in ultrasound imaging," *IEEE Transactions on Ultrasonics, Ferroelectrics, and Frequency Control*, vol. 57, no. 6, pp. 1329–1346, June 2010.

[16] C.-I. C. Nilsen and S. Holm, "Wiener beamforming and the coherence factor in ultrasound imaging," *IEEE Transactions on Ultrasonics, Ferroelectrics, and Frequency Control*, vol. 57, no. 6, pp. 1329–1346, 2010.

[17] T. Chernyakova, D. Cohen, M. Shoham, and Y. C. Eldar, "Imap beamforming for high quality high frame rate imaging," *IEEE Transactions on Ultrasonics, Ferroelectrics, and Frequency Control*, vol. 66, no. 12, pp. 1830–1844, 2019.

[18] Y. C. Eldar and G. Kutyniok, *Compressed Sensing: Theory and Applications*. Cambridge University Press, 2012.

[19] Y. C. Eldar, *Sampling Theory: Beyond Bandlimited Systems*. Cambridge University Press, 2015.

[20] K. Gedalyahu, R. Tur, and Y. C. Eldar, "Multichannel sampling of pulse streams at the rate of innovation," *IEEE Transactions on Signal Processing*, vol. 59, no. 4, pp. 1491–1504, 2011.

[21] R. Tur, Y. C. Eldar, and Z. Friedman, "Innovation rate sampling of pulse streams with application to ultrasound imaging," *IEEE Transactions on Signal Processing*, vol. 59, no. 4, pp. 1827–1842, 2011.

[22] N. Wagner, Y. C. Eldar, A. Feuer, G. Danin, and Z. Friedman, "Xampling in ultrasound imaging," in *Medical Imaging 2011: Ultrasonic Imaging, Tomography, and Therapy*, vol. 7968. International Society for Optics and Photonics, 2011, p. 796 818.

[23] N. Wagner, Y. C. Eldar, and Z. Friedman, "Compressed beamforming in ultrasound imaging," *IEEE Transactions on Signal Processing*, vol. 60, no. 9, pp. 4643–4657, 2012.

[24] M. Mishali, Y. C. Eldar, and A. J. Elron, "Xampling: Signal acquisition and processing in union of subspaces," *IEEE Transactions on Signal Processing*, vol. 59, no. 10, pp. 4719–4734, 2011.

[25] M. Mishali, Y. C. Eldar, O. Dounaevsky, and E. Shoshan, "Xampling: Analog to digital at sub-Nyquist rates," *IET Circuits, Devices and Systems*, vol. 5, no. 1, pp. 8–20, 2011.

[26] T. Michaeli and Y. C. Eldar, "Xampling at the rate of innovation," *IEEE Transactions on Signal Processing*, vol. 60, no. 3, pp. 1121–1133, 2012.

[27] T. Chernyakova, R. Cohen, R. Mulayoff, Y. Sde-Chen, C. Fraschini, J. Bercoff, and Y. C. Eldar, "Fourier-domain beamforming and structure-based reconstruction for plane-wave imaging," *IEEE Transactions on Ultrasonics, Ferroelectrics, and Frequency Control*, vol. 65, no. 10, pp. 1810–1821, 2018.

[28] A. Burshtein, M. Birk, T. Chernyakova, A. Eilam, A. Kempinski, and Y. C. Eldar, "Sub-Nyquist sampling and Fourier domain beamforming in volumetric ultrasound imaging," *IEEE Transactions on Ultrasonics, Ferroelectrics, and Frequency Control*, vol. 63, no. 5, pp. 703–716, 2016.

[29] A. Lahav, T. Chernyakova, and Y. C. Eldar, "Focus: Fourier-based coded ultrasound," *IEEE Transactions on Ultrasonics, Ferroelectrics, and Frequency Control*, vol. 64, no. 12, pp. 1828–1839, 2017.

[30] C.-L. Liu and P. Vaidyanathan, "Maximally economic sparse arrays and cantor arrays," in *Proc. IEEE 7th International Workshop on Computational Advances in Multi-Sensor Adaptive Processing*. IEEE, 2017, pp. 1–5.

[31] R. Cohen and Y. C. Eldar, "Sparse convolutional beamforming for ultrasound imaging," *IEEE Transactions on Ultrasonics, Ferroelectrics, and Frequency Control*, vol. 65, no. 12, pp. 2390–2406, 2018.

[32] ——, "Sparse convolutional beamforming for ultrasound imaging," *arXiv:1805.05101*, 2018.

[33] A. Austeng and S. Holm, "Sparse 2D arrays for 3D phased array imaging-design methods," *IEEE Transactions on Ultrasonics, Ferroelectrics, and Frequency Control*, vol. 49, no. 8, pp. 1073–1086, 2002.

[34] R. Cohen and Y. C. Eldar, "Sparse array design via fractal geometries," *IEEE Transactions on Signal Processing*, vol. 68, pp. 4797–4812, 2020.

[35] A. Besson, R. E. Carrillo, D. Perdios, M. Arditi, O. Bernard, Y. Wiaux, and J.-P. Thiran, "A compressed beamforming framework for *(IUS)*. IEEE, 2016, pp. 1–4.

[36] O. Lorintiu, H. Liebgott, M. Alessandrini, O. Bernard, and D. Friboulet, "Compressed sensing reconstruction of 3D ultrasound data using dictionary learning and line-wise subsampling," *IEEE Transactions on Medical Imaging*, vol. 34, no. 12, pp. 2467–2477, 2015.

[37] I. A. Huijben, B. S. Veeling, K. Janse, M. Mischi, and R. J. van Sloun, "Learning subsampling and signal recovery with applications in ultrasound imaging," *IEEE Transactions on Medical Imaging*, vol. 39, no. 12, pp. 3955–3966, 2020.

[38] A. Mamistvalov and Y. C. Eldar, "Compressed Fourier-domain convolutional beamforming for wireless ultrasound imaging," *arXiv:2010.13171*, 2020.

[39] D. Perdios, A. Besson, M. Arditi, and J.-P. Thiran, "A deep learning approach to ultrasound image recovery," in *Proc. 2017 IEEE International Ultrasonics Symposium*. IEEE, 2017, pp. 1–4.

[40] K. Kulkarni, S. Lohit, P. Turaga, R. Kerviche, and A. Ashok, "Reconnet: Non-iterative reconstruction of images from compressively sensed measurements," in *Proc. IEEE Conference on Computer Vision and Pattern Recognition*, 2016, pp. 449–458.

[41] R. J. Williams, "Simple statistical gradient-following algorithms for connectionist reinforcement learning," *Machine Learning*, vol. 8, nos. 3–4, pp. 229–256, 1992.

[42] R. S. Sutton and A. G. Barto, *Reinforcement Learning: An Introduction*. MIT Press, 2018.

[43] A. C. Luchies and B. C. Byram, "Deep neural networks for ultrasound beamforming," *IEEE Transactions on Medical Imaging*, vol. 37, no. 9, pp. 2010–2021, 2018.

[44] Y. H. Yoon, S. Khan, J. Huh, and J. C. Ye, "Efficient B-mode ultrasound image reconstruction from sub-sampled RF data using deep learning," *IEEE Transactions on Medical Imaging*, 2018.

[45] S. Khan, J. Huh, and J. C. Ye, "Adaptive and compressive beamforming using deep learning for medical ultrasound," *IEEE Transactions on Ultrasonics, Ferroelectrics, and Frequency Control*, vol. 67, no. 8, pp. 1558–1572, 2020.

[46] S. Khan, J. Huh, and J. C. Ye, "Switchable deep beamformer," *arXiv:2008.13646*, 2020.

[47] B. Luijten, R. Cohen, F. J. de Bruijn, H. A. Schmeitz, M. Mischi, Y. C. Eldar, and R. J. van Sloun, "Deep learning for fast adaptive beamforming," in *Proc. IEEE International Conference on Acoustics, Speech and Signal Processing*. IEEE, 2019, pp. 1333–1337.

[48] A. A. Nair, K. N. Washington, T. D. Tran, A. Reiter, and M. A. L. Bell, "Deep learning to obtain simultaneous image and segmentation outputs from a single input of raw ultrasound channel data," *IEEE Transactions on Ultrasonics, Ferroelectrics, and Frequency Control*, vol. 67, no. 12, pp. 2493–2509, 2020.

[49] A. Wiacek, E. González, and M. A. L. Bell, "Coherenet: A deep learning architecture for ultrasound spatial correlation estimation and coherence-based beamforming," *IEEE Transactions on Ultrasonics, Ferroelectrics, and Frequency Control*, vol. 67, no. 12, pp. 2574–2583, 2020.

[50] N. Kessler and Y. C. Eldar, "Deep-learning based adaptive ultrasound imaging from sub-Nyquist channel data," *arXiv:2008.02628*, 2020.

[51] A. Mamistvalov and Y. C. Eldar, "Deep unfolded recovery of sub-Nyquist sampled ultrasound image," *arXiv:2103.01263*, 2021.

[52] D. Hyun, L. L. Brickson, K. T. Looby, and J. J. Dahl, "Beamforming and speckle reduction using neural networks," *IEEE Transactions on Ultrasonics, Ferroelectrics, and Frequency Control*, vol. 66, no. 5, pp. 898–910, 2019.

[53] M. Bell, J. Huang, D. Hyung, Y. Eldar, R. van Sloun, and M. Mischi, "Challenge on ultrasound beamforming by deep learning," 2020 [Online; accessed September 15, 2021].

[54] D. Hyun, A. Wiacek, S. Goudarzi, S. Rothlübbers, A. Asif, K. Eickel, Y. Eldar, J. Huang, M. Mischi, H. Rivaz, D. Sinden, R. van Sloun, H. Strohm, and M. Bell, "Deep learning for ultrasound image formation: Cubdl evaluation framework and open datasets," *IEEE Transactions on Ultrasonics, Ferroelectrics and Frequency Control*, vol. 68, no. 12, pp. 3466–3483, 2021.

[55] S. Khan, J. Huh, and J. C. Ye, "Deep learning-based universal beamformer for ultrasound imaging," in *Proc. Conference on Medical Image Computing and Computer Assisted Intervention*. Springer, 2019, pp. 619–627.

[56] F. Vignon, J. S. Shin, F. C. Meral, I. Apostolakis, S.-W. Huang, and J.-L. Robert, "Resolution improvement with a fully convolutional neural network applied to aligned per-channel data," in *Proc. 2020 IEEE International Ultrasonics Symposium*. IEEE, 2020, pp. 1–4.

[57] B. Luijten, R. Cohen, F. J. de Bruijn, H. A. Schmeitz, M. Mischi, Y. C. Eldar, and R. J. van Sloun, "Deep learning for fast adaptive beamforming," in *Proc. IEEE International Conference on Acoustics, Speech and Signal Processing*. IEEE, 2019, pp. 1333–1337.

[58] F. Chollet, "Antirectifier," online, accessed: 14-03-2021.

[59] I. K. Holfort, F. Gran, and J. A. Jensen, "Broadband minimum variance beamforming for ultrasound imaging," *IEEE Transactions on Ultrasonics, Ferroelectrics, and Frequency Control*, vol. 56, no. 2, pp. 314–325, 2009.

[60] B. Byram, K. Dei, J. Tierney, and D. Dumont, "A model and regularization scheme for ultrasonic beamforming clutter reduction," *IEEE Transactions on Ultrasonics, Ferroelectrics, and Frequency Control*, vol. 62, no. 11, pp. 1913–1927, 2015.

[61] A. C. Luchies and B. C. Byram, "Training improvements for ultrasound beamforming with deep neural networks," *Physics in Medicine and Biology*, vol. 64, no. 4, p. 045018, 2019.

[62] ——, "Assessing the robustness of frequency-domain ultrasound beamforming using deep neural networks," *IEEE Transactions on Ultrasonics, Ferroelectrics, and Frequency Control*, vol. 67, no. 11, pp. 2321–2335, 2020.

[63] J. Lu, F. Millioz, D. Garcia, S. Salles, W. Liu, and D. Friboulet, "Reconstruction for diverging-wave imaging using deep convolutional neural networks," *IEEE Transactions on Ultrasonics, Ferroelectrics, and Frequency Control*, vol. 67, no. 12, pp. 2481–2492, 2020.

[64] N. Chennakeshava, B. Luijten, O. Drori, M. Mischi, Y. C. Eldar, and R. J. van Sloun, "High resolution plane wave compounding through deep proximal learning," in *Proc. 2020 IEEE International Ultrasonics Symposium*. IEEE, 2020, pp. 1–4.

[65] K. Gregor and Y. LeCun, "Learning fast approximations of sparse coding," in *Proc. International Conference on Machine Learning*, 2010, pp. 399–406.

[66] J. A. Jensen, "Simulation of advanced ultrasound systems using field II," in *Proc. 2nd IEEE International Symposium on Biomedical Imaging: Nano to Macro*. IEEE, 2004, pp. 636–639.

[67] B. E. Treeby and B. T. Cox, "k-Wave: Matlab toolbox for the simulation and reconstruction of photoacoustic wave fields," *Journal of Biomedical Optics*, vol. 15, no. 2, p. 021 314, 2010.

[68] J. Youn, M. L. Ommen, M. B. Stuart, E. V. Thomsen, N. B. Larsen, and J. A. Jensen, "Detection and localization of ultrasound scatterers using convolutional neural networks," *IEEE Transactions on Medical Imaging*, vol. 39, no. 12, pp. 3855–3867, 2020.

[69] Z. Wang, A. C. Bovik, H. R. Sheikh, and E. P. Simoncelli, "Image quality assessment: From error visibility to structural similarity," *IEEE Transactions on Image Processing*, vol. 13, no. 4, pp. 600–612, 2004.

[70] X. Huang and S. Belongie, "Arbitrary style transfer in real-time with adaptive instance normalization," in *Proc. IEEE International Conference on Computer Vision*, 2017, pp. 1501–1510.

[71] G. Peyré, M. Cuturi *et al.*, "Computational optimal transport," *Foundations and Trends® in Machine Learning*, vol. 11, nos. 5–6, pp. 355–607, 2019.

[72] C. Villani, *Optimal Transport: Old and New*. Springer Science & Business Media, 2008, vol. 338.

[73] M. H. Jafari, H. Girgis, N. Van Woudenberg, N. Moulson, C. Luong, A. Fung, S. Balthazaar, J. Jue, M. Tsang, P. Nair *et al.*, "Cardiac point-of-care to cart-based ultrasound translation using constrained cycleGAN," *International Journal of Computer Assisted Radiology and Surgery*, vol. 15, no. 5, pp. 877–886, 2020.

[74] S. Khan, J. Huh, and J. C. Ye, "Variational formulation of unsupervised deep learning for ultrasound image artifact removal," *IEEE Transactions on Ultrasonics, Ferroelectrics, and Frequency Control*, vol. 68, no. 6, pp. 2086–2100, 2021.

11 Ultrasound Image Artifact Removal using Deep Neural Networks

Jaeyoung Huh, Shujaat Khan, and Jong Chul Ye

Ultrasound (US) imaging is susceptible to several types of artifacts. Most artifacts appear because of transducer limitations and simplified assumptions on the wave propagation. The artifacts are sometimes a useful component that contains tissue information; however, they often lead to a misinterpretation in the clinical diagnosis. Therefore, to improve the clinical utility of ultrasound in difficult-to-image patients and settings, a number of artifact removal methods have been proposed that aim at boosting image quality. Classical optimization-based methods have severe limitations due to their limited performance and high computational requirements. Furthermore, it is difficult to set the parameters for producing high-quality output. A quick remedy for the aforementioned issues is the deep-learning approach, which offers high performance compared with the traditional methods despite a significantly reduced runtime complexity. Another big advantage is that the parameters learned during training phase can be used to process different input images. This has motivated the scientific community to design deep-neural-network-based approaches for US artifact removal tasks.

11.1 Introduction

Ultrasound (US) images are susceptible to artifacts due to the physical limitations of the imaging mechanisms. These artifacts often interfere with an accurate diagnosis. Since most of the artifacts are generated by physical phenomena, understanding US artifacts requires a knowledge of the basic physics and underlying assumptions in US imaging.

Specifically, US images are generated using some specific and general assumptions. The general assumptions are listed below.

- The ultrasonic pulse is transmitted in a straight line and its return echo also comes back to the transducer in a straight line.
- The speed of sound in a tissue is constant and it is 1540 m/s.
- The depth at which reflection occurs is proportional to the time taken for the echo to return to the transducer.
- The noise generation phenomenon is simple. For example, delay and sum (DAS) is an additive Gaussian model from which an optimal matched filter is derived.
- The transducer has perfect physical characteristics such as ideal bandwidth and ideal geometry.

There are two major approaches used to avoid artifacts. One is mechanical and the other is software-based. The mechanical approach adjusts the probe position or wave emission mode, etc. For example, tissue harmonic imaging (THI) [1] can suppress the clutter and reverberation artifacts. Spatial compounding imaging (SCI) [2] can reduce the speckle noise or side lobes artifact. The software approach converts the US image with artifacts into an artifact-free image using an algorithm. For instance, there are various iterative methods based on a forward model or on applying filters in a post-processing step to reduce the speckle pattern in the image [3, 4].

In the following section we will discuss various artifacts in the context of US imaging. Later, some recently proposed approaches that are designed to overcome the various US artifacts using deep-learning techniques will be presented. The final section provides a summary of the chapter.

11.2 Ultrasound Artifacts

Compared with other medical imaging modalities such as computed tomography (CT) or magnetic resonance imaging (MRI), the anatomical structure in US images is generally not clearly visible. Because of this, the image is difficult to interpret and reliable diagnosis requires a lot of experience.

As mentioned above, US imaging is based on several assumptions. Because of such assumptions, there are several limitations in visualizing the exact tissue structure. In the following, we separate US artifacts into four main categories based on their causes: (1) time–space mismatch, (2) mismatch in the speed of sound, (3) attenuation, and (4) the fundamental limitations of US physics.

11.2.1 Time–Space Mismatch

The assumption that the reflection depth is proportional to the return time of the echo to the transducer sometimes does not necessarily hold owing to variations in the medium density. This tends to change the return echo path into a different path, which leads to the image artifacts during the reconstruction phase in the US image.

Specifically, the basic imaging principle of US imaging is based on the fact that the transmitted wave is reflected at the boundaries between two different tissues whose acoustic impedance is different. The reflected echo comes back to the transducer, and according to the travel time, a boundary can be located using $d = t_{travel} \times c/2$, where d denotes the depth of the boundary, t_{travel} is the time from transmission to acquisition, and c is the speed of sound. Some common artifacts that appear due to the time–space mismatch assumptions are as follows.

Reverberation

During image acquisition, if an interface with high reflectivity is present in parallel between the structure and the transducer interface, the transmitted wave reflects multiple times. The more it is reflected, the longer it takes to return to the transducer.

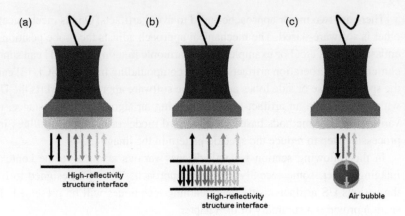

Figure 11.1 The principle of time–space mismatch-related artifacts: (a) reverberation artifacts, (b) comet tail artifacts, (c) ring-down artifacts.

According to the third assumption above, the later-returned echo is mapped behind the original boundary of the object or tissue. Hence multiple parallel lines appear in the reconstructed image. As the intensity of the echo is reduced, so the brightness of the parallel line is also reduced. An example of the reverberation phenomenon is shown in Fig. 11.1(a).

Since reverberation occurs when the reflectivity of the structure is high, it is often used to interpret the nature of imaging structures, typical examples of which include air and metal, etc. Due to this property, the reverberation is considered as evidence of pneumatosis [5]. Similarly, in cardiac US images, reverberation is frequently observed owing to its limited acoustic window.

Comet Tail

The comet tail artifact is special case of reverberation artifacts. As with normal reverberation, it occurs several times between interfaces with high reflectivity. However, reflection usually occurs between two close interfaces that have high reflectivity. Therefore, this artifact appears as several short-length lines in the US image. The principle of comet tail artifacts is shown in Fig. 11.1(b).

Ring-Down

While it is also similar to the reverberation artifact, the ring-down artifact is created by vibration of an air bubble. The bubble creates continuous waves resulting in parallel straight lines in the reconstructed image. Since air bubbles is the main cause of such artifacts, they occurs with pneumobilia [6]. In Fig. 11.1(c), the origin of the ring-down artifact is illustrated.

11.2.2 Speed-of-Sound Mismatch

One major assumption in conventional US imaging is that the speed of sound (SOS) in tissue is a constant 1540 m/s. However, the speed of sound is usually different

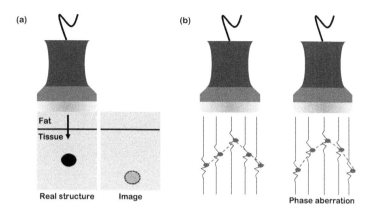

Figure 11.2 The principle of speed-of-sound mismatch-related artifacts: (a) speed displacement, (b) phase aberration.

depending on the properties of tissue. For example, the speed in fat tissue is 1470 m/s and in muscle 1620 m/s. If the wrong speed of sound is used for image formulation, the quality deteriorates considerably and leads to a noticeable loss of contrast and resolution. The SOS-related artifacts usually appear in between layers with high speed difference such, as fatty layers.

Speed Displacement

As mentioned above, there is a difference between the assumed and the actual speed of sound in a tissue. This difference can make the calculation of depth unreliable. That is, if the wave passes through the tissue at low speed, the structure will be imaged far from its original position. On the other hand, if the wave passes through the tissue at high speed, the structure is mapped in front of the original position. The principle of speed displacement is shown in Fig. 11.2(a).

Phase Aberration

A transducer captures the signal using array elements. The return echo signal is calibrated for each delay in each element. Since the speed of sound does not correspond to the exact value, the calibration results in the distortion of the wave signal. This deteriorates the lateral resolution in the US image. The principle of phase aberration is shown in Fig. 11.2(b).

11.2.3 Attenuation

Here, the wave transmitted into the tissue is weakened along the travel path. There are several reasons for the attenuation of the wave amplitude. Among them, refraction, reflection, and absorption are the main reasons for the decrease in signal intensity and are functions of the frequency of the wave. High frequency leads to high attenuation and low frequency leads to low attenuation.

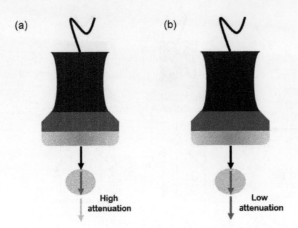

Figure 11.3 The principle of attenuation-related artifacts: (a) acoustic shadowing artifact, (b) acoustic enhancement artifact.

Acoustic Shadowing

When a wave is transmitted into a tissue, depending on the nature of the medium it travels deep into the body. If the wave encounters a structure that has strong reflectivity or high attenuation properties, the amplitude of the wave drops rapidly. For this reason, the echo signal on the rear side of the structure becomes weak in intensity, resulting in a shadow artifact. In the reconstructed image, the low-intensity signal appears as a black region compared with the surrounding area. In Fig. 11.3(a), the principle of acoustic shading is shown. Owing to the characteristic attenuation, high-density objects like a stone can be easily identified. Therefore acoustic shadowing is often used for the detection of gallstones and travertine [5].

Acoustic Enhancement

In contrast with the acoustic shadow artifact, the acoustic enhancement artifact occurs when the wave encounters a structure with a low attenuation factor, e.g., water. When the wave passes through the weak attenuation structure, the signal intensity drop is less than the expected or background intensity. Therefore, the post area of the structure has a high amplitude and it appears with high intensity in the US image. The principle of acoustic enhancement is presented in Fig. 11.3(b). Owing to the characteristic attenuation, low-density objects or substances like body fluids can be easily detected. Therefore, acoustic enhancement can be used to discriminate fluid-filled structures such as the gallbladder, ascites, and cysts, etc. [7].

11.2.4 Fundamental Limitations of US Physics

Resolution refers to the ability of an instrument to distinguish two adjacent points. Depending on the distance between reflective interfaces, two structures can be recognized separately. However, the secondary beams generated from the crystal expansion and contraction in the transducer yield resolution degradation.

Similarly, the contrast in an image is a measure of the intensity-based resolution. If an imaging object reflects a different amount of energy, then it can be visualized in the resulting image. The relative difference in the intensity of the background and the object of interest is called the contrast. Signal interference can lead to a speckle pattern, which induces contrast degradation.

Limited Bandwidth

The bandwidth is one of the most important characteristics of a transducer and has significant effects on the image quality. An ideal transducer should have infinite bandwidth so that it can record all the useful information. However, there is a physical limit to increasing the bandwidth, and in fact a low bandwidth often yields inevitable artifacts in the US image such as blurring. For example, as shown in Fig. 11.4, the spectrum of the signal is limited by the bandwidth of the transducer. Since the bandwidth of the transducer thus does not cover all the spectrum of the original signal, information is lost.

Secondary lobes To generate medical US images, different types of transducers are used. The measurements from the receiver array are combined with the help of a beamformer. Secondary lobe artifacts arise from reflections of unwanted ultrasound energy directed off-axis from the main beam. Owing to this spatial-domain spectral leakage, main-lobe, grating-lobes, and side lobe signals are generated. While the grating lobes are transducer dependent and usually occur at more oblique angles, the side lobes are transducer independent and are frequently generated for all type of transducers. In addition, the grating lobes have strong intensity, whereas the sidelobes have low intensity and usually disappear through the tissues. If the secondary-lobe signal is reflected, for example, owing to a high-reflectivity structure, an artifact appears and can be clinically misinterpreted. The principle of the secondary-lobe artifact is shown in Fig. 11.4(b).

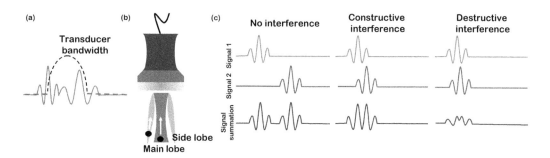

Figure 11.4 The principle of artifacts caused by the fundamental limitations of US physics: (a) limited bandwidth, (b) side lobe artifact, (c) speckle pattern.

Speckle Speckle gives a granular pattern that is easy to see in the US image. It is created on a smaller scale by the interactions of the beam. The wave is reflected with the law of reflections, however, some reflectors scatter the wave in all directions. An example of a speckle pattern in the US image is shown in Fig. 11.4(c). While speckle noise can be used to detect a non-specular structure, it usually degrades the quality of image for anatomical structure identification and is one of the major sources of resolution and contrast loss. Unlike pure random noise, speckle noise is not devoid of information and changes in amplitude in a fully developed speckle do not represent the underlying (anatomical) variations in back-scattering intensity. Therefore, speckle denoising is a challenging task.

11.3 Deep Learning for US Artifact Removal

Deep-learning techniques are receiving a lot of attention in medical imaging community because of their various advantages, including efficient computation and state-of-the-art performance in various tasks such as denoising, super-resolution, and reconstruction problems [8–10]. In particular, the authors of [8] proposed the first supervised deep-learning method for denoising low-dose computed tomography (CT) images. In [11], a supervised learning method was proposed for accelerated MRI imaging. Similarly, for efficient ultrasound imaging, the authors of [10] used a convolutional neural network to interpolate missing radio-frequency data (RF) and achieved high-quality US images with as low as 8 times subsampled channel data.

Motivated by the success of deep learning in various medical imaging modalities, a number of researchers have tried to find mathematical links between conventional iterative methods and deep convolution neural networks [8, 12–15]. These mathematical theories have not only helped in understanding the origin of performance gain but also have provided insights for designing better architectures [16].

There are various deep-learning-based solutions for ultrasound image denoising; however, in this chapter we will cover only five major problems, namely (1) deconvolution, (2) despeckling, (3) removal of reverberation, (4) phase aberration correction, and (5) side lobe artifact removal.

11.3.1 Deconvolution

Owing to the limited bandwidth of the transducer, the spatial resolution of an ultrasound image can be compromised, resulting in blurry images. The blurring effect can be modeled as a linear convolution of a tissue reflectivity function r and a point spread function s. The blurred image can thus be represented as follows:

$$b = r \circledast s + \eta, \qquad (11.1)$$

Here, b is the blurred noisy image due to a Gaussian noise η.

Deconvolution is a reverse of the convolution process through which we can estimate the true reflectivity function r, which is sharpened by the inverse of the point spread function s. The quality of the recovered tissue reflectivity function is linked to the beamforming technique, and finding the beamforming and deconvolution filter matrices that lead to high resolution with good signal-to-noise ratio (SNR) is technically challenging.

In conventional deconvolution approaches, to estimate a high-quality image (the tissue reflectivity function), the inverse problem defined below is solved through some iterative optimization process [17]:

$$\hat{r} = \arg\min_{r} \|b - r \circledast s\|^2 + \lambda R, \qquad (11.2)$$

where $\hat{r}$ is the estimated tissue reflectivity function, and R denotes the constraint terms. For the aforementioned problem, different axial and lateral constraints can be designed. This handcrafted constraint-designing approach shows comparable results; however, it requires detailed physical information to design an accurate model and takes a lot of computational time. To deal with these issues, recently, a number of deep-learning-based solutions have been proposed.

RF-domain approach: Adaptive and compressive beamforming using deep learning for medical ultrasound [18]

- Purpose: The filter matrices for deconvolution and for the adaptive beamformer should be spatially varying and depend on input signal to achieve the best trade-off between noise and resolution. Therefore, an exact calculation is usually computationally expensive, and only approximate forms are used, which limits the accuracy. A quick remedy to reduce the runtime computational complexity would be to store the pre-calculated nonlinear mapping. Unfortunately, as stated before, the mapping depends on the input, so it requires a huge amount of memory for variable input depths.

  To deal with the aforementioned problems, in [18], the authors, using the mathematical link between adaptive beamformer and convolution neural networks [13], designed a deep-learning-based beamforming method. In particular, the proposed adaptive beamforming technique, named as "DeepBF" is based on a convolutional neural network (CNN).
- Method: As shown in Fig. 11.5, after the time-of-flight correction, the RF data cube consisting of three neighboring depth planes is used as an input of the network. The target of the network is to recover the in-phase and quadrature data (the IQ data) of the center depth plane. To design a deconvolution beamformer, the iterative method of sparse reconstruction [17] was used to generate the target IQ data for supervised learning. The network was trained in a supervised manner to minimize the mean square error using the standard gradient descent method. The neural network a trained by the following equation:

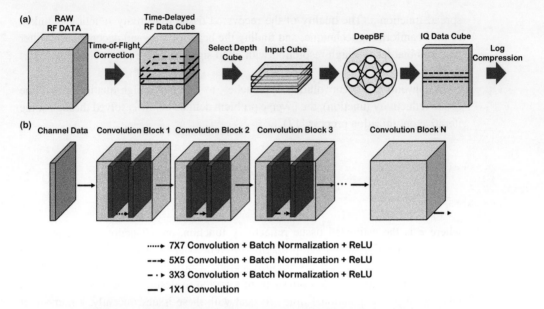

Figure 11.5 (a) The proposed method. From the time-delayed RF data cube, a three-depth cube is extracted. This can be translated into the IQ data of the mid-line of the three-depth cube using DeepBF. The final image is acquired through a log-compression step. (b) The network architecture for DeepBF.

$$\min_{\Theta} \sum_{t=1}^{T} \sum_{n=1}^{N} \|v_n - \tau_{\Theta}(s_n^{(t)})\|_2^2, \qquad (11.3)$$

where $(s_n^{(t)})_{n,t=1}^{N,T}$, $(v_n)_{n,t=1}^{N,T}$ represent a pair of three-input depth planes and the corresponding target IQ data plane, respectively. Here n labels the depth planes, and N is the total number of depth planes; t labels the subsampling patterns, and T is the number of sampling patterns. The quantity τ_{Θ} refers to the neural network model that is trained by minimizing the loss function mentioned above.

Image-domain approach: Unsupervised deconvolution neural network for high-quality ultrasound imaging [19]

- Purpose: One big challenge in machine learning is to gather a large training dataset. Although plenty of ultrasound images are available online, however, for supervised training of a model, a pair of noisy and artifact image or RF channel data is required, which is often infeasible. To deal with this, the authors of [19] suggested an unsupervised learning-based deconvolution method. A major difference in this approach is that, unlike conventional deconvolution methods e.g., [17] and [18], which require RF-channel-domain data, the proposed method works in the image domain and can be trained without paired RF channel data.

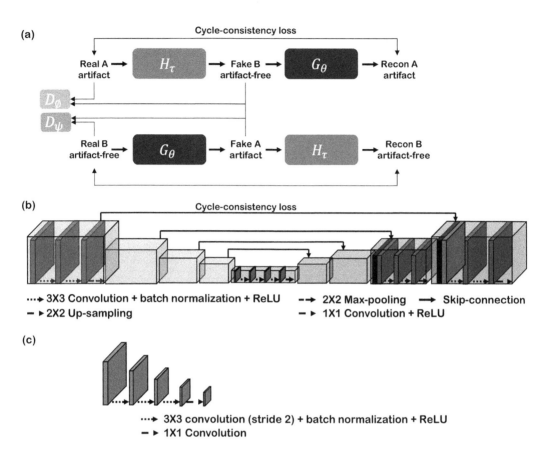

Figure 11.6 (a) The proposed GAN with cycle-consistency loss. The action of H_τ is to generate an artifact-free image from a degraded image. The action of G_θ is to generate a degraded image. To avoid mode-collapse, a cycle-consistency relationship between the input artifact image and the reconstructed artifact image is maintained. (b) The proposed generator architecture. (c) The proposed discriminator architecture.

- Method: As to [18], the proposed method was trained to perform variable imaging, i.e., a single universal model was used to process variable-rate subsampled images. The major differences in the proposed method comes from the type of learning and input-domain data. Unlike DeepBF [18], which processes one depth plane at a time, in the proposed method, the network has access to the full image immediately. For unsupervised learning, a cycle-consistency generative adversarial network (GAN) [20] was used. The input datasets consisted of original DAS images and the target dataset was a collection of deconvoluted images generated from [18]. The model configurations are shown in Fig. 11.6.

  The method may be compared with two supervised training methods: (1) the above-mentioned DeepBF [18], trained using RF-domain data, and (2) the same network architecture trained in the image domain using paired images. The method successfully learned to mimic the deconvolution effect and on both the phantom and *in vivo* scans it demonstrated high contrast and improved spatial resolution. Since the method was designed for compressive deconvolution in addition to fully sampling the DAS image, a low-resolution image created using subsampled channel data can also be processed for high-contrast/high-resolution results.

11.3.2 Despeckle

Speckle patterns occur because of random constructive and destructive interference of the waves. Although speckle is a useful component in diagnosis, as in the speckle tracking procedure, it can interfere with an accurate diagnosis [21], especially in regions of small structures. For this reason, speckle denoising is one of the main research areas in image quality enhancement.

The objective in speckle denoising is to suppress speckle patterns without blurring the structure boundaries. In this regard, a variety of methods have been proposed. Each method exploits different signal characteristics, e.g., smoothness, anisotropy, similarity in adjacent patches, etc. In [22], rank-minimization was utilized to discriminate between noise-free and noisy patterns. In particular, nonlocal patches with high similarity are collected and a low-rank component representing the smoothest and the most representative patch is selected from these. The objective function of the nonlocal low-ranked-based speckle-denoising method is given as follows:

$$\min_{\psi_D, \psi_\eta} \sum_{i=1}^{M} w_i \sigma_i(\psi_D) + \alpha \sum_{g \in \psi_\eta} \|g\|_\infty,$$

Here, w_i is the weight of the singular value σ_i of the low-rank component (ϕ_D); ψ_η is a sparse component satisfying $\psi_I = \psi_D + \psi_\eta$, where ψ_I denotes the raw input. The quantity g represents the patches from ψ_η. This method shows significantly improved results; however, it has a high computational burden and time-consuming problems. Moreover, the results are highly dependent on the parameters, so it cannot provide optimal results robustly. In this regard, the deep-learning approach shows considerable advantages.

RF-domain approach: Beamforming and speckle reduction using neural networks [23]

- Purpose: Conventional beamformers process channel data to estimate the particular statistical components of image, whereas in proposed beamformer the objective is to estimate the echogenicity map of the tissue. There are a number of speckle denoising algorithms that work as a post-processing filter such as wavelet transformation or iterative methods [24, 25]. One major motivation of RF-domain based speckle reduction method is that the conventional method rely on the quality of input images. If conventional beamforming method fail to reconstruct any anatomical feature, the speckle reduction method can not recover it. To address this problem, using a neural network, a speckle-reduced beamforming technique is proposed.

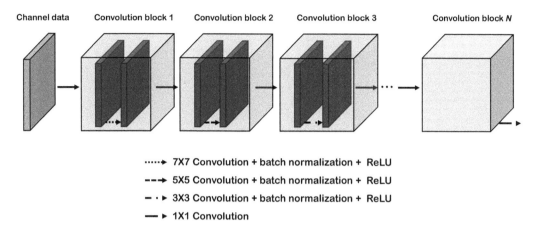

Figure 11.7 Network architecture for the proposed method. The channel data is used as input of the network and the speckle-reduced image is the output of the network.

- Method: To take advantage of the various contrast features, real photographic images such as Places2 [26] and ImageNet [27] were used. The dataset was used to simulate the ultrasound channel data using Field-II [28, 29]. In addition to real photographic images, simulated phantom images were used. To generate speckle-free simulated phantom images, a nonlocal mean value filter [30] was utilized.

  In particular, 128 RF channels were used to simulate the ultrasound imaging. For the model input, the 128 elements were divided into 16 sub-apertures of 8 elements. The sub-aperture data was beamformed using delay-and-sum to produce in-phase and quadrature (IQ) data. The real and the imaginary components of the IQ data were separated and concatenated along the input channel axis. Finally, the input dimension of $H \times W \times 32$ is achieved, where H, W denote the height and width of the image. The model consists of $N = 16$ convolution blocks, and the architecture is presented in Fig. 11.7. The network was trained by minimizing the l_1 loss and the SSIM loss between log-compressed input and the log-compressed and positively weighted output of the network.

  In this study, authors compared their results with conventional DAS, spatial compounding (SC) and optimized Bayesian non-local means (OBNLM) [31] using cyst and harmonic B-mode images of a kidney. The proposed method is found to produce good generalization and achieve superior contrast performance while maintaining the resolution by fine-detail preservation.

Image-domain approach: Ultrasound speckle reduction using generative adversarial networks [32]

- Purpose: In this research, the authors proposed an image-to-image transformation method for speckle denoising in real-time ultrasonography using a generative adversarial network (GAN) model (Fig. 11.4).
- Method: The authors proposed a simple generative adversarial network model composed of a single generator and discriminator. The noisy input is transformed

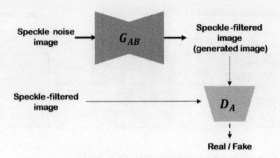

Figure 11.8 The proposed method for ultrasound speckle reduction. The generator generates a speckle-filtered image. The discriminator discriminates which of the generated image and real speckle-filtered image is real rather than fake.

into a noise-free image through the generator. Then the discriminator distinguishes whether the image is real or fake. To train the network, the authors used 600 cardiac B-mode images from 200 patients. The target speckle-free image was generated using a nonlocal low-rank (NLLR) algorithm [22].

The authors compared their method with Gaussian blurring and the target NLLR method. The Gaussian blur makes the image smooth out; however, the structural details also become blurred. On the other hand, the NLLR algorithm suppresses the speckle patterns well and the structural details remain preserved. The proposed method successfully mimics the NLLR and produces indiscernibly different results and with significantly less runtime computational complexity: it generates an artifact-free image in just 286 ms, which is a huge improvement over the 160 seconds taken by the original NLLR.

11.3.3 Reverberation

Reverberation artifacts generally occur at high-reflectivity interfaces. With reverberation artifacts, multiple reflections appear as straight lines in the US image. Recently, a method was proposed for obtaining the error filter parameters. The filter can be applied to the noisy data to acquire the noise-free signal [33]. The proposed method requires the solution of the following objective function:

$$J = \|Ma - d\|_2^2,$$
$$\hat{a} = (M^T M + \mu I)^{-1} M^T d,$$

where $\hat{a}$ is the estimated error filter; M denotes the convolution matrix consisting of the Fourier transforms of channel RF signals and d represents the noisy data. This classical parametric approach can suppress the reverberation artifact successfully by correcting the off-axis signal; however, it requires high computational power for calculating the matrix inversion. Furthermore, owing to its parametric nature, it often produces additional artifacts caused by the unwanted off-axis signal.

RF-domain approach: Reverberation noise suppression in ultrasound channel signals using a 3D fully convolutional neural network [34]

- Purpose: Diffused reverberation artifacts not only degrade image quality, but also hinder the processes of adaptive beamforming and sound velocity estimation, etc. There are several ways to solve these problems. For example, tissue harmonic imagining (THI) [1] is one of the methods. However, it cannot solve the problem perfectly and requires further post-processing. An early attempt to deal with reverberation artifacts using a DNN was made in [35]; however the method could not handle reverberation noise adequately. Since reverberation noise is associated with the lateral and axial components of the image, in this research the authors used a 3D convolutional neural network to robustly suppress the reverberation artifacts.
- Method: Similarly to the aforementioned speckle-reduction method [23], the authors used various photographic images from ImageNet [27] and a Field-II [28, 29] simulator program to generate ultrasonic channel signals from 128 elements. To generate the reverberation noise data, they used Gaussian-distributed random noise and a band pass filter. The reverberation IQ data was used as the input dataset, and the reverberation-free IQ data was used as the target dataset. The input had dimensions $100 \times 100 \times 128 \times 2$, where the numbers denote the axial and lateral lengths, the number of channels, and the real or imaginary axis, respectively. To process this 4D input, 3D neural convolution blocks were used, as shown in Fig. 11.9. The network was trained in a supervised fashion using l_2 loss between the sign-preserved logarithmically compressed signals.

The model was evaluated for variable dynamic ranges, and found to produce better contrast and reverberation-free results compared to the unfiltered image.

Image-domain approach: Deep unfolded robust PCA with application to cluster suppression in ultrasound [15]

- Purpose: Clutter signals constitute an image-quality degradation factor in vascular ultrasound imaging. The latter is often required for the analysis of the blood volume and velocity. In vascular imaging microbubbles are normally used for visual enhancement; therefore, it is important to separate the microbubble signal from the clutter signal. Traditionally, with iterative methods the signals are separated into low-rank and sparse components, where the low-rank components represent the background and the sparse components represent the microbubble signal. Iterative methods are generally time-consuming and require parameter tuning. In this research, the authors proposed a convolutional neural-network-based robust principal component analysis (CORONA) algorithm to acquire the desired signal.
- Method: Two types of datasets were used for the training of the network. First, the low-rank and sparse components were generated by simulating the ultrasound

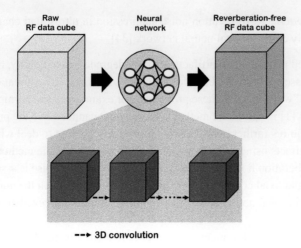

Figure 11.9 Block diagram of the proposed reverberation correction method. The input data is 3D IQ data with reverberation and the output data is reverberation-free 3D IQ data. The network consists of 3D convolution blocks.

contrast agent (UCA) signal and tissue signal. Second *in vivo* data are decomposed into the low-rank and sparse components with the iterative method FISTA [36] and using robust principal component analysis (RPCA). To mimic the robust-PCA method, the unfolding approach was used to design the neural network model. The model was trained in a supervised manner to minimize the following loss function:

$$L(\theta) = \frac{1}{2N} \sum_{i=1}^{K} \| f_S(D_i, \theta) - \hat{S}_i \|_F^2$$

$$+ \frac{1}{2K} \sum_{i=1}^{K} \| f_L(D_i, \theta) - \hat{L}_i \|_F^2.$$

Here θ represents the learnable parameters, f_S, f_L denote the networks for the sparse and low-rank components, respectively, D_i represents the in-phase and quadrature components of the received signal which can be represented as $D_i = L_i + S_i + N$; N is the additive noise. The input cine-lopes are processed by dividing them into fixed-size 3D patches, i.e., image height, width, and time frame.

The proposed method may be compared with the FISTA-based conventional RPCA, wall filtering, and singular-value-decomposition (SVD)-based methods. The proposed method demonstrates superior performance and successfully suppresses the clutter, achieving clearer separation of the blood vessels at significantly reduced computational time compared with the other algorithms.

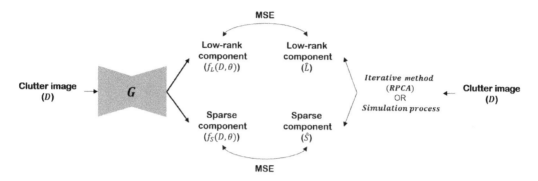

Figure 11.10 The proposed scheme for the clutter suppression method. The generator separates the low-rank component and the sparse component. The network is trained by minimizing the MSE loss between the generated component and the target component, which is generated from the iterative method or a simulation process.

11.3.4 Phase Aberration

Accurate estimation of the speed of sound (SoS) is important. Differences in SoS yield phase aberration which degrades the US image quality. Conventionally, to deal with phase aberration artifacts, coherent-factor-based beamforming methods were used [37, 38]. Coherent-factor (CF)-based post-filtration can be coupled with any beamforming method; the coherence factor is the ratio of coherent to incoherent sums of the received signal:

$$CF[z,x] = \frac{|\sum y_m[z,x]|^2}{M \sum |y_m[z,x]|^2},$$

where y is the time-delayed signal and m is the element index; $[z,x]$ denotes the pixel and M is the number of elements. The state-of-the-art coherent-factor filtration method uses a generalized coherence factor (GCF), which is defined by

$$GCF = \frac{\text{low-frequency energy}}{\text{total energy}} = \frac{\sum_{k \in L} |p(k)|^2}{\sum_{k=0}^{N-1} |p(k)|^2},$$

where $p(k)$ is the spectrum of the received data, L is the set representing the low-frequency region, and N is the total number of points in the spectrum. As can be seen, the method requires optimal parameter selection for the effective suppression of phase aberration artifacts, and so it often degrades the dynamic range of the ultrasound image.

Image-domain approach: Phase aberration correction: a convolutional neural network approach [39]

- Purpose: Phase aberration is one of the factors that degrades the ultrasound image quality. It occurs because of variations in the speed of sound (SoS), which causes a difference between the real and assumed SoS values. Phase aberration blocks the coherent-signal summation and therefore prevents accurate beam focusing. In this

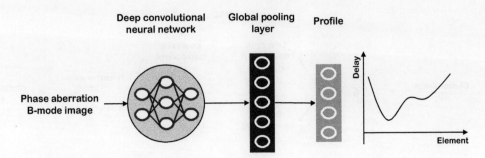

Figure 11.11 A B-mode image with phase aberration artifact is put into a network composed of a deep convolutional neural network and global pooling layer. The output of the network is the phase aberration profile for each channel element.

research, a deep-learning-based solution was proposed for aberration correction. The method utilizes an aberrator-profile-generation network with a convolutional neural network.

- Method: First, various phase aberrator profiles were generated using [40] (see Fig. 11.11). The Field-II simulation program is utilized to generate the B-mode images after the phase aberrator profile has been obtained. The paired datasets are used for the training of the network. The input of the network is a phase-aberrated B-mode image and the target of the network is the phase-aberrator profile, which is a vector that is as long as the number of receiver elements.

The proposed method was compared with conventional DAS beamformer and nearest-neighbor cross-correlation (NNCC) algorithms [41, 42] for weak, moderate, and strong aberration scenarios. The estimated profile from the proposed method in each case was almost the same as the ground-truth profile. Moreover, in the quantitative result, the proposed method also shows better performance than the DAS and NNCC methods.

11.3.5 Side Lobes

When the side lobe signal is reflected by a high-reflectivity structure, the return signal causes some artifacts in the image. It degrades the image quality and hinders accurate diagnosis. To suppress the side lobe artifacts, a number of methods have been proposed. A classical method for side lobe suppression is minimum-variance-based beamforming (MVBF) [43].

The MVBF method can produce state-of-the-art results; however, since calculating the adaptive weights from the received signal requires inverse covariance matrix calculation, it is seldom used for real-time applications. Furthermore, it usually requires a large number of channels, which is another bottle-neck for low-cost systems.

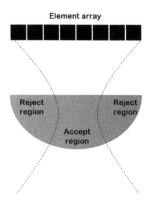

Figure 11.12 The data acquisition area.

RF-domain approach: Deep neural networks for ultrasound beamforming [35]

- Purpose: Ultrasound artifacts hinder the clinical use of US images for reliable diagnosis. To improve the image quality by suppressing the artifacts, the image reconstruction of the aperture-domain model (ADMIRE) was proposed; however, ADMIRE requires an extensive calculation [44, 45] and hence is unsuitable for real-time imaging. To suppress the off-axis scattering artifact in real-time imaging, in [35] the authors proposed a neural-network-based beamformer.
- Method: Since in real scanner data we cannot determine which lobe is the result of off-axis scattering, the authors utilized the Field-II [28, 29] simulator to generate paired phantom data with and without off-axis scattering. A region is selected as the pass region and acoustic reflection from that region is treated as clean data while measurement from the other regions are treated as off-axis scattering. An example of off-axis scattering and the pass region is shown in Fig. 11.12. In particular, the phantom data consist of different point targets placed at different positions, and, depending on their relative position with respect to active channel subset (active aperture), e.g., the acceptance or rejection areas, their RF measurements were multiplied by one or zero respectively. First the channel data was changed to the spectrogram using a short time Fourier transform (STFT) and the individual spectrogram was fed to the neural network. A fully connected multi-layer perceptron neural network was used. The network was trained in a supervised manner. A flowchart of the proposed method is shown in Fig. 11.13.

The proposed method was evaluated on phantom and *in vivo* scanner data. Compared with conventional beamforming methods, e.g., DAS and CF [46], the proposed method shows qualitatively and quantitatively better contrast and reduced off-axis scattering:

RF-domain approach: Adaptive ultrasound beamforming using deep learning [47]

- Purpose: Minimum variance or Capon beamforming (MVBF) is considered to be a state-of-the art adaptive beamforming technique for ultrasound imaging [46, 48].

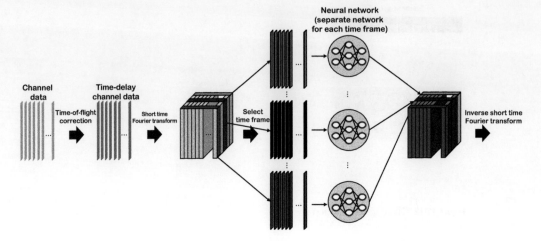

Figure 11.13 A block diagram of the proposed method. After the time-of-flight correction, the STFT step is performed to change the raw channel data into a spectrogram. By using separate neural networks for each time case, new channel data can be created after inverse STFT.

It is a robust technique used for high-quality image formation; however, it requires extensive computation for the calculation of the inverse of the correlation matrix. This is why it is seldom used for real-time ultrasound imaging. To address this problem, the authors of [47] proposed an efficient DNN architecture that can mimic the MVBF output with reduced computational complexity.
- Method: The overall framework of the proposed method is shown in Fig. 11.14. The purpose of this network is to adaptively calculate the apodization weights according to the input channel data. For model design and validation, phantom and real *in vivo* data of plane-wave and synthetic aperture datasets scanned using linear array were used. Additionally, using the Field-II [28, 29] simulator, point scatterer phantom data were generated. The target data was processed using eigen-based minimum-variance (EBMV) beamformer [49]. The network was trained in supervised fashion to minimize the following loss function:

$$L_{total} = \lambda L_{SMSLE} + (1-\lambda) L_{unity}, \tag{11.4}$$

$$L_{SMSLE} = \frac{1}{2} \| \log_{10}(P^+_{ABLE}) - \log_{10}(P^+_{MV}) \|^2_2$$

$$+ \frac{1}{2} \| \log_{10}(P^-_{ABLE}) - \log_{10}(P^-_{MV}) \|^2_2 \tag{11.5}$$

$$L_{unity} = |1^T f_\theta(y[x,z]) - 1|^2, \tag{11.6}$$

where λ is a parameter that determines the ratio of the losses and P^+, P^- denote the positive part and negative part of the beamformed data, where $P_{ABLE}[x,z] = f_\theta(y[x,z])^H y[x,z]$. The quantities x, z represent the pixel index and $y[x,z]$ is the channel data; f_θ is the network to be trained.

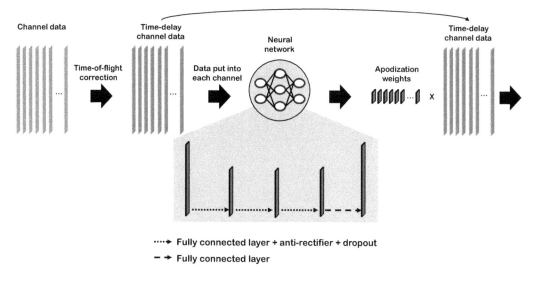

Figure 11.14 A block diagram of the proposed method. The time-delayed channel data is put into the neural network and then the apodization weights are generated for each channel. The apodization weights are then elementwise multiplied to the input channel data to generate side-lobe-suppressed channel data.

The method was evaluated using intravascular synthetic aperture ultrasound (IVUS) and plane wave imaging data for contrast and resolution enhancement and clutter and side lobe artifact suppression.

It shows better contrast and lateral resolution compared with DAS and iterative maximum aposteriori algorithms (iMAPs) [50]. Furthermore, the side lobe artifacts were noticeably suppressed and substructures were clearly visible:

RF-domain approach: Image quality enhancement using a DNN for plane wave medical ultrasound imaging [51]

- Purpose: The plane wave imaging (PWI) provides a high frame rate; however, it suffers from severe side lobe artifacts. The plane wave compounding (PWC) method suppresses these artifacts at the cost of a reduced frame rate. In [51], the authors proposed a method that can suppress those artifacts. In particular, the proposed method generates high-quality images mimicking the focused scan signal from the low-quality PWI scan signal by using a DNN.
- Method: The flow chart of this research is shown in Fig. 11.15. A multi-layer perceptron-based fully connected neural network was utilized. As shown in the flow chart, each sample from the Fast Fourier Transform (FFT) of the channel data is used as the input data of the network. The output of the network is converted into the original domain through the inverse-FFT (IFFT). Since the network was trained in supervised manner, the paired data is generated using the Field-II simulation program.

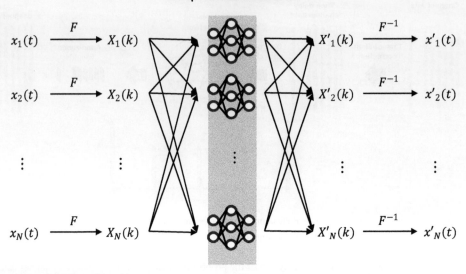

Figure 11.15 The proposed scheme for the side lobe suppression method. The quantity $x_N(t)$ denotes the Nth channel data. After F (FFT) each frequency component is put into a separate neural network. The output of the network is changed into a time-domain signal by F^{-1}, i.e., IFFT.

The method was compared with conventional DAS, CF, and [35] methods. The proposed method successfully mimics the target quality results and is shown to achieve superior reconstruction performance in both the phantom and *in vivo* scan's data.

11.4 Summary and Outlook

Herein, we have discussed source effects and different deep-learning-based solutions for the suppression of various types of ultrasound imaging artifacts. In this chapter, apart from the artifact physics, we discussed motivation, data processing pipelines, target image generation, network architecture, and different training methods used to design artifact removal. For example, blurring artifacts which occur due to limited bandwidth can be suppressed using data in different domains; each domain has its own merits. Similarly, the task of speckle denoising, which is often performed by a post-processing filter, can also be done in the channel domain and can the result in improved de-noising with better structural details.

Unlike conventional artifact-removal methods, which rely on well-defined mathematical relationships and require sophisticated feature engineering, one big advantage of the DNN approach is its exponentially large expressiveness, which can help learn patterns from data without explicit modeling.

Despite being a robust and effective artifact removal technique, a big challenge in deep learning is the need to prepare of sophisticated target data. To deal with this, simulated phantom data are often used, although well-designed simulations remain just an approximation to real measurements. To overcome this limitation of supervised learning approaches, unsupervised learning approaches using cycleGAN can be potentially useful.

In conclusion, a number of strategies have been explored for the removal of various types of ultrasound imaging artifacts. In this regard, deep learning is a promising tool for improved ultrasound imaging.

References

[1] T. S. Desser and R. B. Jeffrey, "Tissue harmonic imaging techniques: Physical principles and clinical applications," in *Seminars in Ultrasound, CT and MRI*, vol. 22, no. 1. Elsevier, 2001, pp. 1–10.

[2] R. R. Entrekin, B. A. Porter, H. H. Sillesen, A. D. Wong, P. L. Cooperberg, and C. H. Fix, "Real-time spatial compound imaging: Application to breast, vascular, and musculoskeletal ultrasound," in *Seminars in Ultrasound, CT and MRI*, vol. 22, no. 1. Elsevier, 2001, pp. 50–64.

[3] R. N. Czerwinski, D. L. Jones, and W. D. O'Brien, "Ultrasound speckle reduction by directional median filtering," in *Proc. International Conference on Image Processing*, vol. 1. IEEE, 1995, pp. 358–361.

[4] T. C. Aysal and K. E. Barner, "Rayleigh-maximum-likelihood filtering for speckle reduction of ultrasound images," *IEEE Transactions on Medical Imaging*, vol. 26, no. 5, pp. 712–727, 2007.

[5] M. Baad, Z. F. Lu, I. Reiser, and D. Paushter, "Clinical significance of US artifacts," *Radiographics*, vol. 37, no. 5, pp. 1408–1423, 2017.

[6] A. Hindi, C. Peterson, and R. G. Barr, "Artifacts in diagnostic ultrasound," *Reports in Medical Imaging*, vol. 6, pp. 29–48, 2013.

[7] F. W. Kremkau and K. Taylor, "Artifacts in ultrasound imaging," *Journal of Ultrasound in Medicine*, vol. 5, no. 4, pp. 227–237, 1986.

[8] E. Kang, J. Min, and J. C. Ye, "A deep convolutional neural network using directional wavelets for low-dose X-ray CT reconstruction," *Medical Physics*, vol. 44, no. 10, pp. e360–e375, 2017.

[9] D. Mahapatra, B. Bozorgtabar, and R. Garnavi, "Image super-resolution using progressive generative adversarial networks for medical image analysis," *Computerized Medical Imaging and Graphics*, vol. 71, pp. 30–39, 2019.

[10] Y. H. Yoon, S. Khan, J. Huh, and J. C. Ye, "Efficient B-mode ultrasound image reconstruction from sub-sampled RF data using deep learning," *IEEE Transactions on Medical Imaging*, 2018.

[11] H. Lee, J. Lee, H. Kim, B. Cho, and S. Cho, "Deep-neural-network-based sinogram synthesis for sparse-view CT image reconstruction," *IEEE Transactions on Radiation and Plasma Medical Sciences*, vol. 3, no. 2, pp. 109–119, 2018.

[12] J. C. Ye, Y. Han, and E. Cha, "Deep convolutional framelets: A general deep learning framework for inverse problems," *SIAM Journal on Imaging Sciences*, vol. 11, no. 2, pp. 991–1048, 2018.

[13] J. C. Ye and W. K. Sung, "Understanding geometry of encoder–decoder CNNS," in *Proc. International Conference on Machine Learning*. PMLR, 2019, pp. 7064–7073.

[14] B. Sim, G. Oh, J. Kim, C. Jung, and J. C. Ye, "Optimal transport driven cycleGAN for unsupervised learning in inverse problems," *SIAM Journal on Imaging Sciences*, vol. 13, no. 4, pp. 2281–2306, 2020.

[15] O. Solomon, R. Cohen, Y. Zhang, Y. Yang, Q. He, J. Luo, R. J. van Sloun, and Y. C. Eldar, "Deep unfolded robust PCA with application to clutter suppression in ultrasound," *IEEE Transactions on Medical Imaging*, vol. 39, no. 4, pp. 1051–1063, 2019.

[16] S. Khan, J. Huh, and J. C. Ye, "Variational formulation of unsupervised deep learning for ultrasound image artifact removal," *IEEE Transactions on Ultrasonics, Ferroelectrics, and Frequency Control*, pp. 1–1, 2021.

[17] J. Duan, H. Zhong, B. Jing, S. Zhang, and M. Wan, "Increasing axial resolution of ultrasonic imaging with a joint sparse representation model," *IEEE Transactions on Ultrasonics, Ferroelectrics, and Frequency Control*, vol. 63, no. 12, pp. 2045–2056, 2016.

[18] S. Khan, J. Huh, and J. C. Ye, "Adaptive and compressive beamforming using deep learning for medical ultrasound," *IEEE Transactions on Ultrasonics, Ferroelectrics, and Frequency Control*, vol. 67, no. 8, pp. 1558–1572, 2020.

[19] S. Khan, J. Huh, and J. C. Ye, "Unsupervised deconvolution neural network for high quality ultrasound imaging," in *Proc. 2020 IEEE International Ultrasonics Symposium*. IEEE, 2020, pp. 1–4.

[20] J.-Y. Zhu, T. Park, P. Isola, and A. A. Efros, "Unpaired image-to-image translation using cycle-consistent adversarial networks," in *Proc. IEEE International Conference on Computer Vision*, 2017, pp. 2223–2232.

[21] L. Bohs, B. Geiman, M. Anderson, S. Gebhart, and G. Trahey, "Speckle tracking for multi-dimensional flow estimation," *Ultrasonics*, vol. 38, nos. 1–8, pp. 369–375, 2000.

[22] L. Zhu, C.-W. Fu, M. S. Brown, and P.-A. Heng, "A non-local low-rank framework for ultrasound speckle reduction," in *Proc. IEEE Conference on Computer Vision and Pattern Recognition*, 2017, pp. 5650–5658.

[23] D. Hyun, L. L. Brickson, K. T. Looby, and J. J. Dahl, "Beamforming and speckle reduction using neural networks," *IEEE Transactions on Ultrasonics, Ferroelectrics, and Frequency Control*, vol. 66, no. 5, pp. 898–910, 2019.

[24] A. Khare, M. Khare, Y. Jeong, H. Kim, and M. Jeon, "Despeckling of medical ultrasound images using Daubechies' complex wavelet transform," *Signal Processing*, vol. 90, no. 2, pp. 428–439, 2010.

[25] P. Coupé, P. Hellier, C. Kervrann, and C. Barillot, "Nonlocal means-based speckle filtering for ultrasound images," *IEEE Transactions on Image Processing*, vol. 18, no. 10, pp. 2221–2229, 2009.

[26] B. Zhou, A. Lapedriza, A. Khosla, A. Oliva, and A. Torralba, "Places: A 10 million image database for scene recognition," *IEEE Transactions on Pattern Analysis and Machine Intelligence*, vol. 40, no. 6, pp. 1452–1464, 2017.

[27] O. Russakovsky, J. Deng, H. Su, J. Krause, S. Satheesh, S. Ma, Z. Huang, A. Karpathy, A. Khosla, M. Bernstein *et al.*, "Imagenet large scale visual recognition challenge," *International Journal of Computer Vision*, vol. 115, no. 3, pp. 211–252, 2015.

[28] J. A. Jensen, "Field: A program for simulating ultrasound systems," in *Proc. 10th Nord Baltic Conference on Biomedical Imaging*, vol. 4, supplement 1, part 1. Citeseer, 1996, pp. 351–353.

[29] J. A. Jensen and N. B. Svendsen, "Calculation of pressure fields from arbitrarily shaped, apodized, and excited ultrasound transducers," *IEEE Transactions on Ultrasonics, Ferroelectrics and Frequency Control*, vol. 39, no. 2, pp. 262–267, 1992.

[30] J. Yang, J. Fan, D. Ai, X. Wang, Y. Zheng, S. Tang, and Y. Wang, "Local statistics and non-local mean filter for speckle noise reduction in medical ultrasound image," *Neurocomputing*, vol. 195, pp. 88–95, 2016.

[31] P. Coupé, P. Hellier, C. Kervrann, and C. Barillot, "Bayesian non local means-based speckle filtering," in *Proc. 5th IEEE International Symposium on Biomedical Imaging: From Nano to Macro*. IEEE, 2008, pp. 1291–1294.

[32] F. Dietrichson, E. Smistad, A. Ostvik, and L. Lovstakken, "Ultrasound speckle reduction using generative adversial networks," in *Proc. IEEE International Ultrasonics Symposium*. IEEE, 2018, pp. 1–4.

[33] J. Shin, L. Huang, and J. T. Yen, "Spatial prediction filtering for medical ultrasound in aberration and random noise," *IEEE Transactions on Ultrasonics, Ferroelectrics, and Frequency Control*, vol. 65, no. 10, pp. 1845–1856, 2018.

[34] L. L. Brickson, D. Hyun, M. Jakovljevic, and J. J. Dahl, "Reverberation noise suppression in ultrasound channel signals using a 3D fully convolutional neural network," *IEEE Transactions on Medical Imaging*, vol. 40, no. 4, pp. 1184–1195, 2021.

[35] A. C. Luchies and B. C. Byram, "Deep neural networks for ultrasound beamforming," *IEEE Transactions on Medical Imaging*, vol. 37, no. 9, pp. 2010–2021, 2018.

[36] A. Beck and M. Teboulle, "A fast Iterative shrinkage-thresholding algorithm for linear inverse problems," *SIAM Journal on Imaging Sciences*, vol. 2, no. 1, pp. 183–202, 2009.

[37] R. Mallart and M. Fink, "Adaptive focusing in scattering media through sound-speed inhomogeneities: The van Cittert Zernike approach and focusing criterion," *Journal of the Acoustical Society of America*, vol. 96, no. 6, pp. 3721–3732, 1994.

[38] C.-I. C. Nilsen and S. Holm, "Wiener beamforming and the coherence factor in ultrasound imaging," *IEEE Transactions on Ultrasonics, Ferroelectrics, and Frequency Control*, vol. 57, no. 6, pp. 1329–1346, 2010.

[39] M. Sharifzadeh, H. Benali, and H. Rivaz, "Phase aberration correction: A convolutional neural network approach," *IEEE Access*, vol. 8, pp. 162 252–162 260, 2020.

[40] J. J. Dahl, D. A. Guenther, and G. E. Trahey, "Adaptive imaging and spatial compounding in the presence of aberration," *IEEE Transactions on Ultrasonics, Ferroelectrics, and Frequency Control*, vol. 52, no. 7, pp. 1131–1144, 2005.

[41] M. O'donnell and S. Flax, "Phase aberration measurements in medical ultrasound: Human studies," *Ultrasonic Imaging*, vol. 10, no. 1, pp. 1–11, 1988.

[42] M. O'Donnell and W. E. Engeler, "Correlation-based aberration correction in the presence of inoperable elements," *IEEE Transactions on Ultrasonics, Ferroelectrics, and Frequency Control*, vol. 39, no. 6, pp. 700–707, 1992.

[43] J. Capon, "High-resolution frequency–wavenumber spectrum analysis," *Proceedings of the IEEE*, vol. 57, no. 8, pp. 1408–1418, 1969.

[44] B. Byram and M. Jakovljevic, "Ultrasonic multipath and beamforming clutter reduction: A chirp model approach," *IEEE Transactions on Ultrasonics, Ferroelectrics, and Frequency Control*, vol. 61, no. 3, pp. 428–440, 2014.

[45] B. Byram, K. Dei, J. Tierney, and D. Dumont, "A model and regularization scheme for ultrasonic beamforming clutter reduction," *IEEE Transactions on Ultrasonics, Ferroelectrics, and Frequency Control*, vol. 62, no. 11, pp. 1913–1927, 2015.

[46] F. Vignon and M. R. Burcher, "Capon beamforming in medical ultrasound imaging with focused beams," *IEEE Transactions on Ultrasonics, Ferroelectrics, and Frequency Control*, vol. 55, no. 3, pp. 619–628, 2008.

[47] B. Luijten, R. Cohen, F. J. De Bruijn, H. A. Schmeitz, M. Mischi, Y. C. Eldar, and R. J. Van Sloun, "Adaptive ultrasound beamforming using deep learning," *IEEE Transactions on Medical Imaging*, vol. 39, no. 12, pp. 3967–3978, 2020.

[48] J. F. Synnevag, A. Austeng, and S. Holm, "Adaptive beamforming applied to medical ultrasound imaging," *IEEE Transactions on Ultrasonics, Ferroelectrics, and Frequency Control*, vol. 54, no. 8, pp. 1606–1613, 2007.

[49] A. M. Deylami, J. A. Jensen, and B. M. Asl, "An improved minimum variance beamforming applied to plane-wave imaging in medical ultrasound," in *Proc. IEEE International Ultrasonics Symposium*. IEEE, 2016, pp. 1–4.

[50] T. Chernyakova, D. Cohen, M. Shoham, and Y. C. Eldar, "Imap beamforming for high-quality high frame rate imaging," *IEEE Transactions on Ultrasonics, Ferroelectrics, and Frequency Control*, vol. 66, no. 12, pp. 1830–1844, 2019.

[51] Y. Qi, Y. Guo, and Y. Wang, "Image quality enhancement using a deep neural network for plane wave medical ultrasound imaging," *IEEE Transactions on Ultrasonics, Ferroelectrics, and Frequency Control*, vol. 68, no. 4, pp. 926–934, 2020.

Part III

Generative Models for Biomedical Imaging

12 Image Synthesis in Multi-Contrast MRI with Generative Adversarial Networks

Tolga Çukur, Mahmut Yurt, Salman Ul Hassan Dar, Hyungjin Chung, and Jong Chul Ye

12.1 Introduction

The remarkable level and diversity of soft tissue contrasts in magnetic resonance imaging (MRI) has rendered it a preferred imaging modality for diagnostic imaging. Yet, prolonged examinations and associated healthcare costs often prohibit the acquisition of comprehensive multi-contrast protocols. Even when such protocols are viable, image quality might be compromised in a subset of acquisitions owing to system imperfections and/or uncooperative patients. As a result, there is emerging interest in the synthesis of MR images that are of poor quality or completely absent from the protocol [1–5]. Learning-based methods based on deep neural networks (DNNs) are gaining immense traction in this domain by their ability to capture joint distributions of multi-contrast MR images and to identify nonlinear mappings between separate contrasts [6–10].

Two main classes of synthesis approaches have come forth for multi-contrast MRI synthesis, namely unconditional and conditional methods. In unconditional synthesis, the aim is to generate new, independent, samples from a target image distribution. The most common base neural architecture basis for unconditional MRI synthesis has been generative adversarial networks (GANs). Such networks involve a game-theoretic competition between two subnetworks: a generator that tries to generate synthetic data samples, and a discriminator that tries to distinguish the fake samples from the generator using real samples of the desired data distribution. When the generator grows sufficiently strong to deceive the discriminator, it can generate realistic data samples. Given their unparalleled ability to learn data distributions, GANs have been adopted for synthesizing high-quality MR images given simply random noise samples as input [11, 12].

In conditional synthesis, the aim is to generate samples from a target distribution that are consistent with prior information from a source distribution. For MRI synthesis, the goal is to create an image from a target contrast given input images of the same anatomy from different source contrasts. As such, the structural content of the source images are used as prior information to conditionally improve the quality of the target samples generated. As in the unconditional case, GAN models have proven their exceptional performance for conditional MR image synthesis, as they characteristically offer much better capture of structural details compared with conventional learning-based models [8–10, 13–15].

Generative adversarial network models have enabled a leap in the realism and quality of multi-contrast MRI synthesis, powering it to improve the diagnostic value of MRI examinations. In this chapter, we discuss recent GAN-based synthesis approaches and their applications in MRI synthesis. We start with a quick overview of the physics underlying tissue contrasts in MRI. We then give a review of GAN basics, followed by an overview of existing unconditional and conditional models. We highlight collaborative GANs as a representative case of unified synthesis models for multiple distinct target contrasts. Finally, we close the chapter with a summary and outlook for future research in this area.

12.2 Physics for MR Contrast

Owing to the abundance of hydrogen in the body's water, it is the most widely imaged nucleus in MRI. Single protons within the hydrogen nuclei spin around a central axis along the direction of its magnetic moment. The spin axes are incoherently aligned in the absence of external inputs, resulting in zero net magnetization. However, when the spins are inserted into a static external magnetic field (B_o), they precess around the applied field at the Larmor frequency and a net magnetic moment builds up in the direction of B_o, i.e., the longitudinal direction. A radio-frequency (RF) pulse exerted on the spins then tips their orientation to the transverse plane, with the flip angle of the pulse determining the degree of tip. Once the RF pulse is turned off, the magnetization gradually returns to equilibrium by back-aligning itself with B_o. This restoration results in a decay of the transverse component of magnetization, with time constant T_2, and growth of the longitudinal component, with time constant T_1. The transverse component of the tissue magnetization is then sensed through flux changes on dedicated receiver coils. Meanwhile, additional magnetic field gradients are applied along Cartesian axes for spatial encoding. The resulting signal recorded at the receiver coil is given as

$$s(t) = \int_x \int_y \int_z M(x,y,z) e^{-t/T_2(\mathbf{r})} \exp\left(-i\gamma \int_0^t \mathbf{G}(\tau).\mathbf{r}\, d\tau\right) dxdydz \quad (12.1)$$

where $M(x,y,z)$ is the spatial distribution of the transverse magnetization, $\mathbf{r} = [x,y,z]$, $\mathbf{G} = [G_x, G_y, G_z]$, T_2 is the relaxation constant, and γ denotes the nucleus-specific gyromagnetic ratio. In the k-space formulation, the gradient terms are expressed with spatial frequency variables along each axis, k_x, k_y, and k_z. Setting $k_x(t) = (\gamma/2\pi) \int_0^t G_x(\tau) d\tau$, $k_y(t) = (\gamma/2\pi) \int_0^t G_y(\tau) d\tau$, and $k_z(t) = (\gamma/2\pi) \int_0^t G_z(\tau) d\tau$, the signal equation modifies to

$$s(t) = \int_x \int_y \int_z M(x,y,z) e^{-t/T_2(\mathbf{r})} e^{-i2\pi[k_x(t)+k_y(t)+k_z(t)]} dxdydz. \quad (12.2)$$

Note that Eq. (12.2) corresponds to the 3D Fourier transform of the magnetization, and k-space is simply the Fourier transform of the imaged volume. The periodic execution of the RF and gradient fields determines the signal levels of various tissues in MRI, which are dependent on the tissue-specific relaxation parameters (T_1 and T_2).

Therefore, the tissue contrasts in MRI can be manipulated by varying the RF pulse strength and pulse sequence timing. A mainstream sequence in MRI is the spin-echo sequence, whose signal is expressed as:

$$S_o = K\rho(\mathbf{r})[1 - e^{-TR/T_1(\mathbf{r})}]e^{-TE/T_2(\mathbf{r})} \tag{12.3}$$

where $\rho(\mathbf{r})$ is the proton density, K is the instrumental scaling constant, TR is the repetition time of the pulse sequence, TE is the echo time, and T_1 and T_2 are relaxation parameters. In spin-echo sequences, the tissue contrast can be adjusted to weight either the T_1 or T_2 values by altering TE/TR. For instance, T_1-weighted contrast can be obtained by choosing low TE and moderate TR, whereas T_2-weighted contrast can be obtained by choosing moderate TE and long TR. Through multiple acquisitions with different sequence parameters, the same anatomy can be imaged under different tissue contrasts, increasing the breadth of diagnostic information captured.

12.3 Brief Review of Generative Adversarial Networks (GANs)

Generative models are of great interest thanks to their ability to model high-dimensional probability distributions. In supervised learning, pairs of input–labels are given to the model, and the model learns the mapping between the pairs. On the other hand, the purpose of generative models, or unsupervised learning methods in general, is to approximate the distribution $p(\cdot)$ over the data x. Generative models can be divided into two large categories: explicit-density models and implicit-density models. While the former try to maximize the likelihood either directly or indirectly, the latter implicitly model the distribution, leveraging the property of being able to sample from the distribution.

Concretely, GANs define a prior $p(z)$ of input-noise variables which will be used to sample the data from the modeled distribution. An input variable then passes through a generator G, which maps the input variable to the distribution of the target high-dimensional data (e.g., the image). Another component that consists of a GAN is the discriminator D. The discriminator learns to represent the probability $D(x) \in [0, 1]$ that the given data x belongs to the real data distribution. Values closer to 0 mean that the discriminator has decided that it is unlikely that the given data was sampled from the real distribution, and values closer to 1 mean that the given data was realistic. The two components, G and D are pitted against each other in a mini–max game, where D tries to maximize the probability of correctly classifying the real and fake data, while G tries to minimize the probability of D being correct by generating realistic data to fool D [16]. Often, this competitive learning strategy is described with an analogy to a police officer (D) and a counterfeiter (G), where the counterfeiter gets better and better at deceiving the police officer, while the police officer gets consistently better at catching the counterfeiter.

The actual implementation of a GAN, as depicted in Fig. 12.1, models G and D with differentiable deep neural networks, and the network parameters are updated by optimizing the following loss function:

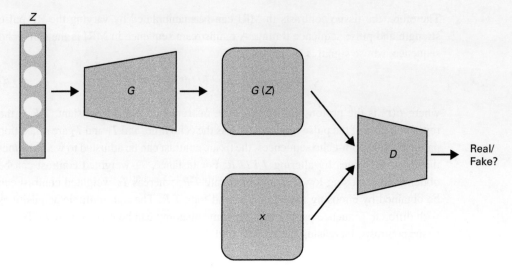

Figure 12.1 Illustration of a generative adversarial network

$$\min_{G} \max_{D} V(D, G), \qquad (12.4)$$

where

$$V(D, G) = \mathbb{E}_x[\log D(x)] + \mathbb{E}_z[\log(1 - D(G(z)))]. \qquad (12.5)$$

By iterating the update steps between G and D, the objective of GAN is to reach a Nash equilibrium, which means that each component is at its optimal state in relation to the opposite component. When the training is properly performed, the generator is able to produce very realistic data that are hard to distinguish from the real data.

While one acknowledges the strong representational power of GANs, they are also notorious for being hard to optimize. Often, optimization fails and the model either stagnates in bad local minima or ends up with mode collapse [17]. Therefore, the design of GAN loss functions has been an area of great interest [18–21]. When Eq. (12.5) is decoupled into two separate components, it can be written as

$$\begin{aligned} V^{(D)} &= -\mathbb{E}_x[\log D(x)] - \mathbb{E}_z[\log(1 - D(G(z)))], \\ V^{(G)} &= \mathbb{E}_z[\log(1 - D(G(z)))]. \end{aligned} \qquad (12.6)$$

Although the vanilla (i.e., basic) GAN loss described in Eq. (12.6) is mathematically sound, it does not perform particularly well in practice. A simple heuristic fix is to change the cost function for the generator to

$$V^{(G)} = -\mathbb{E}_z[\log D(G(z))]. \qquad (12.7)$$

By changing the cost function to Eq. (12.7), the generator receives a stronger gradient at the early stage of training, and thus is able to escape from bad local minima:

$$V^{(D)} = \mathbb{E}_x[D(x)] - \mathbb{E}_z[D(G(z))]$$
$$V^{(G)} = -\mathbb{E}_z[D(G(z))] \quad (12.8)$$

Later, the authors of the Wasserstein GAN (WGAN) [18] proposed to use the "earth mover's" (EM) distance as the metric between the generated distribution and the target distribution, as in Eq. (12.8). In contrast to minimizing the Kullback–Leibler (KL) divergence in the vanilla GAN [16], it was shown that the WGAN loss is superior at driving the cost to a global optimum both theoretically and experimentally. This idea was further extended in WGAN-gp [19], where the authors imposed a gradient penalty on top of the WGAN loss function to match a Lipschitz constraint.

Least squares GAN (LSGAN) [20] is another widely used GAN objective, described in Eq. (12.9):

$$V^{(D)} = \mathbb{E}_x[(D(x) - b)^2] + \mathbb{E}_z[(D(G(z)) - a)^2],$$
$$V^{(G)} = \mathbb{E}_z[(D(G(z)) - c)^2]. \quad (12.9)$$

The main motivation of LSGAN is to pull the distribution of the generated data to the real-data manifold; using this simple loss formulation has proved to be efficient in many applications including image-to-image translation [22, 23].

Thanks to the great advances in training strategy and the development of efficient neural architectures, GANs have revolutionized deep learning and generative modeling over the past few years. Interestingly, although the first intuitive use of GANs was to generate images out of random noise, GANs have been applied to various fields, such as text-to-image generation [24], photo blending [25], and image-to-image translation [22, 23]. Consequently, GANs are nowadays the main workhorse of unsupervised deep learning, achieving the state-of-the-art in many areas including the field of MR image synthesis.

12.4 MR Contrast Conversion using GAN

Generative adversarial models with convolutional neural network (CNN) backbones are being widely adopted for MR image synthesis with a high degree of realism and structural detail. Depending on whether prior information from source contrasts is available, two main classes of model emerge: unconditional and conditional GANs. In this section, we will overview the basics of the two types of model and discuss their existing applications.

12.4.1 Unconditional GANs

Unconditional GANs learn to generate samples from a random noise vector without any extra information regarding the target MR images [11, 12, 26–31]. The generator is trained with the aim of generating realistic-looking MR images, and the discriminator is trained to distinguish real images from synthetic images. Classical GANs are trained to minimize the following adversarial loss function:

$$\ell_{adv} = \mathbb{E}_{x_t}[\log(D(x_t))] + \mathbb{E}_z[\log(1 - D(G(z)))] \tag{12.10}$$

where z is the noise vector, x_t denotes an image from the target distribution, G denotes the generator, and D is the discriminator; G is trained to minimize ℓ_{adv} while D is trained to maximize ℓ_{adv}.

Unconditional GANs are highly suited for data augmentation, to help improve the training of learning-based methods. As such, they have found broad use in image analysis tasks including segmentation and classification. For instance, synthetic cardiac MRI samples have been shown to improve classification performance [12], and volumetric brain MRI samples have been leveraged for enhanced segmentation [27].

As discussed before, a main limitation of vanilla GANs is their training instabilities, which could compromise the realism and diversity of generated images. Enhanced divergence measures have been proposed to alleviate this limitation, including the earth mover's distance in Wasserstein GAN (WGAN) [18]; this has been demonstrated for multi-contrast brain MRI synthesis [28, 31, 32]. Another prominent approach rests on the Pearson χ^2 divergence in least squares GAN (LSGAN) [20]; this approach has been demonstrated in prostate MR image synthesis for subsequent classification [32].

Another limitation of vanilla GANs involves the difficulty of training them on full-resolution images directly. Progressively growing GANs (PCGANs) that sequentially increase the image resolution during synthesis have been successfully demonstrated for brain image synthesis in glioma patients [29] and later were adopted for segmentation [30].

12.4.2 Conditional GANs

Conditional GANs learn to synthesize images in a target domain, given input images from a separate source domain as prior information. In the case of MRI, the source and target can be images of the same anatomy under separate MR contrasts.

In cross-contrast image synthesis, the aim is to learn the nonlinear mapping among images of distinct MRI tissue contrasts. A pioneering study in this field introduced pixel-GAN (pGAN) and cycleGAN (cGAN) methods for paired and unpaired MRI synthesis, respectively [8]. In the presence of training data consisting of co-registered images of source and target contrasts, pGAN leverages the pixel-wise and perceptual losses between the synthesized and real target images in conjunction with an adversarial loss (Fig. 12.2).

The pixel-wise loss can be expressed as

$$\ell_{pix} = \mathbb{E}_{x_t,x_s}[\|G(x_s) - x_t\|_1] \tag{12.11}$$

where x_s denotes the image of the source contrast. The perceptual loss function can be expressed as:

$$\ell_{perc} = \mathbb{E}_{x_t,x_s}[\|V(G(x_s)) - V(x_t)\|_1] \tag{12.12}$$

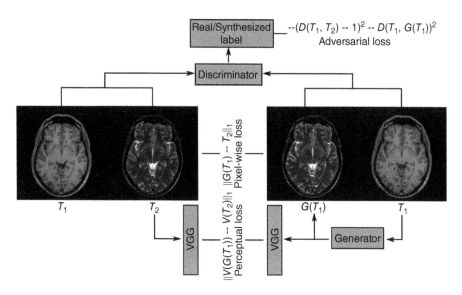

Figure 12.2 pGAN is a conditional GAN trained using co-registered images of source and target contrasts [8]. pGAN is trained to minimize an adversarial loss, a pixel-wise loss, and a perceptual loss. Note that, in the figure, T_1 and T_2 correspond respectively to x_s and x_t in Eq. (12.13). © 2019 IEEE. Reprinted, with permission, from [8]

where $V(\cdot)$ denotes a pre-trained computer vision model. The pGAN method enables a leap in accuracy of synthetic MRI over both conventional and prior deep-learning methods in the brain, as illustrated in Fig. 12.3.

For cases when paired images of source and target contrasts from the same set of subjects are unavailable, cGAN can be used to allow unpaired training with cycle-consistency loss (Fig. 12.4). Cycle-consistency enforces self-supervision on the GAN model:

$$\ell_{cc} = \mathbb{E}_{x_t, x_s}[\|G_{t \mapsto s}(G_{s \mapsto t}(x_s)) - x_s\|_1 + \|G_{s \mapsto t}(G_{t \mapsto s}(x_t)) - x_t\|_1] \quad (12.13)$$

where $G_{s \mapsto t}$ is trained to recover an image of the target contrast given an image of the source contrast, and $G_{t \mapsto s}$ is trained to synthesize an image of the source contrast given an image of the target contrast.

On the basis of the conditional GAN framework for multi-contrast MRI synthesis introduced by [8], later studies have either adopted it for other multi-contrast MRI synthesis applications [33, 34] or proposed additional loss terms to enforce enhanced priors [35, 36], and there have been many-to-one variants that aggregate information from multiple source contrasts [35, 37–39]. The authors of [37] incorporated multiple source contrasts by concatenating them at the input level in the form of a many-to-one GAN, and showed that this can enhance the overall recovery performance. In [38] it was demonstrated that traditional many-to-one GANs based on concatenation at the input level might be less sensitive to the unique information present within each source contrast, and this could lead to sub-optimal performance. The authors of [38] further

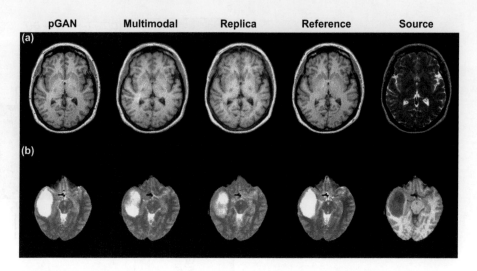

Figure 12.3 Representative T_1-weighted images of a healthy subject, and T_2-weighted images of a glioma patient recovered via pGAN: multimodal based on a traditional convolutional neural networks, and replica based on random forests are shown [8]. Compared with the basic the multimodal and replica, pGAN shows a remarkable recovery performance in both the healthy subject and the glioma patient. © 2019 IEEE. Reprinted, with permission, from [8]

showed that fusing features from multiple one-to-one GANs dedicated for each source contrast and a many-to-one GAN can lead to efficient recovery of both complementary and shared information across the source contrasts.

Several fundamental advances have been introduced to the main conditional GAN framework in recent studies. First, multi-tasking has been suggested as a means to improve the quality of MRI synthesis [40–44]. In [42] the authors performed synergistic synthesis and reconstruction by providing highly undersampled images of the target contrasts as additional priors. Similarly, [44] jointly super-resolved and synthesized the target contrast images.

Second, training instabilities were investigated particularly for 3D models whose complexity renders the learning process suboptimal for inevitably limited medical datasets. Spectral normalization and feature matching were proposed as two common methods to improve GAN training [45]. Self-attention modules were incorporated to reduce residual errors in focal, important, image regions [45]. Despite these advances, the need for massive datasets for training 3D models remains. Recently, the authors of [46] proposed the progressive decomposition of volumetric mapping into 2D mappings in a multi-planar fashion. The progressive model enforces spatial context to prevent the incoherence and artifacts commonly encountered in 2D models and can train accurate models for datasets that are several orders of magnitude smaller.

Most GAN-based models for MRI synthesis assume the availability of either paired or unpaired samples of multi-contrast MRI images from source and target domains. In practice, however, scan time limitations render it difficult to acquire fully sampled images. A recent study addresses this vital gap in the literature by introducing a semi-supervised GAN (ssGAN) model that can directly learn to synthesize MR images

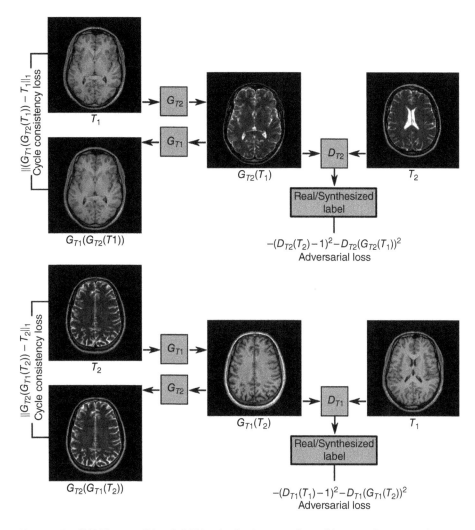

Figure 12.4 cGAN is a conditional GAN trained using unregistered images of source and target contrasts [8]. cGAN is trained to minimize an adversarial loss and a cycle-consistency loss. Note that, in the figure, T_1 and T_2 correspond respectively to x_s and x_t in Eq. (12.13). © 2019 IEEE. Reprinted, with permission, from [8]

from undersampled source and target contrasts [47]. A selective k-space loss function in ssGAN delivers a performance on a par with state-of-the-art supervised methods; it has highly accelerated datasets that are easier to collect.

12.5   Collaborative GAN for MR Contrast Conversion

Magnetic resonance contrast conversion is related to a topic that is broadly investigated in the field of statistics and machine learning: *missing data imputation*, or, more simply, *imputation*. Formally, the goal of imputation is to estimate the missing element in the dataset using the remainder of the data and statistical modeling. Based on different modeling assumptions and theories, there are several well-known methods for imputation: regression imputation, nonnegative matrix factorization, stochastic imputation, etc. [48, 49]. However, these conventional methods have limitations when dealing with images. Unlike low-dimensional tabular data, image data exist in a high-dimensional manifold, which makes them very hard to model directly. Fortunately, owing to the recent development of deep learning, image imputation has now become relevant. One typical method to tackle image imputation in the field of deep learning is to consider the problem as image-to-image translation [9, 10, 22, 23, 50].

Among many studies, cycleGAN [23] has been the main workhorse owing to its capability to train a network even in the case of unpaired settings. While in a typical GAN there exists a single generator–discriminator set, cycleGAN exploits two sets of generator–discriminator pairs to formally address the mapping between two different image spaces,

$$\hat{x}_t = G_{i \mapsto t}(x_i), \tag{12.14}$$

$$\hat{x}_i = G_{t \mapsto i}(x_t), \tag{12.15}$$

where i, t are the input domain and the target domain, respectively, and $G_{i \mapsto t}, G_{t \mapsto i}$ are the generators that execute the mapping. However, a major shortcoming of using cycleGAN is that it cannot be used in situations where we have N different domains of interest. If we were to use cycleGAN to deal with all N domains, then we would need $N(N-1)$ different generators, which would be prohibitive.

To consider multiple domains effectively, starGAN [50] was proposed. In starGAN, a *single* generator–discriminator set is used. Formally put, the mapping in starGAN is described as

$$\hat{x}_t = G(x_i; t), \tag{12.16}$$

where the single generator G is conditioned with t, and the subscript from Eq. (12.14) is dropped. Here, the generator knows the target domain t to be estimated by a mask vector that is concatenated with the input image. However, starGAN is also not a perfect fit for image imputation. Although it is able to exploit multiple domains with a single generator, it uses a *single* input to estimate the missing component, when it would be much more plausible to use multiple information from different domains synergistically.

12.5.1 Collaborative GAN

Accordingly, it would be beneficial to design a mapping to consider all other input images for missing data imputation. This idea was first proposed in collaborativeGAN (collaGAN) [9, 10]. A conceptual diagram of collaGAN is depicted in Fig. 12.5, where an example of synthesizing MR contrast using collaGAN is shown.

In this subsection, we formally address the collaGAN framework, handling multiple input domains to generate more feasible image imputation. For simplicity in expla-

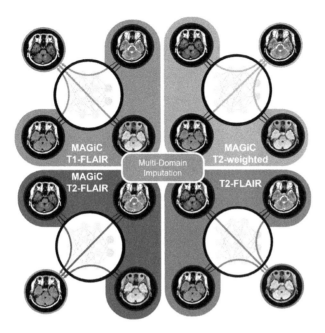

Figure 12.5 Conceptual diagram of collaGAN. © [2020] *Nature Machine Intelligence*, vol. 2, no. 1, pp. 34–42. Springer Nature. CollaGAN enables flexible multi-domain imputation within a single GAN architecture.

nation, we will assume that we have $N = 4$ domains: $a, b, c,$ and d. When the target domain is a, and a target image x_a exists, collaGAN tries to estimate a collaborative mapping from the set of images that exist in different domains: $\{x_a\}^C = \{x_b, x_c, x_d\}$. Here, the superscript C denotes the complementary set. More properly, we have

$$\hat{x}_t = G(\{x_t\}^C; t), \qquad (12.17)$$

where $t \in \{a, b, c, d\}$ is the corresponding target domain for each missing image. Notice that there are N different combinations of multiple inputs and single outputs. These combinations are randomly chosen at the training stage so that the generator learns to synergistically combine multiple images from different domains to synthesize the missing image. In the following, we review the overall flow of collaGAN and the training methodology.

Multiple Cycle-Consistency Loss

In cycleGAN, the cycle-consistency loss is used to impose a constraint that the mapping between the two domains should be the inverse of each other. In collaGAN, this cycle-consistency loss is redefined so that the inverse mapping can be achieved in *any* other domains. Specifically, let us assume that we have generated an output in domain a, denoted as $\hat{x}_a$, with the generator G, as shown in the middle panel of Fig. 12.6. Subsequently, we can define three different inverse mappings:

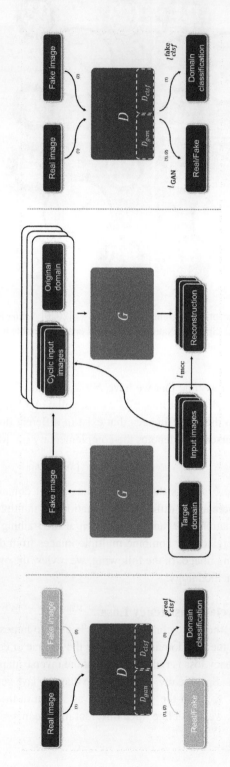

Figure 12.6 Training pipeline for collaGAN.

$$\tilde{x}_{b|a} = G(\{\hat{x}_a, x_c, x_d\}; b) \tag{12.18}$$

$$\tilde{x}_{c|a} = G(\{\hat{x}_a, x_b, x_d\}; c) \tag{12.19}$$

$$\tilde{x}_{d|a} = G(\{\hat{x}_a, x_b, x_c\}; d). \tag{12.20}$$

Then, we can calculate the *multiple cycle-consistency loss* for domain a as the following:

$$\ell_{mcc,a} = \|x_b - \tilde{x}_{b|a}\|_1 + \|x_c - \tilde{x}_{c|a}\| + \|x_d - \tilde{x}_{d|a}\|_1. \tag{12.21}$$

This is illustrated in the middle panel of Fig. 12.6 indexed with ℓ_{mcc} and a bidirectional arrow. Generally, ℓ_{mcc} reads

$$\ell_{mcc,t} = \sum_{t' \neq t} \|x_{t'} - \tilde{x}_{t'|t}\|_1. \tag{12.22}$$

In the original paper [9], the authors claim to use a more sophisticated cycle-consistency loss on top of the ℓ_1 loss, coined the structural similarity index loss. The new quantity directly minimizes the structural similarity index (SSIM), which leads to a more feasible output:

$$\tilde{x}_{t'|t} = G(\{\hat{x}_t\}^C; t'). \tag{12.23}$$

Interested readers are referred to [9].

Discriminator Loss

Now that the cycle-consistency loss has been defined, we are ready to define the discriminator loss designed for multi-domain image imputation. In most GAN architectures, the role of the discriminator is to learn to tell how realistic the given image is. However, there is one more role of the discriminator in collaGAN. It also has to classify to which domain the given image belongs. A schematic diagram of the colla-GAN discriminator is depicted in Fig. 12.6 (left) and (right). Within the discriminator D we see two components: D_{gan} and D_{clsf}. Here, D_{gan} refers to the adversarial loss, defined analogously to the other GAN architectures, and D_{clsf} refers to the domain classification loss. Before jointly training the generator and the discriminator, the domain classifier part of the discriminator, D_{clsf}, is trained with only real images:

$$\ell_{clsf}^{real}(D_{clsf}) = \mathbb{E}_{x_t}[-\log(D_{clsf}(t; x_t))], \tag{12.24}$$

where $(D_{clsf}(t; x_t))$ outputs a probability value of x_t belonging to the class t. Training D_{clsf} prior to jointly training G and D stabilizes training by building a strong classifier to provide a better guide to the generator. At the actual joint training stage, the generated fake images are also used to update the parameters in G with the following loss function:

$$\ell_{clsf}^{fake}(G) = \mathbb{E}_{\hat{x}_{t|t}}[-\log(D_{clsf}(t; \hat{x}_{t|t}))]. \tag{12.25}$$

The other part of the discriminator minimizes the LSGAN loss [20], known to stabilize the training process while preventing mode collapse. The parameters of D_{gan} are updated by minimizing

$$\ell_{gan}(D_{gan}) = \mathbb{E}_{x_t}[(D_{gan}(x_t) - 1)^2] + \mathbb{E}_{\tilde{x}_{t|t}}[(D_{gan}(\tilde{x}_{t|t}))^2], \qquad (12.26)$$

whereas the parameters of G are updated by minimizing

$$\ell_{gan}(G) = \mathbb{E}_{\tilde{x}_{t|t}}[(D_{gan}(\tilde{x}_{t|t}) - 1)^2]. \qquad (12.27)$$

12.5.2 MR Contrast Synthesis using CollaGAN

As discussed in earlier sections, different combinations of MR contrast images deliver diverse information about the patient being scanned. Among them, T_1-weighted (T1), T_2-weighted (T2), gadolinium-contrast-enhanced T_1-weighted (T1Gd), T_2 fluid-attenuated inversion recovery (T2F) are canonical examples of MR contrasts that are widely used in clinical situations. Full acquisition of different MR contrast images would be the most beneficial, but in most cases this is hard to achieve: acquiring a complete set of MR contrast images requires a painfully long scanning time, and the protocols among different medical centers vary. Even if a complete set has been acquired, systematic and operational errors have often corrupted one of the images. Subsequently, such images cannot be used, which eventually leads to statistical errors that hinder exact analysis [48].

Recently, a synthetic MRI technique called magnetic resonance image compilation (MAGiC, GE Healthcare) [51] has grown in popularity owing to its ability to generate multiple contrast MR images using the newly developed multidynamic multiecho (MDME) scan. With MAGiC, one can generate different contrast images such as T_1, T_2, and T_2-FLAIR, but it is known to generate substantial artifacts [51–53]. For example, it is known that the SNR is degraded with MAGiC FLAIR in comparison with conventional FLAIR, and the flow-related artifacts are enhanced. Hence, even after the acquisition of the MAGiC sequence, additional MR scans need to be incorporated to confirm the diagnosis, which eventually increases time and cost.

Instead, when we have erroneous acquisition among one of the contrast images, we can directly use collaGAN to impute the missing data, indicated by the question marks in Fig. 12.7. In the figure we can see that collaGAN is indeed able to impute the missing data in any given domain: MAGiC T_1-FLAIR, T_2, MAGiC T_2-FLAIR, and T_2-FLAIR. It is notable that by jointly combining the information from multiple domains, collaGAN is able to provide more accurate results compared with methods such as cycleGAN or starGAN which utilize a single-input domain.

The versatility of collaGAN readily extends to the imputation of pathological data. For example, in Fig. 12.8, we see the results of reconstruction with visible lesions. In (a), collaGAN is able to perform reconstruction even with an erroneous MAGiC T_2-FLAIR image, showing a hyperintensity signal of the cerebrospinal fluid (CSF). Abnormal signals are also well reconstructed in (b). In (c), while the MAGiC T_2-FLAIR image does not capture the hyperintensity signal, indicated by the arrow, collaGAN is able to reconstruct the signal. In (d), we see that collaGAN again corrects for the error induced in MAGiC T_2-FLAIR. From the reconstruction results, it is clear that collaGAN can accurately reconstruct missing MR contrasts.

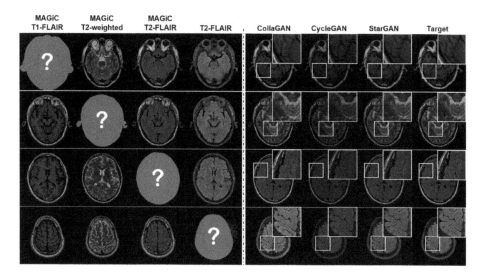

Figure 12.7 MR contrast imputation results using different methods. © [2020] *Nature Machine Intelligence*, vol. 2, no. 1, pp. 34–42. Springer Nature.

Assessing the Importance of MR Contrast

Interesting enough, this also leads to a fundamental question – can collaGAN synthesize *any* contrast? If not, what are the conditions required for collaGAN to perform well? In fact, this question was rigorously investigated in [10]. Specifically, by estimating specific contrasts one by one, it is possible to verify which contrasts can or cannot be generated through the imputation process, and ultimately which contrasts are irreplaceable. To ponder this question, the present authors performed a quantitative study by comparing the segmentation performance of the synthesized MR contrasts, replacing real contrast images with the synthesized images, one by one. If the segmentation performance did not drop even with the synthesized image, then it would mean that the contrast was replaceable. On the other hand, if there were a sufficient performance drop, that would mean that such contrasts were irreplaceable. From the results, the authors concluded that T1, T2, and T2F images are replaceable with synthetic images from collaGAN, which shows almost no difference in performance. However, a clear distinction was seen for a gadolinium contrast agent injected T_1-weighted image. From the quantitative results, we can again verify that Gd-enhanced contrasts are crucial for segmentation performance, and they are not replaceable with collaGAN imputation.

12.6 Summary and Outlook

For both unconditional synthesis and conditional synthesis, GANs have shown to be the perfect fit to such means thanks to their ability to learn probability distributions.

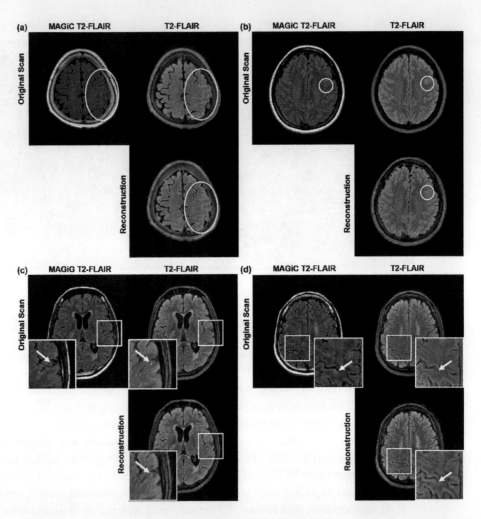

Figure 12.8 MR contrast imputation results on pathological data using collaGAN. © [2020] *Nature Machine Intelligence*, vol. 2, no. 1, pp. 34–42. Springer Nature.

For unconditional synthesis, the objective is to stochastically generate MR images of target contrast. Conditional synthesis refers to the case where the model is to learn a nonlinear mapping to the different MR tissue contrasts without altering the physiological information. Furthermore, by merging collaborative information about multiple contrast images, missing data imputation among many different domains is also effectively solved with GANs. Although promising results have been shown, developments in the area are still at an early stage. Interesting research directions have been proposed in previous work, which include application of the more advanced methods and rigorous validation in clinical settings. In effect, MRI image synthesis techniques should be able to reduce the burden of costly MR scans, benefiting both patients and hospitals.

References

[1] S. Roy, A. Carass, and J. Prince, "A compressed sensing approach for MR tissue contrast synthesis," in *Proc. Conference on Information Processing in Medical Imaging*, vol. 22, 2011, pp. 371–383.

[2] ——, "Magnetic resonance image example-based contrast synthesis," *IEEE Transactions on Medical Imaging*, vol. 32, no. 12, pp. 2348–2363, 2013.

[3] Y. Huang, L. Beltrachini, L. Shao, and A. F. Frangi, "Geometry regularized joint dictionary learning for cross-modality image synthesis in magnetic resonance imaging," in *Proc. International Workshop on Simulation and Synthesis in Medical Imaging*. Springer, 2016, pp. 118–126.

[4] Y. Huang, L. Shao, and A. F. Frangi, "Cross-modality image synthesis via weakly-coupled and geometry co-regularized joint dictionary learning," *IEEE Transactions on Medical Imaging*, vol. 37, no. 3, pp. 815–827, 2018.

[5] A. Jog, A. Carass, S. Roy, D. L. Pham, and J. L. Prince, "Random forest regression for magnetic resonance image synthesis," *Medical Image Analysis*, vol. 35, pp. 475–488, 2017.

[6] H. Van Nguyen, K. Zhou, and R. Vemulapalli, "Cross-domain synthesis of medical images using efficient location-sensitive deep network," in *Proc. Conference on Medical Image Computing and Computer-Assisted Intervention*. Springer, 2015, pp. 677–684.

[7] A. Chartsias, T. Joyce, M. Valerio Giuffrida, and S. Tsaftaris, "Multimodal MR synthesis via modality-invariant latent representation," *IEEE Transactions on Medical Imaging*, vol. 37, no. 3, pp. 803–814, 2018.

[8] S. U. H. Dar, M. Yurt, L. Karacan, A. Erdem, E. Erdem, and T. Çukur, "Image synthesis in multi-contrast MRI with conditional generative adversarial networks," *IEEE Transactions on Medical Imaging*, vol. 38, no. 10, pp. 2375–2388, 2019.

[9] D. Lee, J. Kim, W.-J. Moon, and J. C. Ye, "Collagan: Collaborative GAN for missing image data imputation," in *Proc. IEEE/CVF Conference on Computer Vision and Pattern Recognition*, 2019, pp. 2487–2496.

[10] D. Lee, W.-J. Moon, and J. C. Ye, "Assessing the importance of magnetic resonance contrasts using collaborative generative adversarial networks," *Nature Machine Intelligence*, vol. 2, no. 1, pp. 34–42, 2020.

[11] F. Calimeri, A. Marzullo, C. Stamile, and G. Terracina, "Biomedical data augmentation using generative adversarial neural networks," in *Proc. Conference on Artificial Neural Networks and Machine Learning*. Springer, 2017, pp. 626–634.

[12] L. Zhang, A. Gooya, and A. F. Frangi, "Semi-supervised assessment of incomplete LV coverage in cardiac MRI using generative adversarial nets," in *Proc. International Workshop on Simulation and Synthesis in Medical Imaging*. Springer, 2017, pp. 61–68.

[13] B. Yu, L. Zhou, L. Wang, J. Fripp, and P. Bourgeat, "3D cGAN based cross-modality MR image synthesis for brain tumor segmentation," in *Proc. IEEE 15th International Symposium on Biomedical Imaging*, 2018, pp. 626–630.

[14] F. Liu, "SUSAN: Segment unannotated image structure using adversarial network," *Magnetic Resonance in Medicine*, vol. 81, no. 5, pp. 3330–3345, 2018.

[15] S. Olut, Y. H. Sahin, U. Demir, and G. Unal, "Generative adversarial training for MRA image synthesis using multi-contrast MRI," in *Proc. Conference on Predictive Intelligence in Medicine*. Springer, 2018, pp. 147–154.

[16] I. J. Goodfellow, J. Pouget-Abadie, M. Mirza, B. Xu, D. Warde-Farley, S. Ozair, A. Courville, and Y. Bengio, "Generative adversarial networks," in *Advances in Neural Information Processing Systems*, 2014, pp. 2672–2680.

[17] I. Goodfellow, "Nips 2016 tutorial: Generative adversarial networks," *arXiv:1701.00160*, 2016.

[18] M. Arjovsky, S. Chintala, and L. Bottou, "Wasserstein generative adversarial networks," in *Proc. International Conference on Machine Learning*, 2017, pp. 214–223.

[19] I. Gulrajani, F. Ahmed, M. Arjovsky, V. Dumoulin, and A. C. Courville, "Improved training of wasserstein GANs," in *Advances in Neural Information Processing Systems*, 2017, pp. 5767–5777.

[20] X. Mao, Q. Li, H. Xie, R. Y. Lau, Z. Wang, and S. Paul Smolley, "Least squares generative adversarial networks," in *Proc. IEEE International Conference on Computer Vision*, 2017, pp. 2794–2802.

[21] J. H. Lim and J. C. Ye, "Geometric GAN," *arXiv:1705.02894*, 2017.

[22] P. Isola, J.-Y. Zhu, T. Zhou, and A. A. Efros, "Image-to-image translation with conditional adversarial networks," in *Proc. IEEE Conference on Computer Vision and Pattern recognition*, 2017, pp. 1125–1134.

[23] J.-Y. Zhu, T. Park, P. Isola, and A. A. Efros, "Unpaired image-to-image translation using cycle-consistent adversarial networks," in *Proc. IEEE International Conference on Computer Vision*, 2017, pp. 2223–2232.

[24] H. Zhang, T. Xu, H. Li, S. Zhang, X. Wang, X. Huang, and D. N. Metaxas, "StackGAN: Text to photo-realistic image synthesis with stacked generative adversarial networks," in *Proc. International Conference on Computer Vision*, 2017, pp. 5907–5915.

[25] H. Wu, S. Zheng, J. Zhang, and K. Huang, "Gp-GAN: Towards realistic high-resolution image blending," in *Proc. 27th ACM International Conference on Multimedia*, 2019, pp. 2487–2495.

[26] C. Bermudez, A. J. Plassard, L. T. Davis, A. T. Newton, S. M. Resnick, and B. A. Landman, "Learning implicit brain MRI manifolds with deep learning," in *Proc. Conference on Medical Imaging 2018: Image Processing*, vol. 10 574, SPIE, 2018, pp. 408–414.

[27] A. K. Mondal, J. Dolz, and C. Desrosiers, "Few-shot 3D multi-modal medical image segmentation using generative adversarial learning," *arXiv*, 2018.

[28] Y. Han and J. C. Ye, "Framing U-Net via deep convolutional framelets: Application to sparse-view CT," *IEEE Transactions on Medical Imaging*, vol. 37, no. 6, pp. 1418–1429, 2018.

[29] A. Beers, J. Brown, K. Chang, J. P. Campbell, S. Ostmo, M. F. Chiang, and J. Kalpathy-Cramer, "High-resolution medical image synthesis using progressively grown generative adversarial networks," *arXiv:1805.03144*, 2018.

[30] C. Bowles, L. Chen, R. Guerrero, P. Bentley, R. Gunn, A. Hammers, D. A. Dickie, M. V. Hernández, J. Wardlaw, and D. Rueckert, "GAN augmentation: Augmenting training data using generative adversarial networks," *arXiv:1810.10863*, 2018.

[31] G. Kwon, C. Han, and D. Shik Kim, "Generation of 3D brain MRI using auto-encoding generative adversarial networks," in *Proc. Conference on Medical Image Computing and Computer Assisted Intervention*, vol. 11 766. Springer, 2019, pp. 118–126.

[32] H. Yu and X. Zhang, "Synthesis of prostate MR images for classification using capsule network-based GAN model," *Sensors*, vol. 20, no. 20, p. 5736, 2020.

[33] D. Abramian and A. Eklund, "Refacing: Reconstructing anonymized facial features using GANs," in *2019 IEEE 16th International Symposium on Biomedical Imaging*, 2019, pp. 1104–1108.

[34] X. Gu, H. Knutsson, M. Nilsson, and A. Eklund, "Generating diffusion MRI scalar maps from T1 weighted images using generative adversarial networks," in *Image Analysis*. Springer, 2019, pp. 489–498.

[35] B. Yu, L. Zhou, L. Wang, Y. Shi, J. Fripp, and P. Bourgeat, "Ea-GANs: Edge-aware generative adversarial networks for cross-modality MR image synthesis," *IEEE Transactions on Medical Imaging*, vol. 38, no. 7, pp. 1750–1762, 2019.

[36] X. Liu, F. Xing, J. L. Prince, A. Carass, M. Stone, G. E. Fakhri, and J. Woo, "Dual-cycle constrained bijective vae-GAN for tagged-to-cine magnetic resonance image synthesis," *arXiv:2101.05439*, 2021.

[37] A. Sharma and G. Hamarneh, "Missing MRI pulse sequence synthesis using multi-modal generative adversarial network," *IEEE Transactions on Medical Imaging*, vol. 39, no. 4, pp. 1170–1183, 2020.

[38] M. Yurt, S. U. Dar, A. Erdem, E. Erdem, K. K. Oguz, and T. Çukur, "mustGAN: Multi-stream generative adversarial networks for MR image synthesis," *Medical Image Analysis*, vol. 70, p. 101 944, 2021.

[39] B. E. Dewey, C. Zhao, J. C. Reinhold, A. Carass, K. C. Fitzgerald, E. S. Sotirchos, S. Saidha, J. Oh, D. L. Pham, P. A. Calabresi, P. C. van Zijl, and J. L. Prince, "DeepHarmony: A deep learning approach to contrast harmonization across scanner changes," *Magnetic Resonance Imaging*, vol. 64, pp. 160–170, 2019.

[40] L. Xiang, Y. Chen, W. Chang, Y. Zhan, W. Lin, Q. Wang, and D. Shen, "Deep-learning-based multi-modal fusion for fast MR reconstruction," *IEEE Transactions on Biomedical Engineering*, vol. 66, no. 7, pp. 2105–2114, 2019.

[41] A. Falvo, D. Comminiello, S. Scardapane, M. Scarpiniti, and A. Uncini, "A multimodal dense U-Net for accelerating multiple sclerosis MRI," in *Proc. IEEE 29th (MLSP)*. IEEE, 2019.

[42] S. Dar, M. Yurt, M. Shahdloo, M. Ildiz, B. Tinaz, and T. Cukur, "Prior-guided image reconstruction for accelerated multi-contrast MRI via generative adversarial networks," *IEEE Journal on Selected Topics in Signal Processing*, vol. 14, no. 6, pp. 1072–1087, 2020.

[43] W.-J. Do, S. Seo, Y. Han, J. C. Ye, S. H. Choi, and S.-H. Park, "Reconstruction of multicontrast MR images through deep learning," *Medical Physics*, vol. 47, no. 3, pp. 983–997, 2020.

[44] K. H. Kim, W. J. Do, and S. H. Park, "Improving resolution of MR images with an adversarial network incorporating images with different contrast," *Medical Physics*, vol. 45, no. 7, pp. 3120–3131, 2018.

[45] H. Lan, A. W. Toga, and F. Sepehrband, "SC-GAN: 3D self-attention conditional GAN with spectral normalization for multi-modal neuroimaging synthesis," *bioRxiv*, 2020.

[46] M. Yurt, M. Özbey, S. U. H. Dar, B. Tınaz, K. K. Oğuz, and T. Çukur, "Progressively volumetrized deep generative models for data-efficient contextual learning of MR image recovery," *arXiv preprint arXiv:2011.13913*, 2020.

[47] M. Yurt, S. U. H. Dar, B. Tınaz, M. Özbey, and T. Çukur, "Semi-supervised learning of mutually accelerated multi-contrast MRI synthesis without fully-sampled ground-truths," *arXiv:2011.14347*, 2020.

[48] A. N. Baraldi and C. K. Enders, "An introduction to modern missing data analyses," *Journal of School Psychology*, vol. 48, no. 1, pp. 5–37, 2010.

[49] C. K. Enders, *Applied Missing Data Analysis*. Guilford Press, 2010.

[50] Y. Choi, M. Choi, M. Kim, J.-W. Ha, S. Kim, and J. Choo, "StarGAN: Unified generative adversarial networks for multi-domain image-to-image translation," in *Proc. IEEE Conference on Computer Vision and Pattern Recognition*, 2018, pp. 8789–8797.

[51] L. N. Tanenbaum, A. J. Tsiouris, A. N. Johnson, T. P. Naidich, M. C. DeLano, E. R. Melhem, P. Quarterman, S. Parameswaran, A. Shankaranarayanan, M. Goyen *et al.*, "Synthetic MRI for clinical neuroimaging: Results of the magnetic resonance image compilation (magic) prospective, multicenter, multireader trial," *American Journal of Neuroradiology*, vol. 38, no. 6, pp. 1103–1110, 2017.

[52] A. Hagiwara, M. Warntjes, M. Hori, C. Andica, M. Nakazawa, K. K. Kumamaru, O. Abe, and S. Aoki, "SyMRI of the brain: Rapid quantification of relaxation rates and proton density, with synthetic MRI, automatic brain segmentation, and myelin measurement," *Investigative Radiology*, vol. 52, no. 10, p. 647, 2017.

[53] A. Hagiwara, M. Hori, K. Yokoyama, M. Takemura, C. Andica, T. Tabata, K. Kamagata, M. Suzuki, K. Kumamaru, M. Nakazawa *et al.*, "Synthetic MRI in the detection of multiple sclerosis plaques," *American Journal of Neuroradiology*, vol. 38, no. 2, pp. 257–263, 2017.

13 Regularizing Deep-Neural-Network Paradigm for the Reconstruction of Dynamic Magnetic Resonance Images

Jaejun Yoo and Michael Unser

In this chapter, we provide an overview of a recent image-reconstruction method that uses a deep generative algorithm for dynamic magnetic resonance imaging (dMRI). We begin by briefly introducing the imaging modality of dMRI, the associated image-reconstruction problem, and existing reconstruction approaches. Next, we introduce a time-dependent deep image prior (TD-DIP), which exploits the structure of convolutional neural networks (CNNs) as a regularizing prior. We show some representative results and discuss the pros and cons of this regularizing paradigm. Finally, we discuss a few potential remaining limitations.

13.1 Introduction

Magnetic resonance imaging (MRI) is a noninvasive and high-fidelity imaging technology that plays a central role in medical research and diagnosis [1]. By leveraging the nuclear magnetic resonance (NMR) phenomenon, an MRI scanner collects k-space measurements which are samples of the spatial Fourier transform of the spin distribution within a specimen (Fig. 13.1). Given these samples, the MR image can be decoded by the inverse Fourier transform of the measured data. Hence, to obtain a high-quality image without aliasing artifacts, the k-space (or frequency domain) must be sampled densely enough to satisfy the Nyquist sampling criterion.

13.1.1 Challenges of dMRI

Unlike static MRI, which captures a single image, the aim of dMRI is to monitor dynamic processes such as structural and functional changes in the heart, lung, or liver. Now, the difficulty with dynamic imaging is that sampling must be also performed along the time axis while satisfying the sampling criterion (i.e., (k, t)-space, Fig. 13.1). However, owing to physical constraints, MR scanners have a fixed sampling rate, which makes it difficult to keep up with fast organ motions. The resulting undersampling causes signal aliasing and distortions in the reconstructed images; to overcome this calls for sophisticated reconstruction procedures.

13.1.2 Image-Formation Model of dMRI

Before we introduce MRI reconstruction approaches, we begin with a brief discussion of the MR forward model with radial sampling pattern (Fig. 13.1). This pattern is often used in dMRI due to its robustness to movement artifacts [2–4].

Let $\mathbf{x} \in \mathbb{C}^N$ be the vectorized two-dimensional (2D) image of the underlying specimen, where N is the number of pixels, and let $\mathbf{y}(t) \in \mathbb{C}^M$ be its k-space measurements acquired from the MR scanner at time t. This is called an angular spoke. It consists of M uniform samples of a radial line in k-space at orientation $\vartheta(t) \in [0, \pi)$. By invoking the central-slice theorem, we can interpret a spoke as the 1D Fourier transform of the Radon transform of the image at angle ϑ. This measurement process can be described by the linear relation

$$\mathbf{y}(t) = \mathbf{H}(\vartheta)\mathbf{x}(t), \tag{13.1}$$

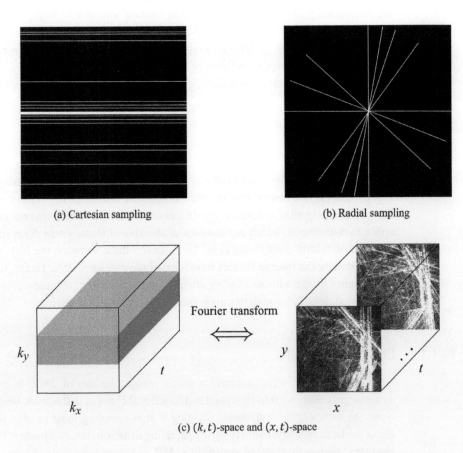

Figure 13.1 Canonical k-space sampling patterns and measurement space of dMRI. (a) k-space with 1D Cartesian sampling pattern; (b) k-space with radial sampling pattern; (c) (k, t)-space; the navigator data are denoted by the gray region, which is the low-frequency part of (k, t)-space.

where $\mathbf{H}(\vartheta)$ is the $M \times N$ system matrix that represents the combined effect of taking the 2D Fourier transform of $\mathbf{x}$ and resampling along a radial line of direction ϑ.

In practice, we acquire a series of S spokes taken at regularly spaced time points $t_s = t_0 + s\Delta t$, $s = 0, \ldots, S-1$ with step size Δt. The orientations of the spokes follow the uniform-angle or golden-angle strategy [3, 4]

$$\vartheta_s = \vartheta_0 + \omega_0 s \Delta t, \qquad (13.2)$$

where ϑ_s gives the orientation of a spoke at time $t_s = t_0 + s\Delta t$, with ω_0 its angular velocity. The golden-angle strategy is simply a specific way to add irrationality into the sampling process by setting $\omega_0 \Delta t / \pi \notin \mathbb{Q}$. This is approximated by setting $\omega_0 \Delta t \approx 111.25°$, which allows one to sample radial lines in k-space with minimal information overlap. This results in the forward imaging model

$$\mathbf{y}_s = \mathbf{H}_s \mathbf{x}_s, \qquad (13.3)$$

where the matrix $\mathbf{H}_s$ encodes the acquisition in direction θ_s at time t_s. Our task is to reconstruct the image sequence $\{\mathbf{x}_s\}_{s=0}^{S-1}$ or matrix $\mathbf{X} = [\mathbf{x}_0 \cdots \mathbf{x}_{S-1}] \in \mathbb{C}^{N \times S}$ from the measurement sequence $\{\mathbf{y}_s\}_{s=0}^{S-1}$ or matrix $\mathbf{Y} \in \mathbb{C}^{M \times S}$.

However, the reconstruction problem associated with Eq. (13.3) is severely ill-posed because $M \ll N$. The difficulty lies in the limited number of k-space samples that can be acquired within a unit of time. This causes a trade-off between spatial and temporal resolutions, which hinders the detection of fast events or structures concentrated in small areas.

13.2 Reconstruction Approaches in dMRI

Giving a limited budget of samples, a typical way to address the undersampling problem is to choose a smart acquisition of the samples in (k,t)-space. Then the loss of information is compensated by assuming data redundancy in either the spatial or the temporal domain, or both [1]. In this section, we introduce a few representative approaches. Note that our list is far from being exhaustive; we refer the interested readers to [1] and references therein for a more extensive survey.

13.2.1 Sparsity and Low-Rank-Based Methods

Compressed sensing (CS) is one of the most representative approaches that exploit data redundancy. Assuming that the underlying spatiotemporal patterns can be compactly represented in some transform domain, CS dynamic imaging solves the following regularized optimization problem using the forward model (13.3):

$$\mathbf{X}^* = \arg\min_{\mathbf{X}} \|\mathbf{Y} - \mathbf{H}\mathbf{X}\|_2^2 + \lambda \mathcal{R}\{\mathbf{X}\}, \qquad (13.4)$$

where $\mathcal{R}(\cdot)$ is a sparsity-inducing regularization functional and $\mathbf{H} = [\mathbf{H}_0 \ldots \mathbf{H}_{S-1}]$. To encourage temporal dependency, a finite-difference operator $\mathbf{D}$ is typcally used as a sparsifying transform along the temporal domain, leading to $\mathcal{R}\{\mathbf{X}\} = \|\mathbf{D}\mathbf{X}\|_1$. Since its introduction by Lustig et al. [5], compressed sensing has been the main workhorse in modern accelerated MRI research [3, 4, 6–12].

Closely related to the concept of sparsity is the idea of low-rank regularization, which has also been actively explored in dMRI [13–15]. Instead of using a sparsifying transform with an ℓ_1 norm, one can use a nuclear norm as a convex surrogate of the rank function, leading to $\mathcal{R}\{\mathbf{X}\} = \|\mathbf{X}\|_*$, as demonstrated by Candès and Recht [16]. In fact, there exists a close relationship between compressed sensing and low-rank algorithms. Jin and his colleagues [17–20] found a duality between the CS and low-rank Hankel-matrix approaches. On the basis of this observation, Jin et al. proposed a unified framework for compressed sensing and parallel MRI in terms of low-rank Hankel-matrix approaches [17]. They also provided theoretical performance guarantees in [21].

Nevertheless, these schemes seem to be less effective when it comes to dMRI with extensive interframe motion. To address this in cardiac applications, motion-resolved reconstruction strategies have been proposed. By using electrocardiograms or self-gating techniques, the idea is to rebin the data to a few cardiac and respiratory phases and to use this augmented dataset to recover images [22, 23].

13.2.2 Manifold Learning

Manifold learning is another way to introduce prior information into the reconstruction process [24, 25]. It captures nonlinear and nonlocal dependencies between images in the time series. The concept is to assume that each image frame in the dataset is a point on some smooth low-dimensional manifold in high-dimensional space. This assumption becomes reasonable in many real applications, especially when the dynamic movements depend on only a few physiological parameters such as cardiac and respiratory movements [26–29]. By exploiting such nonlinear and nonlocal redundancies in the dataset, manifold-learning methods solve Eq. (13.4) with the manifold-smoothness regularization $\mathcal{R}\{\mathbf{X}\} = \|\mathcal{M}\{\mathbf{X}\}\|_2$, where $\mathcal{M}\{\cdot\}$ is a manifold-mapping function. Because this strategy requires a knowledge of the manifold, most methods use a "navigator acquisition scheme" that approximates the manifold's structure (or, equivalently, the associated graph-Laplacian matrix) using pre-calculated or learned relationships among a set of k-space measurements. For example, in Cartesian sampling, a set of densely scanned low-frequency measurements is used as navigator signals and, in radial sampling, a set of nearby radial measurements are typically used to estimate the relationship (Figure 13.1). Then, the estimated Laplacian matrix is used as guidance for the inpainting of the missing k-space information.

However, all the above-mentioned methods are limited by constraints over the signal-to-noise ratio (SNR), restrictions in the scanner design (e.g., the coils), hand-picked priors, multiple processing steps, and lack of efficient algorithms for the convex optimization of unstructured data.

13.2.3 Supervised Deep Learning

More recently, along with the development of deep-learning techniques in various biomedical imaging modalities [30–33], supervised learning approaches have also been applied to address the fast and accurate reconstruction of partially sampled MRI [34–43]. However, to fully exploit the advantages of supervised learning, one needs to acquire a large set of ground-truth data. In dMRI, this would require acquisition of the full (k, t)-space at a high temporal rate, which is impossible because of fundamental physical limitations such as relaxation times. This makes it difficult to employ supervised learning approaches for dMRI reconstruction.

13.3 Regularizing Deep-Neural-Network Paradigm

The regularization paradigm in the context of deep neural networks is a unique unsupervised approach that combines the benefits of deep-learning approaches and model-based optimization methods. In contrast with supervised-learning methods, this approach requires neither prior training nor a paired set of training data [44, 45].

13.3.1 Deep Image Prior (DIP)

Deep image prior (DIP) [44] is one of the most representative and most successful examples in this approach. Instead of learning a direct mapping from the data to the ground-truth image, DIP deploys a random latent variable input in an attempt to fit its output to a (noisy) measurement, by training a deep generative neural network. More specifically, similarly to the canonical model-based optimization methods, DIP finds the parameters θ of a neural network f_θ by iteratively optimizing over the cost function:

$$\theta^* = \arg\min_{\theta} \| \mathbf{y} - \mathbf{H}(f_\theta(\mathbf{z})) \|_2^2, \tag{13.5}$$

where $\mathbf{z} \in \mathbb{R}^L$ is a fixed but random input, $\mathbf{H} \in \mathbb{R}^{M \times N}$ encodes the forward model, and $\mathbf{y} \in \mathbb{R}^M$ is the measurement. For example, in the image-super-resolution problem, $\mathbf{y}$ is a noisy low-resolution image and $\mathbf{H}$ is a downsampling operator. The output of the optimized network $\mathbf{x}^* = f_{\theta^*}(\mathbf{z})$ then yields a reconstructed high-resolution image. The key intuition behind DIP is that the structure of the deep network itself, which could be a series of convolutional layers with nonlinear operations, already imposes a strong inductive prior on the search space. Ulyanov *et al.* [44] found that, with enough model capacity and sufficiently many iterations, the model can eventually overfit the data, thus even fitting the noise, but it does so very reluctantly. As a result, the network favors smooth and natural-looking images over unstructured noise. It is found empirically that DIP achieves surprisingly good reconstruction quality when applied with appropriate care such as an early stopping of the optimization procedure [44, 45].

Owing to this implicit algorithm, which is agnostic to data, DIP is applicable to a variety of linear inverse problems such as image denoising, inpainting, and super-resolution, by simply adjusting the forward model. However, as its application has been restricted to static images with a relatively simple linear forward model, more sophisticated methods for the reconstruction of dMRI images are needed.

13.3.2 Time-Dependent Deep Image Prior (TD-DIP)

To solve the challenging dMRI reconstruction problem, we need to extend the concept of DIP. First, instead of taking a single input, the time-dependent deep image prior (TD-DIP) takes a sequence $\{\mathbf{z}_s\}_{s=0}^{S-1}$ of inputs, where $\mathbf{z}_s \in \mathbb{R}^L$ (Fig. 13.2). Then, we optimize a simple common neural network f_θ to map this input sequence to the sequence $\{\mathbf{y}_s\}_{s=0}^{S-1}$ of spokes. The weights of the network are adjusted (i.e., trained) to finally take the value

$$\theta^* = \arg\min_\theta \frac{1}{S} \sum_{s=0}^{S-1} \|\mathbf{y}_s - \mathbf{H}_s f_\theta(\mathbf{z}_s)\|^2. \tag{13.6}$$

Note that the optimization is done in the frequency domain, unlike the original DIP, which works in the image domain. This naturally enforces the output image sequence to be consistent with the k-space measurements. After the training is finished, one then reconstructs the dynamic image sequence $\{\mathbf{x}_s^*\}_{s=0}^{S-1}$ as $\mathbf{x}_s^* = f_{\theta^*}(\mathbf{z}_s)$.

Manifold Design

When successful, and given enough information, the network is able to catch the temporal redundancy of data even with a set of randomly sampled inputs, i.e., $\{\mathbf{z}_s\}_{s=0}^{S-1}$ contains S i.i.d. realizations. However, when the reconstruction involves complicated physics and there are too few measurements, as is the case in dMRI, we need to devise a stronger prior to regularize the search space of the underlying model. To this end, we explicitly designed a manifold $\mathcal{Z}$, thereby effectively injecting another structural prior into the network. For a quasi-periodic signal such as cardiac motion, we can encode the expected behavior by letting the manifold take the structure of a helix – we sample an ordered sequence $\{\mathbf{z}_s\}_{s=0}^{S-1}$ from the helix, where $\mathbf{z}_s \in \mathbb{R}^3$. This guides the network to associate spatial closeness of the input variables with temporal closeness of the images. This helps the network to understand the temporal variation of the measurements and to reconstruct an image sequence accordingly.

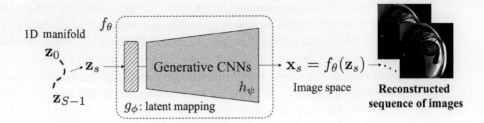

Figure 13.2 Image restoration with TD-DIP.

Algorithm 13.1 Time-dependent DIP for dMRI

Input: Set $\{\mathbf{y}_s\}_{s=0}^{S-1}$ as the number of measurements, number of iterations n_{iter}, batch size B, and number of cycles p.
Output: Reconstructed images $\{\mathbf{x}_s\}_{s=0}^{S-1}$.

1: Select a manifold $\mathcal{Z}$.
2: Sample $\{\mathbf{z}_s\}_{s=0}^{S-1}$ from $\mathcal{Z}$.
3: **for** n_{iter} iterations **do**
4:     Randomly sample a batch $\{s_0, \ldots, s_{B-1}\}$ of size B from $\{0, \ldots, S-1\}$.
5:     Compute the batch loss of Eq. (13.7).
6:     Update $\boldsymbol{\theta}$ with gradient $\nabla_{\boldsymbol{\theta}} L_B(\boldsymbol{\theta})$.
7: **end for**
8: Reconstruct $\{\mathbf{x}_s\}_{s=0}^{S-1} = \{(h \circ g)_{\boldsymbol{\theta}^*}(\mathbf{z}_s)\}_{s=0}^{S-1}$.

Mapping Network (MapNet)

Although a helix represents cardiac motion well, the heartbeats may lack regularity. These small variations may confuse the model and prevent it from achieving the best reconstruction. To address this, we introduce a mapping network (MapNet) inspired by a recent study on generative models [46, 47], which is known to provide richer representation power to the network by adding flexibility to the fixed latent space. Here, we used a few fully connected layers with nonlinear operations to design the MapNet g_ϕ. It learns to map the fixed manifold (the helix) into a more expressive latent space $\mathcal{W} = g_\phi(\mathcal{Z})$ of higher dimensions, for instance, $g_\phi : \mathbb{R}^3 \to \mathbb{R}^{64}$. Specifically, our model $f_{\boldsymbol{\theta}}$ now has a hierarchical architecture that consists of the MapNet g_ϕ followed by the CNN h_ψ, so that $f_{\boldsymbol{\theta}} = h_\psi \circ g_\phi$ and $\boldsymbol{\theta} = \{\phi, \psi\}$ (Fig. 13.2). This leads us to replace Eq. (13.6) by

$$L_S(\boldsymbol{\theta}) = \frac{1}{S} \sum_{s=0}^{S-1} \|\mathbf{y}_s - \mathbf{H}_s (h \circ g)_{\boldsymbol{\theta}}(\mathbf{z}_s)\|^2. \quad (13.7)$$

The role of g_ϕ is to learn a flexible latent space that helps h_ψ to reconstruct the true dynamics.

Final Algorithm

Our optimization scheme is given in Algorithm 13.1. We minimize the loss function (13.7) using standard gradient descent methods [48] for n_{iter} iterations. At each iteration, instead of Eq. (13.7) a batch loss $L_B(\boldsymbol{\theta})$ is updated, where a batch $\{s_0, \ldots, s_{B-1}\}$ of size B is randomly sampled from the index set $\{0, \ldots, S-1\}$. The corresponding latent variables $\{\mathbf{z}_{s_b}\}_{b=0}^{B-1}$ are fed to the network and its parameters are updated using the gradient with respect to $\boldsymbol{\theta}$.

Table 13.1. Performance on the retrospective dataset for multiple heart cycles: averaged RSNR over three runs and their standard deviations for several CNN latent-space designs.

Method	RSNR (dB)
GRASP [3]	24.2123
Helix	27.78 ± 0.07
Helix + MapNet	**28.05 ± 0.04**

13.4 Results

To illustrate the performance of TD-DIP and evaluate it quantitatively, we tested our method on a downsampled real cardiac cine dataset. This dataset originally consisted of full k-space measurements that were acquired using a cardiac retrospective gating with a Cartesian sampling pattern. To simulate the undersampling scenario, we sampled the k-space measurements by mimicking the golden-angle radial-sampling procedure.

For comparison, we used the images from the original k-space measurements as the ground truth and GRASP [3] as the baseline method that uses compressed-sensing techniques. We used the regressed signal-to-noise ratio (RSNR) as a quantitative metric. For ground truth $\mathbf{x}$ and reconstructed image $\mathbf{x}^*$, the RSNR is calculated as

$$\text{RSNR} = \max_{a,b \in \mathbb{R}} 20 \log \frac{\|\mathbf{x}\|_2}{\|\mathbf{x} - a\,\mathbf{x}^* + b\|_2}, \tag{13.8}$$

where a higher RSNR corresponds to a better reconstruction.

To demonstrate the effect of MapNet, we compared the final performance with and without it (Table 13.1). Even without MapNet, TD-DIP outperforms the baseline performance by 3.57 dB, which validates the effectiveness of the algorithm. MapNet further enhances the performance over the basic model (TD-DIP without MapNet). This can also be seen from qualitative comparisons (Fig. 13.3). This figure the variation of cardiac movements from the diastolic (relaxed) phase to systolic (contracted) phase. The residual images reveal that the compressed-sensing algorithm has artifacts in the heart areas that undergo dynamic movements, while ours reconstructs both static and dynamic areas. For more extensive results, see [49].

13.5 Discussion

The TD-DIP algorithm takes the best of both the model-based optimization approaches and the learning-based approaches. Unlike many learning-based methods, TD-DIP does not require a large dataset to train the model. Thus, it does not suffer from train–test inconsistency, which allows for more general application to various data

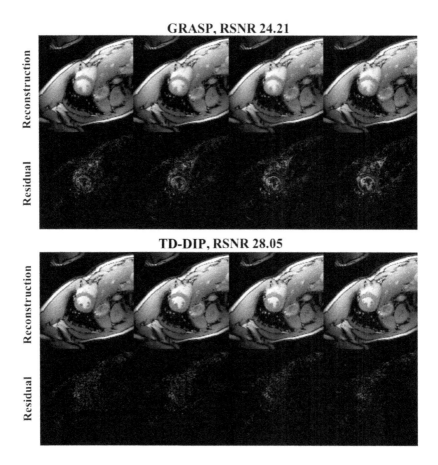

Figure 13.3 Qualitative comparison of the reconstructed cardiac images using GRASP and using TD-DIP. Cardiac movements from the diastole phase to systole phase are displayed. The absolute residual images are shown to highlight the difference between the reconstructed outputs and the ground truth.

from different devices. Meanwhile, it benefits from the rich representation and the regularizing power of deep neural networks so that it can outperform analytically designed priors such as sparsity.

Another major benefit of TD-DIP is its ability to reconstruct temporally continuous dynamic images. It even allows one to recover subframe images by sampling a new input from the helix manifold. Because TD-DIP represents images by means of a learned parametric function f_{θ^*} and a sequence of samples from the helix, it can perform an interpolation in the input manifold to find an unseen, intermediate image between two consecutive frames. This is a unique property of TD-DIP that is impossible in other standard algorithms.

Despite these advantages of our framework, there still remain several theoretical and practical challenges. At present, the main limitation is that we do not clearly understand what image priors the network learns by using this "deep image prior" framework. Although the results are remarkable, what we know is that the deep network somehow works as a denoiser that favors structured or smooth patterns. This needs more investigations and theoretical analysis, which is, we think, an important research direction.

The major practical bottleneck of TD-DIP is that, similarly to other model-based optimization approaches, it requires separate optimization for each case. Unlike the supervised-learning scheme, which moves the learning cost from the online optimization to the offline training, this test–time optimization is inevitable in TD-DIP since it only assumes unpaired data for a single case. Its running time, however, is comparable with or only slightly worse than that of standard CS algorithms, which is acceptable considering the huge improvement in quality of the reconstructed images. As the computational hardware development is expected to continue improving, we believe that this gap will reduce since the parallelization of the training step of our algorithm is easier than that of standard CS algorithms.

13.6 Conclusion

We have described an unsupervised deep-learning-based algorithm for dynamic MRI, called time-dependent deep image prior (TD-DIP). By using an explicit input manifold and a mapping network, TD-DIP is able to exploit the data redundancy and to achieve a high spatial and temporal resolution. The experimental results show that TD-DIP can successfully reconstruct the dynamic movements of a severely undersampled cardiac cine MRI dataset. Since TD-DIP inherits the benefits of the regularizing deep-neural-network paradigm, it can be trained without a paired dataset, while still outperforming other model-based algorithms.

Acknowledgements

We thank Dr. Kyong Hwan Jin, Dr. Harshit Gupta, Dr. Jérome Yerly, and Dr. Matthias Stuber for their contribution to the development of TD-DIP. We also thank Professor Jong Chul Ye at KAIST for providing the bSSFP cardiac MRI k-space dataset. This work was funded partly by the European Research Council (ERC) Grant 692726 (H2020-ERC Project GlobalBioIm).

References

[1] J. C. Ye, "Compressed sensing MRI: a review from signal processing perspective," *BMC Biomedical Engineering*, vol. 1, no. 1, pp. 1–17, 2019.

[2] H. Chandarana, T. K. Block, A. B. Rosenkrantz, R. P. Lim, D. Kim, D. J. Mossa, J. S. Babb, B. Kiefer, and V. S. Lee, "Free-breathing radial 3d fat-suppressed T1-weighted gradient echo sequence: A viable alternative for contrast-enhanced liver imaging in

patients unable to suspend respiration," *Investigative Radiology*, vol. 46, no. 10, pp. 648–653, 2011.

[3] L. Feng, R. Grimm, K. T. Block, H. Chandarana, S. Kim, J. Xu, L. Axel, D. K. Sodickson, and R. Otazo, "Golden-angle radial sparse parallel MRI: Combination of compressed sensing, parallel imaging, and golden-angle radial sampling for fast and flexible dynamic volumetric MRI," *Magnetic Resonance in Medicine*, vol. 72, no. 3, pp. 707–717, 2014.

[4] L. Feng, L. Axel, H. Chandarana, K. Block, D. Sodickson, and R. Otazo, "XD-grasp: Golden-angle radial MRI with reconstruction of extra motion-state dimensions using compressed sensing," *Magnetic Resonance in Medicine*, vol. 75, no. 2, pp. 775–788, 2016.

[5] M. Lustig, D. Donoho, and J. M. Pauly, "Sparse MRI: The application of compressed sensing for rapid MR imaging," *Magnetic Resonance in Medicine*, vol. 58, no. 6, pp. 1182–1195, 2007.

[6] H. Jung, J. C. Ye, and E. Y. Kim, "Improved k–t blast and k–t sense using FOCUSS," *Physics in Medicine and Biology*, vol. 52, no. 11, p. 3201, 2007.

[7] U. Gamper, P. Boesiger, and S. Kozerke, "Compressed sensing in dynamic mri," *Magnetic Resonance in Medicine*, vol. 59, no. 2, pp. 365–373, 2008.

[8] J. Ji and T. Lang, "Dynamic MRI with compressed sensing imaging using temporal correlations," in *Proc. 5th IEEE International Symposium on Biomedical Imaging: From Nano to Macro*, 2008, pp. 1613–1616.

[9] H. Jung, K. Sung, K. Nayak, E. Kim, and J. Ye, "k–t FOCUSS: A general compressed sensing framework for high resolution dynamic MRI," *Magnetic Resonance in Medicine*, vol. 61, no. 1, pp. 103–116, 2009.

[10] R. Otazo, D. Kim, L. Axel, and D. K. Sodickson, "Combination of compressed sensing and parallel imaging for highly accelerated first-pass cardiac perfusion MRI," *Magnetic Resonance in Medicine*, vol. 64, no. 3, pp. 767–776, 2010.

[11] Y. Wang and L. Ying, "Compressed sensing dynamic cardiac cine MRI using learned spatiotemporal dictionary," *IEEE Transactions on Biomedical Engineering*, vol. 61, no. 4, pp. 1109–1120, 2013.

[12] L. Feng, M. B. Srichai, R. P. Lim, A. Harrison, W. King, G. Adluru, E. V. Dibella, D. K. Sodickson, R. Otazo, and D. Kim, "Highly accelerated real-time cardiac cine MRI using k–t sparse-sense," *Magnetic Resonance in Medicine*, vol. 70, no. 1, pp. 64–74, 2013.

[13] S. G. Lingala, Y. Hu, E. Dibella, and M. Jacob, "Accelerated first pass cardiac perfusion MRI using improved k–t SLR," in *Proc. IEEE International Symposium on Biological Imaging*. IEEE, 2011, pp. 1280–1283.

[14] R. Otazo, E. Candès, and D. K. Sodickson, "Low-rank plus sparse matrix decomposition for accelerated dynamic MRI with separation of background and dynamic components," *Magnetic Resonance in Medicine*, vol. 73, no. 3, pp. 1125–1136, 2015.

[15] S. Ravishankar, B. E. Moore, R. R. Nadakuditi, and J. A. Fessler, "Low-rank and adaptive sparse signal (LASSI) models for highly accelerated dynamic imaging," *IEEE Transactions on Medical Imaging*, vol. 36, no. 5, pp. 1116–1128, 2017.

[16] E. J. Candès and B. Recht, "Exact matrix completion via convex optimization," *Foundations of Computational Mathematics*, vol. 9, no. 6, p. 717, 2009.

[17] K. H. Jin, D. Lee, and J. C. Ye, "A general framework for compressed sensing and parallel MRI using annihilating filter based low-rank Hankel matrix," *IEEE Transactions on Computational Imaging*, vol. 2, no. 4, pp. 480–495, 2016.

[18] K. H. Jin, J.-Y. Um, D. Lee, J. Lee, S.-H. Park, and J. C. Ye, "MRI artifact correction using sparse+ low-rank decomposition of annihilating filter-based Hankel matrix," *Magnetic Resonance in Medicine*, vol. 78, no. 1, pp. 327–340, 2017.

[19] D. Lee, K. H. Jin, E. Y. Kim, S.-H. Park, and J. C. Ye, "Acceleration of MR parameter mapping using annihilating filter-based low rank Hankel matrix (ALOHA)," *Magnetic Resonance in Medicine*, vol. 76, no. 6, pp. 1848–1864, 2016.

[20] J. Lee, K. H. Jin, and J. C. Ye, "Reference-free single-pass EPI Nyquist ghost correction using annihilating filter-based low rank Hankel matrix (ALOHA)," *Magnetic Resonance in Medicine*, vol. 76, no. 6, pp. 1775–1789, 2016.

[21] J. C. Ye, J. M. Kim, K. H. Jin, and K. Lee, "Compressive sampling using annihilating filter-based low-rank interpolation," *IEEE Transactions on Information Theory*, vol. 63, no. 2, pp. 777–801, 2016.

[22] J. Yerly, G. Ginami, G. Nordio, A. J. Coristine, S. Coppo, P. Monney, and M. Stuber, "Coronary endothelial function assessment using self-gated cardiac cine MRI and k–t sparse sense," *Magnetic Resonance in Medicine*, vol. 76, no. 5, pp. 1443–1454, 2016.

[23] J. Chaptinel, J. Yerly, Y. Mivelaz, M. Prsa, L. Alamo, Y. Vial, G. Berchier, C. Rohner, F. Gudinchet, and M. Stuber, "Fetal cardiac cine magnetic resonance imaging in utero," *Scientific Reports*, vol. 7, no. 15 540, pp. 1–10, 2017.

[24] M. Belkin, "Problems of learning on manifolds," Ph. D., University of Chicago. 2003.

[25] M. Belkin, P. Niyogi, and V. Sindhwani, "Manifold regularization: A geometric framework for learning from labeled and unlabeled examples," *Journal of Machine Learning Research*, vol. 7, pp. 2399–2434, 2006.

[26] S. Poddar and M. Jacob, "Dynamic MRI using smoothness regularization on manifolds (STORM)," *IEEE Transactions on Medical Imaging*, vol. 35, no. 4, pp. 1106–1115, 2015.

[27] U. Nakarmi, Y. Wang, J. Lyu, D. Liang, and L. Ying, "A kernel-based low-rank (KLR) model for low-dimensional manifold recovery in highly accelerated dynamic MRI," *IEEE Transactions on Medical Imaging*, vol. 36, no. 11, pp. 2297–2307, 2017.

[28] U. Nakarmi, K. Slavakis, J. Lyu, and L. Ying, "M-MRI: A manifold-based framework to highly accelerated dynamic magnetic resonance imaging," in *Proc. IEEE 14th International Symposium on Biomedical Imaging*, 2017, pp. 19–22.

[29] U. Nakarmi, K. Slavakis, and L. Ying, "Mls: Joint manifold-learning and sparsity-aware framework for highly accelerated dynamic magnetic resonance imaging," in *Proc. IEEE 15th International Symposium on Biomedical Imaging*, 2018, pp. 1213–1216.

[30] H. Gupta, K. H. Jin, H. Q. Nguyen, M. T. McCann, and M. Unser, "CNN-based projected gradient descent for consistent CT image reconstruction," *IEEE Transactions on Medical Imaging*, vol. 37, no. 6, pp. 1440–1453, 2018.

[31] E. Kang, W. Chang, J. Yoo, and J. C. Ye, "Deep convolutional framelet denoising for low-dose CT via wavelet residual network," *Transactions on Medical Imaging*, vol. 37, no. 6, pp. 1358–1369, 2018.

[32] J. Yoo, A. Wahab, and J. C. Ye, "A mathematical framework for deep learning in elastic source imaging," *SIAM Journal on Applied Mathematics*, vol. 78, no. 5, pp. 2791–2818, 2018.

[33] J. Yoo, S. Sabir, D. Heo, K. H. Kim, A. Wahab, Y. Choi, S.-I. Lee, E. Y. Chae, H. H. Kim, Y. M. Bae, Y.-W. Choi, and S. Cho, "Deep learning diffuse optical tomography," *IEEE Transactions on Medical Imaging*, vol. 39, no. 4, pp. 877–887, 2019.

[34] Y. Yang, J. Sun, H. Li, and Z. Xu, "Deep ADMM-Net for compressive sensing MRI," in *Advances in Neural Information Processing Systems*, 2016, pp. 10–18.

[35] S. Wang, Z. Su, L. Ying, X. Peng, S. Zhu, F. Liang, D. Feng, and D. Liang, "Accelerating magnetic resonance imaging via deep learning," in *Proc. 13th International Symposium on Biomedical Imaging*. IEEE, 2016, pp. 514–517.

[36] K. H. Jin, M. T. McCann, E. Froustey, and M. Unser, "Deep convolutional neural network for inverse problems in imaging," *IEEE Transactions on Image Processing*, vol. 26, no. 9, pp. 4509–4522, 2017.

[37] K. C. Tezcan, C. F. Baumgartner, R. Luechinger, K. P. Pruessmann, and E. Konukoglu, "MR image reconstruction using deep density priors," *IEEE Transactions on Medical Imaging*, vol. 38, no. 7, pp. 1633–1642, 2019.

[38] K. Hammernik, T. Klatzer, E. Kobler, M. P. Recht, D. K. Sodickson, T. Pock, and F. Knoll, "Learning a variational network for reconstruction of accelerated MRI data," *Magnetic Resonance in Medicine*, vol. 79, no. 6, pp. 3055–3071, 2018.

[39] Y. Han, J. Kang, and J. C. Ye, "Deep learning reconstruction for 9-view dual energy CT baggage scanner," *arXiv:1801.01258*, 2018.

[40] J. Schlemper, J. Caballero, J. V. Hajnal, A. N. Price, and D. Rueckert, "A deep cascade of convolutional neural networks for dynamic MR image reconstruction," *IEEE Transactions on Medical Imaging*, vol. 37, no. 2, pp. 491–503, 2018.

[41] C. Qin, J. Schlemper, J. Caballero, A. Price, J. Hajnal, and D. Rueckert, "Convolutional recurrent neural networks for dynamic MR image reconstruction," *IEEE Transactions on Medical Imaging*, vol. 38, no. 1, pp. 280–290, 2019.

[42] A. Hauptmann, S. Arridge, F. Lucka, V. Muthurangu, and J. Steeden, "Real-time cardiovascular MR with spatio-temporal artifact suppression using deep learning – proof of concept in congenital heart disease," *Magnetic Resonance in Medicine*, vol. 81, no. 2, pp. 1143–1156, 2019.

[43] M. Mardani, E. Gong, J. Y. Cheng, S. S. Vasanawala, G. Zaharchuk, L. Xing, and J. M. Pauly, "Deep generative adversarial neural networks for compressive sensing MRI," *IEEE Transactions on Medical Imaging*, vol. 38, no. 1, pp. 167–179, 2019.

[44] D. Ulyanov, A. Vedaldi, and V. Lempitsky, "Deep image prior," in *Proc. IEEE Conference on Computer Vision and Pattern Recognition*, 2018, pp. 9446–9454.

[45] R. Heckel and P. Hand, "Deep decoder: Concise image representations from untrained non-convolutional networks," in *Proc. International Conference on Learning Representations*, 2019.

[46] T. Karras, S. Laine, and T. Aila, "A style-based generator architecture for generative adversarial networks," in *Proc. IEEE Conference on Computer Vision and Pattern Recognition*, 2019, pp. 4401–4410.

[47] Y. Choi, Y. Uh, J. Yoo, and J.-W. Ha, "StarGAN v2: Diverse image synthesis for multiple domains," in *Proc. IEEE/CVF Conference on Computer Vision and Pattern Recognition*, 2020, pp. 8188–8197.

[48] D. Kingma and J. Ba, "Adam: A method for stochastic optimization," *arXiv:1412.6980*, 2014.

[49] J. Yoo, K. H. Jin, H. Gupta, J. Yerly, M. Stuber, and M. Unser, "Time-dependent deep image prior for dynamic MRI," *arXiv:1910.01684*, 2019.

14 Regularizing Neural Network for Phase Unwrapping

Thanh-an Pham, Fangshu Yang, and Michael Unser

Quantitative phase imaging (QPI) refers to label-free techniques that produce images containing morphological information. In this chapter, we focus on two-dimensional (2D) phase imaging with a holographic setup. In such a setting, the complex-valued measurements contain both intensity and phase information. The phase is related to the distribution of the refractive index of the underlying specimen. In practice, the collected phase happens to be wrapped (i.e., modulo 2π of the original phase) and one gains quantitative information on the sample only once the measurements are unwrapped. The process of phase unwrapping relies on the solution of an inverse problem, for which numerous methods exist. However, it is challenging to unwrap the phase of particularly complex or thick specimens such as organoids. Under such extreme conditions, classical methods often exhibit unwrapping errors. In the following sections, we first formulate the problem of phase unwrapping and review the existing methods for solving it. Then we present an application of regularizing neural networks to phase unwrapping, which allows us to outline the advantages of a training-free approach, i.e., using a deep image prior, over classical methods or supervised learning.

14.1 Problem Formulation

In 2D phase unwrapping, one wants to recover the phase image $\boldsymbol{\Phi} = (\phi_n) \in \mathbb{R}^N$ from the observed and wrapped phase image $\boldsymbol{\Psi} = (\psi_n) \in [-\pi, \pi)^N$ (e.g., acquired with a digital holographic setup [1]), where both images are represented by vectors, by simply stacking the pixels. The two quantities are related by

$$\boldsymbol{\Phi} = \boldsymbol{\Psi} + 2\pi \mathbf{k}, \tag{14.1}$$

where the integer vector $\mathbf{k} \in \mathbb{Z}^N$ accounts for the integer multiples of 2π usually referred to as "wrap-count" [2]. The pixel-wise observation model is then

$$\psi_n = \mathcal{W}(\phi_n) = \big((\phi_n + \pi) \bmod (2\pi)\big) - \pi, \tag{14.2}$$

where $\mathcal{W}: \mathbb{R} \to [-\pi, \pi)$ represents the wrapping process. There exists another useful relation between the phases $\boldsymbol{\Phi}$ and $\boldsymbol{\Psi}$, given by

$$\mathcal{W}([\boldsymbol{\Delta\Phi}]_{n,*}) = \mathcal{W}([\boldsymbol{\Delta\Psi}]_{n,*}), \quad n \in [1, \ldots, N]. \tag{14.3}$$

In Eq. (14.3), the operator $\boldsymbol{\Delta} : \mathbb{R}^N \mapsto \mathbb{R}^{N \times 2}$ is the discrete gradient

$$\boldsymbol{\Delta \Phi} = \begin{bmatrix} \boldsymbol{\Delta}_x \boldsymbol{\Phi} & \boldsymbol{\Delta}_y \boldsymbol{\Phi} \end{bmatrix}, \tag{14.4}$$

where $\boldsymbol{\Delta}_x : \mathbb{R}^N \mapsto \mathbb{R}^N$ and $\boldsymbol{\Delta}_y : \mathbb{R}^N \mapsto \mathbb{R}^N$ denote horizontal and vertical finite-difference operations, respectively. One takes advantage of Eq. (14.3) when the phase $\boldsymbol{\Phi}$ satisfies the so-called Itoh continuity condition [3]

$$\|[\boldsymbol{\Delta \Phi}]_{n,*}\|_2^2 \leq \pi^2, \quad n \in [1, \ldots, N], \tag{14.5}$$

where $\|\cdot\|_2$ denotes the L_2-norm and $[\boldsymbol{\Delta \Phi}]_{n,*} \triangleq ([\boldsymbol{\Delta}_x \boldsymbol{\Phi}]_n, [\boldsymbol{\Delta}_y \boldsymbol{\Phi}]_n)$ represents the nth component of the vector of the discrete gradient (i.e., the nth row of $\boldsymbol{\Delta \Phi}$). If Eq. (14.5) is satisfied, then Eq. (14.3) simplifies as

$$[\boldsymbol{\Delta \Phi}]_{n,*} = \mathcal{W}([\boldsymbol{\Delta \Psi}]_{n,*}), \quad n \in [1, \ldots, N]. \tag{14.6}$$

This seminal relation forms the basis of numerous recovery methods, because it yields a differentiable observation model $[\boldsymbol{\Delta \Phi}]_{n,*}$ for the modified measurements $\mathcal{W}([\boldsymbol{\Delta \Psi}]_{n,*})$.

14.1.1 Classical Methods

There are two main categories of phase-unwrapping algorithms: path-following [4, 5], and minimum-norm [6–8].

Because wrapping events would artificially affect the phase difference between adjacent pixels, path-following algorithms aim to minimize those values. They perform a line integration along some path established by techniques such as a quality-guided algorithm [9] or a branch-cut algorithm [4]. In consequence, the unwrapped phase might depend on the chosen path.

Minimum-norm methods, by contrast, are global. They estimate the unwrapped phase by optimizing

$$\boldsymbol{\Phi}^* = \arg \min_{\boldsymbol{\Phi} \in \mathbb{R}^N} \|[\boldsymbol{\Delta \Phi} - \mathcal{W}(\boldsymbol{\Delta \Psi})]_{*,1}\|_p^p + \|[\boldsymbol{\Delta \Phi} - \mathcal{W}(\boldsymbol{\Delta \Psi})]_{*,2}\|_p^p, \tag{14.7}$$

where $\|\cdot\|_p$ denoted the discrete ℓ_p norm for some $p > 0$. For $p = 2$ (least squares methods) [10], Eq. (14.7) admits a direct solution that may be obtained by fast Fourier or discrete cosine transformations [7]. However, the L_2 norm tends to produce smooth image edges, especially in the vicinity of the wrapped events [6]. This drawback is reduced by setting $0 \leq p \leq 1$, at the expense of an increased computational cost.

The authors of [11] proposed an iteratively reweighted $L_{2,1}$ mixed norm for the data-fidelity term combined with a Hessian–Schatten norm regularization [12]. They solves the minimization problem with an iterative algorithm (IRTV) based on the alternating-direction method of multipliers [13].

Bioucas-Dias and Valadao [14] designed an energy-minimization framework for phase unwrapping (PUMA) that recovers the wrap-count by computing a series of binary optimizations. Condat et al. [15] recovered the wrap-count in the same spirit;

in effect, they implemented a convex relaxation of the original and nonconvex integer-optimization problem.

14.1.2 Deep-Learning-Based Approaches

Several authors have proposed solutions based on deep learning to address the 2D phase-unwrapping problem. In [16], the wrapped phase was mapped to its unwrapped counterpart by a residual neural network [17]. In [18], a convolutional-neural-network- (CNN)-based framework inspired by semantic segmentation, termed PhaseNet, was used to predict the wrap-count. This was followed by some clustering-based post-processing to further improve the results. More recently, the same team proposed a refined version for extremely noisy conditions [19]; several variants of these techniques have emerged in recent years [20–24].

The aforementioned works rely on supervised learning to map the input–output data pairs. While efficient, these methods require a large training dataset made of pairs of wrapped and ground-truth phase images, which may not be available in some cases. In addition, direct feedforward networks might provide solutions inconsistent with the measurements because they lack a feedback mechanism [25–27]. Nevertheless, these works still suggest that deep learning is an appealing solution for phase unwrapping.

In the following sections, we describe more thoroughly a training-free approach which exploits the remarkable ability of CNNs while ensuring consistency with measurements.

14.2 Phase Unwrapping with Deep Image Prior

14.2.1 Problem Formulation

In phase images, most pixels in Φ satisfy the Itoh continuity condition (14.5). One can then reconstruct the unwrapped phase by computing a weighted energy function and minimizing it as [11]

$$\hat{\Phi} = \arg\min_{\Phi \in \mathbb{R}^N} \sum_{n=1}^{N} w_n(\Phi) \|[\Delta\Phi - \mathcal{W}(\Delta\Psi)]_{n,*}\|_2, \quad (14.8)$$

where $w_n(\Phi) \in \mathbb{R}_{\geq 0}$ is the adaptive nonnegative weight for the nth component of the cost. It is defined as

$$w_n(\Phi) = \begin{cases} \dfrac{1}{\|[\epsilon]_{n,*}\|_2}, & \epsilon_{\min} \leq \|[\epsilon]_{n,*}\|_2 \leq \epsilon_{\max} \\ \dfrac{1}{\epsilon_{\max}}, & \|[\epsilon]_{n,*}\|_2 \geq \epsilon_{\max} \\ \dfrac{1}{\epsilon_{\min}}, & \|[\epsilon]_{n,*}\|_2 \leq \epsilon_{\min} \end{cases}, \quad (14.9)$$

where $\epsilon = (\Delta\Phi - \mathcal{W}(\Delta\Psi))$ enforces sparsity in the loss, and where $\epsilon_{\min}$ and $\epsilon_{\max}$ are the user-defined minimum and maximum boundary weights, respectively. There might be few pixels that violate the Itoh continuity condition, for which one can compensate by imposing prior knowledge (i.e., a regularization term) such as the total variation (TV) [28] or a Hessian–Schatten norm (HS) [12].

Deep Image Prior
In the method to be described next, we add a deep image prior (DIP) [29] to the weighted energy function in Eq. (14.8). The concept of DIP is that the unwrapped phase is generated by the CNN given by

$$\boldsymbol{\Phi} = f_{\boldsymbol{\theta}}(\mathbf{z}), \tag{14.10}$$

where f denotes the neural network and $\boldsymbol{\theta}$ stands for the network parameters to be optimized. The input $\mathbf{z} \in \mathbb{R}^{C \times N}$ to the generative network with C channels is randomly initialized and remains fixed during the optimization. Ulyanov et al. [29] showed that such a prior does regularize the reconstructed image, as a result of the peculiar architecture of CNNs. This structure favors smooth and natural images while making the noise difficult to fit.

Plugging Eq. (14.10) into Eq. (14.8) leads to the optimization solution

$$\hat{\boldsymbol{\theta}} = \arg\min_{\boldsymbol{\theta}} \sum_{n=1}^{N} w_n\big(f_{\boldsymbol{\theta}}(\mathbf{z})\big) \|[\boldsymbol{\Delta}\big(f_{\boldsymbol{\theta}}(\mathbf{z})\big) - \mathcal{W}(\boldsymbol{\Delta}\boldsymbol{\Psi})]_{n,*}\|_2. \tag{14.11}$$

We illustrate in Fig. 14.1 the phase unwrapping achieved with a deep image prior (PUDIP). It is worth underlining that DIP does not require any prior training. The parameters $\boldsymbol{\theta}$ of the network are optimized for each wrapped measurement, which means that the training stage required in supervised learning is absent here.

Note that Eq. (14.8) relies on continuous optimization to solve the discrete-optimization problem (14.1). The only downside of the approach is that the solution that is obtained offers no guarantee regarding consistency between the rewrapped phase $\mathcal{W}(\hat{\boldsymbol{\Phi}})$ and the wrapped phase $\boldsymbol{\Psi}$ [11]. We can overcome this limitation and ensure congruence by the single post-processing step [30]

$$\tilde{\boldsymbol{\Phi}} = f_{\hat{\boldsymbol{\theta}}}(\mathbf{z}) + \mathcal{W}\big(\boldsymbol{\Psi} - f_{\hat{\boldsymbol{\theta}}}(\mathbf{z})\big). \tag{14.12}$$

14.2.2 Architecture and Optimization Strategy

In the above illustrative application, we used an architecture close to a standard U-Net [31] with skip connections which include convolution (four channels) and concatenation. In all the other convolutional layers, a constant number of channels (i.e., 128) was used. The nonlinear activation function is the parametric rectified linear unit [32]. Downsampling to the coarser level is achieved by means of a convolutional module with strides of 2, so that the size of the feature map is halved in the contracting path. The upsampling operation doubles the size through bilinear interpolation.

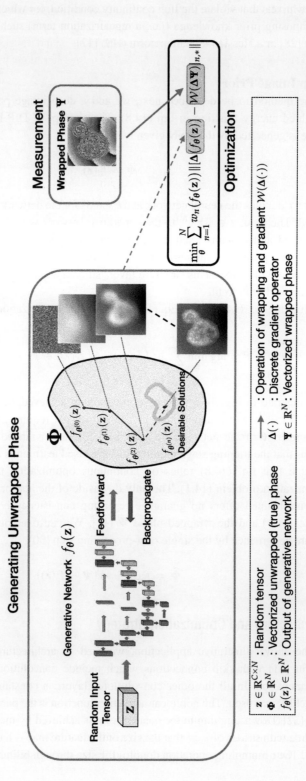

Figure 14.1 Proposed PUDIP for 2D phase unwrapping.

The last layer subtracts an appropriate scalar value from the image. This takes care of the bias intrinsic to phase unwrapping, which can recover phases only up to a global constant. Equation (14.11) was minimized using standard gradient descent methods [33] and the weights w_n were updated every N_w iterations following Eq. (14.9).

14.3 Experiments

In our experiments, we adopted the following strategy. The input variable **z** was a random vector filled with the uniform noise $\mathcal{U}(0, 0.1)$. In practice, the adaptive weights w_n were updated every N_w iterations. We optimized Eq. (14.11) by using an adaptive moment-estimation algorithm (Adam, with $\beta_1 = 0.9$ and $\beta_2 = 0.999$) [33]. The optimization was performed on a desktop workstation (Nvidia Titan X GPU, Ubuntu operating system) and implemented on PyTorch [34]. The reconstruction was repeated four times for each case. The random initialization of **z** did not impact the performance or the time of computation in the experiments. In Algorithm 14.1 we summarize the global optimization scheme.

Algorithm 14.1 Phase unwrapping with deep image prior

Require: $\Psi \in [-\pi, \pi)^N$, N_{outer}, N_w
1: $t \leftarrow 0$
2: $w_n^{(0)} = 1$ for $n = 1, \ldots, N$
3: **for** N_{outer} **do**
4:      Solve with Adam for N_w iterations
5:      $\theta^{(t+1)} = \arg\min_\theta \sum_{n=1}^N w_n^{(t)} \|[\Delta(f_\theta(\mathbf{z})) - \mathcal{W}(\Delta \Psi)]_{n,*}\|_2$
6:      Update the weights with Eq. (15.9) for $n = 1, \ldots, N$: $w_n^{(t+1)} = w_n(\theta^{(t+1)})$
7:      $t \leftarrow t + 1$
8:      Reconstruct $\tilde{\Phi} = f_{\theta^{(t)}}(\mathbf{z}) + \mathcal{W}(\Psi - f_{\theta^{(t)}}(\mathbf{z}))$.
9: **end for**

14.3.1 Simulated Data

Dataset
We simulated samples with controllable difficulties. The samples of type I illustrated in Fig. 14.2 consist of cropped Gaussian blobs similar to [11, 14]. The angle of cropping is increased in such way that phase unwrapping becomes increasingly challenging: the edges in those cropped areas produce phase jumps with an increasing number of pixels that do not respect the Itoh condition. Their presence is sufficient to impact the performance of most methods.

In a second set of samples (type II), we added different levels of speckle noise – a structured noise that occurs in coherent microscopy [35] – to one of the samples of type I. Our motivation there was twofold. First, we wanted to assess the robustness

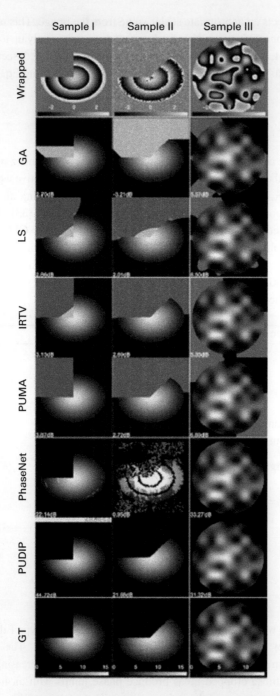

Figure 14.2 Unwrapped phases of three simulated samples. From top to bottom: the wrapped phase; the results obtained by GS, LS, IRTV, PUMA, PhaseNet, and PUDIP; the ground-truth images. From left to right: samples of type I (cropping angle 90 degrees); type II (SNR: 15.7 dB); and type III (matrix size (11 × 11)). The corresponding RSNR in dB is shown at the bottom left of each subfigure.

of the methods to the presence of structured noise. Second, we wanted to probe conventional regularization methods such as the Hessian–Schatten norm which lose their power when the prior is not properly matched to the sample.

Finally, the last set of samples (type III) was directly inspired from [21]. Starting from a random matrix with controllable size, the obtained samples are upsampled via bicubic interpolation to the size of the image of interest. We kept only the central part, as illustrated in Fig. 14.2. The larger the matrix size, the more detailed the image becomes.

Baseline Methods

We compared PUDIP quantitatively with four baseline methods: GA [4], LS [7], PUMA [14], and IRTV [11]. In addition, we performed comparisons with a supervised deep-learning method, PhaseNet [18]. We trained it only once on samples of type III.

Metrics

We used two metrics to assess the quality of the reconstructed phase $\tilde{\Phi}$. The first metric was the regressed signal-to-noise ratio (RSNR), defined as

$$\text{RSNR}(\tilde{\Phi}, \Phi) = \max_{b \in \mathbb{R}^+} \left(20 \log_{10} \left(\frac{\|\Phi\|_2}{\|(\tilde{\Phi} + b) - \Phi\|_2} \right) \right), \tag{14.13}$$

where Φ is the ground truth and b adjusts for a potential global offset. Because phase unwrapping can only recover the phase up to a constant, this adjustment is used to avoid unfair comparisons. Despite correct unwrapping, the recovered phase image sometimes differs from the ground truth because of numerical imprecision. We therefore set the corresponding RSNR to infinity when the RSNR was more than 100 dB. Note that an error in unwrapping can lead to a large difference in the RSNR since it means there is at least a 2π difference between the estimated and true values. To alleviate this effect, we also computed the structural similarity (SSIM), defined as

$$\text{SSIM}(\tilde{\Phi}, \Phi) = \frac{(2\mu_\Phi \mu_{\tilde{\Phi}} + c_1)(2\sigma_{\Phi \tilde{\Phi}} + c_2)}{(\mu_\Phi^2 + \mu_{\tilde{\Phi}}^2 + c_1)(\sigma_\Phi^2 + \sigma_{\tilde{\Phi}}^2 + c_2)}, \tag{14.14}$$

where μ_Φ, $\mu_{\tilde{\Phi}}$, σ_Φ, $\sigma_{\tilde{\Phi}}$, and $\sigma_{\Phi \tilde{\Phi}}$ are the local means, standard deviations, and cross-covariances for images Φ, $\tilde{\Phi}$, respectively. We set the regularization constants to $c_1 = 10^{-4}$ and $c_2 = 9 \times 10^{-4}$.

Results

As shown in Tables 14.1 and 14.2, PUDIP quantitatively outperforms the other methods for all sample types. Let us focus on the samples of type III, shown in Table 14.3. In this case, the training set of PhaseNet is consistent with the testing set, which is expected to perform best. Here, the results show the valuable role of supervised learning when a training set is available. However, the remaining testing sets (types I and II), which mismatch the training set, show a significant drop in performance. This illustrates a powerful use of training-free techniques such as PUDIP. Indeed, the

Table 14.1. RSNS/SSIM, where RSNS is in dB, for the reconstructed-phase images versus the angle of cropping. The RSNR and SSIM of PUDIP are the averages of five experiments.

Angle	GA	LS	IRTV	PUMA	PhaseNet	PUDIP
0°	∞/1.000	∞/1.000	∞/1.000	∞/1.000	24.79/0.979	∞/**1.0000**
45°	6.80/0.897	5.15/0.834	8.61/0.942	10.20/0.959	14.10/0.968	**15.99/0.986**
90°	2.70/0.907	2.86/0.718	3.15/0.733	3.87/0.741	22.14/0.977	**37.75/0.999**
135°	−0.56/0.836	1.32/0.571	2.46/0.651	2.06/0.557	22.01/0.976	**43.52/1.000**
180°	−5.15/0.486	−0.13/0.477	0.84/0.489	∞/1.000	19.33/0.977	∞/**1.000**
225°	−6.70/0.426	−0.43/0.341	−0.24/0.322	2.21/0.118	19.96/0.985	**41.44/1.000**
270°	−8.00/0.365	−1.85/0.239	−1.66/0.224	2.01/0.083	21.23/0.990	∞/**1.000**

Table 14.2. RSNS/SSIM, where RSNS is in dB, for the reconstructed-phase images versus the noise level. The RSNR and SSIM of PUDIP are the averages of five experiments.

Noise level (dB)	GA	LS	IRTV	PUMA	PhaseNet	PUDIP
22.80	−3.58/0.007	1.67/0.148	2.32/0.161	2.34/0.163	3.24/−0.294	**20.51/0.989**
15.70	−3.21/0.008	2.01/0.126	2.84/0.149	2.72/0.150	0.95/−0.140	**20.94/0.991**
11.82	2.81/0.002	2.36/0.102	3.13/0.119	3.13/0.131	2.45/0.204	**20.80/0.990**

Table 14.3. RSNS/SSIM, where RSNS is in dB, for the reconstructed-phase images versus the size of the random matrix. The metric is averaged over four samples for each size. For each sample, we repeated the reconstructions of PUDIP five times. The reported RSNR of PUDIP is then the average of 20 experiments for each size.

Matrix size	GA	LS	IRTV	PUMA	PhaseNet	PUDIP
(3 × 3)	4.18/0.736	3.84/0.722	3.60/0.725	3.72/0.706	**36.30/0.992**	21.94/0.969
(5 × 5)	5.57/0.682	5.55/0.650	4.87/0.659	5.39/0.647	**31.89**/0.956	21.51/**0.958**
(7 × 7)	5.32/0.663	6.28/0.649	5.72/0.634	5.41/0.640	**21.97/0.957**	19.98/0.929
(9 × 9)	5.53/0.651	6.19/0.702	5.47/0.657	5.99/0.687	**39.71/0.963**	20.80/0.929
(11 × 11)	5.71/0.653	6.88/0.648	6.82/0.657	6.88/0.655	**23.63**/0.923	18.65/**0.934**

results show its stable and excellent performance over all the different settings. In Fig. 14.2 three examples (types I, II, and III) of reconstructions are shown. Note that some isolated pixels are wrongly estimated, which explains why the RSNR does not reach "infinity" for cases with seemingly perfect recovery.

14.3.2 Phase Images of Organoids

We applied PUDIP to phase images of organoids [36]. These biological specimens act as mini-organs of large three-dimensional (3D) size, which is challenging for phase unwrapping. The phase images were acquired with an off-axis digital holography microscope. As shown in Fig. 14.3, PUDIP preserves the integrity of the boundaries

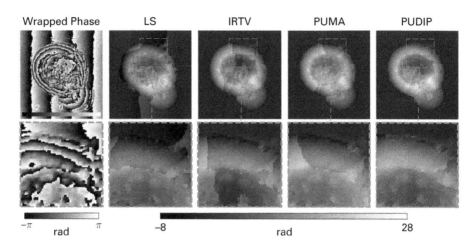

Figure 14.3 Reconstructed phase images of organoids. First column: measured (wrapped) phase image. Second to fifth columns: algorithms using LS, IRTV, PUMA, and PUDIP. First row: reconstructed phase. Second row: zoomed inset. The size of the unwrapped phase image is 260 × 250. For the sake of clarity, we have removed the non-flat (smooth) background of each unwrapped phase. Data courtesy of Professor Matthias Lutolf.

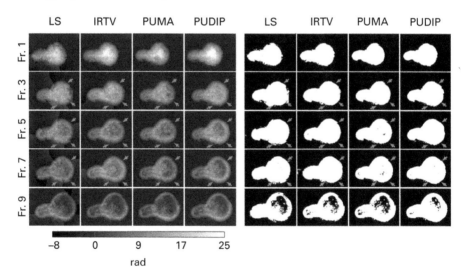

Figure 14.4 Time-lapse reconstructions and segmentation. The frame numbers are indicated at the left-hand side of the figure. The reconstructed images were saturated for visualization purposes. The size of the unwrapped phase image is 280 × 390. For the sake of clarity, we removed the non-flat (smooth) background of each unwrapped phase. We thresholded at 20% of the maximum value of the image. Data courtesy of Professor Matthias Lutolf.

better than the baseline methods do. Since it is made of epithelium, the contour is expected to be somewhat smooth.

In addition, time-lapse measurements of growing organoids were acquired (see Fig. 14.4). We have illustrated the advantage of accurate unwrapping for a typical task: we segmented the unwrapped phase images with a simple threshold (see Fig. 14.4). One observes that the masks are more precise for PUDIP, especially at the border, where it can be seen that the unwrapping could be inaccurate for the baseline methods.

14.4 Discussion

We have presented an application of the deep image prior method for phase unwrapping, with a focus on 2D phase imaging. Deep image prior was able to outperform several baseline methods on simulated and real data. Training-free techniques remove the burden of creating a large training set. While being an adequate approach for biomedical imaging, deep image prior also yields a versatile method that performs well for diverse samples. It is worth mentioning that phase unwrapping is required as well for the 3D variant [37, 38]. Its impact on the reconstructed volume could be of interest for future work.

References

[1] M. K. Kim, "Principles and techniques of digital holographic microscopy," *SPIE Reviews*, vol. 1, no. 1, p. 018 005, 2010.

[2] D. C. Ghiglia and M. D. Pritt, *Two-Dimensional Phase Unwrapping: Theory, Algorithms, and Software*. Wiley, 1998, vol. 4.

[3] K. Itoh, "Analysis of the phase unwrapping algorithm," *Applied Optics*, vol. 21, no. 14, pp. 2470–2470, 1982.

[4] R. M. Goldstein, H. A. Zebker, and C. L. Werner, "Satellite radar interferometry: Two-dimensional phase unwrapping," *Radio Science*, vol. 23, no. 4, pp. 713–720, 1988.

[5] X. Su and W. Chen, "Reliability-guided phase unwrapping algorithm: A review," *Optics and Lasers in Engineering*, vol. 42, no. 3, pp. 245–261, 2004.

[6] D. C. Ghiglia and L. A. Romero, "Minimum L^p-norm two-dimensional phase unwrapping," *Journal of the Optical Society of America A*, vol. 13, no. 10, pp. 1999–2013, 1996.

[7] ——, "Robust two-dimensional weighted and unweighted phase unwrapping that uses fast transforms and iterative methods," *Journal of the Optical Society of America A*, vol. 11, no. 1, pp. 107–117, 1994.

[8] W. He, L. Xia, and F. Liu, "Sparse-representation-based direct minimum L^p-norm algorithm for MRI phase unwrapping," *Computational and Mathematical Methods in Medicine*, vol. 2014, pp. 1–11, 2014.

[9] M. A. Herráez, D. R. Burton, M. J. Lalor, and M. A. Gdeisat, "Fast two-dimensional phase-unwrapping algorithm based on sorting by reliability following a noncontinuous path," *Applied Optics*, vol. 41, no. 35, pp. 7437–7444, 2002.

[10] H. Takajo and T. Takahashi, "Least-squares phase estimation from the phase difference," *Journal of the Optical Society of America A*, vol. 5, no. 3, pp. 416–425, 1988.

[11] U. S. Kamilov, I. P. Papadopoulos, M. H. Shoreh, D. Psaltis, and M. Unser, "Isotropic inverse-problem approach for two-dimensional phase unwrapping," *Journal of the Optical Society of America A*, vol. 32, no. 6, pp. 1092–1100, 2015.

[12] S. Lefkimmiatis, J. P. Ward, and M. Unser, "Hessian Schatten-norm regularization for linear inverse problems," *IEEE Transactions on Image Processing*, vol. 22, no. 5, pp. 1873–1888, 2013.

[13] S. Boyd, N. Parikh, E. Chu, B. Peleato, and J. Eckstein, "Distributed optimization and statistical learning via the alternating direction method of multipliers," *Foundations and Trends in Machine Learning*, vol. 3, no. 1, pp. 1–122, 2011.

[14] J. M. Bioucas-Dias and G. Valadao, "Phase unwrapping via graph cuts," *IEEE Transactions on Image Processing*, vol. 16, no. 3, pp. 698–709, 2007.

[15] L. Condat, D. Kitahara, and A. Hirabayashi, "A convex lifting approach to image phase unwrapping," in *Proc. IEEE International Conference on Acoustics, Speech, and Signal Processing*. IEEE, 2019, pp. 1852–1856.

[16] G. Dardikman and N. T. Shaked, "Phase unwrapping using residual neural networks," in *Proc. Conference on Computational Optical Sensing and Imaging*, 2018, pp. CW3B–5.

[17] K. He, X. Zhang, S. Ren, and J. Sun, "Deep residual learning for image recognition," in *Proc. Conference on Computer Vision and Pattern Recognition*, 2016, pp. 770–778.

[18] G. Spoorthi, S. Gorthi, and R. K. S. S. Gorthi, "PhaseNet: A deep convolutional neural network for two-dimensional phase unwrapping," *IEEE Signal Processing Letters*, vol. 26, no. 1, pp. 54–58, 2018.

[19] G. E. Spoorthi, R. K. S. S. Gorthi, and S. Gorthi, "Phasenet 2.0: Phase unwrapping of noisy data based on deep learning approach," *IEEE Transactions on Image Processing*, vol. 29, pp. 4862–4872, 2020.

[20] T. Zhang, S. Jiang, Z. Zhao, K. Dixit, X. Zhou, J. Hou, Y. Zhang, and C. Yan, "Rapid and robust two-dimensional phase unwrapping via deep learning," *Optics Express*, vol. 27, no. 16, pp. 23 173–23 185, 2019.

[21] K. Wang, Y. Li, Q. Kemao, J. Di, and J. Zhao, "One-step robust deep learning phase unwrapping," *Optics Express*, vol. 27, no. 10, pp. 15 100–15 115, 2019.

[22] J. Zhang, X. Tian, J. Shao, H. Luo, and R. Liang, "Phase unwrapping in optical metrology via denoised and convolutional segmentation networks," *Optics Express*, vol. 27, no. 10, pp. 14 903–14 912, 2019.

[23] C. Li, Y. Tian, and J. Tian, "A method for single image phase unwrapping based on generative adversarial networks," in *Proc. 11th International Conference on Digital Image Processing*, 2019, pp. 272–278.

[24] G. Dardikman-Yoffe, D. Roitshtain, S. K. Mirsky, N. A. Turko, M. Habaza, and N. T. Shaked, "PhUn-Net: Ready-to-use neural network for unwrapping quantitative phase images of biological cells," *Biomedical Optics Express*, vol. 11, no. 2, pp. 1107–1121, 2020.

[25] J. H. Rick Chang, C.-L. Li, B. Poczos, B. V. K. Vijaya Kumar, and A. C. Sankaranarayanan, "One network to solve them all – Solving linear inverse problems using deep projection models," in *Proc. IEEE International Conference on Computer Vision*, 2017, pp. 5888–5897.

[26] H. Gupta, K. H. Jin, H. Q. Nguyen, M. T. McCann, and M. Unser, "CNN-based projected gradient descent for consistent CT image reconstruction," *IEEE Transactions on Medical Imaging*, vol. 37, no. 6, pp. 1440–1453, 2018.

[27] F. Yang, T.-A. Pham, H. Gupta, M. Unser, and J. Ma, "Deep-learning projector for optical diffraction tomography," *Optics Express*, vol. 28, no. 3, pp. 3905–3921, 2020.

[28] L. I. Rudin, S. Osher, and E. Fatemi, "Nonlinear total variation based noise removal algorithms," *Physica D*, vol. 60, no. 1, pp. 259–268, 1992.

[29] D. Ulyanov, A. Vedaldi, and V. Lempitsky, "Deep image prior," in *Proc. IEEE Conference on Computer Vision and Pattern Recognition*, 2018, pp. 9446–9454.

[30] M. D. Pritt, "Congruence in least-squares phase unwrapping," in *Proc. IEEE International Geoscience and Remote Sensing Symposium*, 1997, pp. 875–877.

[31] O. Ronneberger, P. Fischer, and T. Brox, "U-net: Convolutional networks for biomedical image segmentation," in *Proc. International Conference on Medical Image Computing and Computer-Assisted Intervention*. Springer, 2015, pp. 234–241.

[32] K. He, X. Zhang, S. Ren, and J. Sun, "Delving deep into rectifiers: Surpassing human-level performance on imagenet classification," in *Proc. IEEE international Conference on Computer Vision*, 2015, pp. 1026–1034.

[33] D. P. Kingma and J. Ba, "Adam: A method for stochastic optimization," in *Proc. 3rd International Conference on Learning Representations*, 2015.

[34] N. Ketkar, "Introduction to Pytorch," in *Deep Learning with PyThon*. Springer, 2017, pp. 195–208.

[35] J. W. Goodman, *Speckle Phenomena in Optics: Theory and Applications*. Roberts and Co. Publishers, 2007.

[36] N. Brandenberg, S. Hoehnel, F. Kuttler, K. Homicsko, C. Ceroni, T. Ringel, N. Gjorevski, G. Schwank, G. Coukos, G. Turcatti, and M. P. Lutolf, "High-throughput automated organoid culture via stem-cell aggregation in microcavity arrays," *Nature Biomedical Engineering*, vol. 4, no. 9, pp. 863–874, 2020.

[37] U. S. Kamilov, I. P. Papadopoulos, M. H. Shoreh, A. Goy, C. Vonesch, M. Unser, and D. Psaltis, "Learning approach to optical tomography," *Optica*, vol. 2, no. 6, pp. 517–522, 2015.

[38] T.-A. Pham, E. Soubies, A. Ayoub, J. Lim, D. Psaltis, and M. Unser, "Three-dimensional optical diffraction tomography with Lippmann–Schwinger model," *IEEE Transactions on Computational Imaging*, vol. 6, pp. 727–738, 2020.

15 CryoGAN: A Deep Generative Adversarial Approach to Single-Particle Cryo-EM

Michael T. McCann, Laurène Donati, Harshit Gupta, and Michael Unser

CryoGAN [1] uses ideas from deep generative adversarial learning to perform image reconstruction in single-particle cryo-electron microscopy (cryo-EM). In this chapter, we begin by introducing single-particle cryo-EM. We then formulate the associated image-reconstruction problem and discuss the main solutions found in the literature. Next, we describe the CryoGAN algorithm and show some representative results. Finally, we discuss what our experiences with CryoGAN suggest about the advantages and disadvantages of such deep generative adversarial methods in single-particle cryo-EM and beyond.

15.1 The Reconstruction Problem in Single-Particle Cryo-EM

15.1.1 The Quest for Protein Structures

Cryo-EM encompasses a broad range of imaging methods that exploit the wave-like behavior of electrons in a vacuum to produce a high-resolution visualization of biological structures. At the heart of all cryo-EM disciplines lies the use of a transmission electron microscope (TEM) to image radiation-sensitive samples under cryogenic conditions.

Among the various cryo-EM operating modes, a particularly powerful and popular one is the tomographic setup known as single-particle cryo-EM [2]. Its final goal is to characterize the atomic model of proteins, in other words, the three-dimensional (3D) position of each atom in an amino acid chain. The motivation for this is that proteins carry out crucial functions in cells, and alterations in their structure usually affect their ability to perform these functions. Hence, single-particle cryo-EM plays a major role in structural biology and pharmaceutical research.

15.1.2 Single-Particle Cryo-EM

The key to appreciating the specificity of single-particle cryo-EM is to first understand that the conventional tomographic approach, which rotates a single object and takes measurements at every stage of the rotation, is not a viable option for the imaging of single proteins. The reason is that their sensitivity to radiation is so high that they

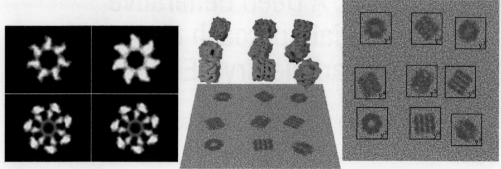

(a) Slices of a simulated protein structure
(b) Imaging procedure
(c) Micrograph showing projections

Figure 15.1 Single-particle cryo-EM. The contrast and signal-to-noise ratio (SNR) of the measurements in the micrograph have been exaggerated for visualization purposes. In practice, the particles are barely visible.

cannot withstand the repetitive exposure to electron beams that would otherwise be necessary in the standard setup.

The twist to get around this is to work with numerous 3D copies of the same protein, called "particles," which are assumed to be structurally identical. The idea is to let these particles adopt random positions and orientations in a thin layer of water, freeze them at cryogenic temperatures, and image all particles simultaneously with parallel electron beams in as few exposures as possible (Fig. 15.1(a)). The collected measurements are called micrographs (Fig. 15.1(b)). They contain thousands of 2D projections of the protein under different orientations. A series of preprocessing steps are then applied to localize, extract, and normalize the individual 2D projections from their micrographs. Finally, sophisticated algorithmic schemes are used to combine the large set of measurements, the ultimate goal being the production of a high-resolution 3D reconstruction (i.e., better than 4 Å). The popularity of this method has rocketed in recent years, culminating in 2017 with Nobel Prizes awarded to Jacques Dubochet, Richard Henderson, and Joachim Frank.

15.1.3 The Image-Formation Model

It is standard in the field [3–5] to model the acquisition of the nth projection through the linear relationship

$$\mathbf{y}^n = \mathbf{H}_{\varphi^n}\mathbf{x} + \mathbf{n}^n, \qquad (15.1)$$

where $\mathbf{y}^n \in \mathbb{R}^M$ is the nth 2D projection image, $\mathbf{x} \in \mathbb{R}^V$ is the 3D structure of interest, and $\mathbf{H}_{\varphi^n} \in \mathbb{R}^{M \times V}$ is the forward operator with imaging parameters $\varphi^n = (\boldsymbol{\theta}^n, \mathbf{t}^n, \mathbf{c}^n)$. These include the projection (Euler) angles $\boldsymbol{\theta}^n = (\theta_1^n, \theta_2^n, \theta_3^n)$, the projection shifts $\mathbf{t}^n = (t_1^n, t_2^n)$, and the contrast transfer function (CTF) parameters $\mathbf{c} = (d_1, d_2, \alpha_{\text{ast}})$, where d_1 is the defocus-major, d_2 the defocus-minor, and α_{ast} the angle of astigmatism. The additive noise $\mathbf{n}^n \in \mathbb{R}^M$ follows a distribution $p_\mathbf{n}$.

The forward operator $\mathbf{H}_\varphi$ is traditionally decomposed as

$$\mathbf{H}_\varphi = \mathbf{C}_\mathbf{c} \mathbf{S}_\mathbf{t} \mathbf{P}_\theta. \qquad (15.2)$$

The projection operator $\mathbf{P}_\theta : \mathbb{R}^V \to \mathbb{R}^M$ (mathematically speaking, the X-ray transform [6]) models the interaction between straight electron beams and the sample. The shift operator $\mathbf{S}_\mathbf{t} : \mathbb{R}^M \to \mathbb{R}^M$ accounts for the fact that particle projections may not be exactly in the center of the corresponding projection image (see Fig. 15.1). The convolution operator $\mathbf{C}_\mathbf{c} : \mathbb{R}^M \to \mathbb{R}^M$ models additional optical effects.

Two major types of noise affect cryo-EM measurements undesirably. They originate from the interaction of electrons with ice and from electron-counting by the TEM detectors [7, 8]. The former leads to a noise that can be modeled as white Gaussian [9]. The latter, a counting process that is by nature discrete, leads to a noise that is usually dominated by Poisson statistics. It is common in the field to model the combination of these types of noise as an additive Gaussian noise.

15.1.4 A Challenging Reconstruction Procedure

The reconstruction task in single-particle cryo-EM consists in the recovery of the 3D structure of interest $\mathbf{x}$ from its set $\{\mathbf{y}^n\}_{n=1}^N$ of 2D measurements, where each measurement is acquired through Eq. (15.1) with unknown imaging parameters.

While the CTF and shift parameters are relatively simple to estimate, the main challenge is the determination of the orientation θ^n for each projection image $\mathbf{y}$, for $n \in 1, \ldots, N$. Indeed, the precise knowledge of the 3D pose of each particle is essential for the deployment of standard tomographic reconstruction. In addition, interactions with the ice surface can sometimes drive particles to favor certain orientations, which leads to a nonuniform distribution of angles. Moreover, as we have just seen, the acquired projections are always extremely numerous, heavily degraded by noise, and affected by complex optical effects.

This collection of problems makes the reconstruction task in single-particle cryo-EM an enduring technical challenge: scientists have spent the better part of the last 30 years designing a solid pipeline that can reliably deliver 3D structures with atomic resolution. The result is an intricate multistep procedure that permits the regular discovery of new structures but which can still be prone to overfitting and lack of reproducibility [10].

15.2 Reconstruction Approaches in Single-Particle Cryo-EM

Several reconstruction approaches have been developed to handle the strongly incomplete data in single-particle cryo-EM. In this section, we give a brief overview of the three main classes of reconstruction algorithms.

15.2.1 Projection Matching

Projection matching approaches refine an initial volume (i.e., 3D image) by alternating between estimation of the orientations and 3D reconstruction from the projections, given the current (however inaccurate) parameter estimates [11, 12]. Speaking very generally, projection matching solves the problem

$$\arg\min_{\mathbf{x},\varphi^1,\ldots,\varphi^N} \sum_{n=1}^{N} D(\mathbf{H}_{\varphi^n}\mathbf{x}, \mathbf{y}^n), \tag{15.3}$$

where D measures the mismatch between the simulated and real projections; the pixel-wise sum of squared errors is a common choice.

The first rough 3D structure is often computed from high-SNR class averages – a complicated task in itself given the challenging imaging conditions. From this first volume, one produces a finite number of synthetic projections with uniformly distributed orientations. These projection templates are used to predict the relative orientation of every 2D projection in the experimental dataset through some appropriate angular-assignment method [13]. This process is then repeated with an increasing number of distinct synthetic projection templates until the optimization fulfills some convergence criterion.

15.2.2 Maximum-Likelihood Estimation

Most modern reconstruction software packages for single-particle cryo-EM rely on a maximum-likelihood (ML) formulation [14]; an early work discussing this method is [15]. The critical refinement over the projection matching (15.3) is that the unknown imaging parameters are marginalized away rather than estimated. The ML problem can be written as

$$\arg\max_{\mathbf{x}} \prod_{n=1}^{N} \int p_{\mathbf{n}}(\mathbf{y}^n - \mathbf{H}_\varphi \mathbf{x}) p_\varphi d\varphi, \tag{15.4}$$

where $p_{\mathbf{n}}$ is the noise distribution and p_φ is the distribution of the unknown imaging parameters. (Often, only the orientations are truly unknown, while the distribution may be considered uniform.) Here, the product arises from the assumption that the projections are statistically independent. The integration corresponds to marginalization (typically performed via numerical integration). Equation (15.4) can be solved using an expectation-maximization (ML-EM) algorithm or using gradient descent. The former is preferred for iterative refinement [16], while a stochastic version of the latter was used for generating initial volume estimates in [17]. In ML-EM, explicit marginalization is sidestepped in the expectation step (E-step) by computing the conditional distribution on the poses for each projection given the volume estimate. This conditional distribution is then used to update the volume in the marginalization step (M-step). The bottom line is that all ML techniques require, either implicitly or explicitly, computations over a large number of poses for each projection.

15.2.3 Method of Moments

The method of moments avoids estimation of the imaging parameters (which is needed in projection matching) or their marginalization (which is needed in ML estimation). Instead, it computes features of the set of measurement projections that can be related to features of the volume and then used for reconstruction. A natural choice for these features is statistical moments, hence the name. For example, we can formulate this as

$$\arg\min_{\mathbf{x}} D(f(\mathbf{x}), g(\{\mathbf{y}^n\}_{n=1}^N)), \qquad (15.5)$$

where g estimates the first and second moments of the projection figures, f computes the same moments for the projections of $\mathbf{x}$ (which can be done analytically from $\mathbf{x}$ without simulating projections), and D measures the discrepancies between them [18]. The main benefit of the approach is that the estimate of the moments gets more and more accurate as the number N of projections grows; this helps to counteract the low SNR of the individual measurements. The method of moments for single-particle cryo-EM was pioneered in [19] and continues to be an area of active research [18].

15.3 The CryoGAN Framework

A key concept in CryoGAN is to cast the cryo-EM reconstruction problem as one of distribution matching: a problem that aims at finding a volume that generates projections that match the data not in a one-to-one correspondence but as distributions. An advantage of this approach as compared to ML estimation (Section 15.2.2) is that it avoids computations over a large number of poses for each projection. In addition, it can be viewed as an extension of the method of moments (Section 15.2.3) wherein we attempt to match *all* the moments rather than just the first few. We now make the distribution-matching concept mathematically precise, thereby developing the theoretical framework of CryoGAN.

15.3.1 Distribution Matching

Mathematically, we consider a distribution $p_{\text{data}}(\mathbf{y})$, where $\mathbf{y}$ represents any single real-world projection image, and a distribution $p_{\text{model}}(\mathbf{y}|\mathbf{x})$ for a 2D projection produced from a model of image formation and a specific reconstruction $\mathbf{x}$. Distribution matching can then be written as

$$\arg\min_{\mathbf{x}} D(p_{\text{data}}(\mathbf{y}), p_{\text{model}}(\mathbf{y}|\mathbf{x})). \qquad (15.6)$$

The immediate problem with Eq. (15.6) is that we have access to neither distribution in closed form but, rather, have samples from each; thus, some empirical estimation must be performed. Another question is whether solving Eq. (15.6) will actually recover the correct reconstruction in cryo-EM; luckily, there exist theoretical results that point to an affirmative answer [1, 20].

As an illustrative example, we show that the ML formulation (15.4) can be viewed as a form of distribution matching. Choosing the Kullback–Leibler (KL) divergence as D in Eq. (15.6), we have that

$$\arg\min_{\mathbf{x}} \mathrm{KL}(p_{\text{data}}(\mathbf{y}), p_{\text{model}}(\mathbf{y}|\mathbf{x})) = \arg\min_{\mathbf{x}} \mathbb{E}_{\mathbf{y} \sim p_{\text{data}}} \left[\log \frac{p_{\text{data}}(\mathbf{y})}{p_{\text{model}}(\mathbf{y}|\mathbf{x})} \right]$$

$$= \arg\min_{\mathbf{x}} \mathbb{E}_{\mathbf{y} \sim p_{\text{data}}} \left[-\log p_{\text{model}}(\mathbf{y}|\mathbf{x}) \right]$$

$$\approx \arg\max_{\mathbf{x}} \sum_{n=1}^{N} \log p_{\text{model}}(\mathbf{y}^n|\mathbf{x})$$

$$= \arg\max_{\mathbf{x}} \prod_{n=1}^{N} \int p_{\text{model}}(\mathbf{y}^n|\mathbf{x}, \boldsymbol{\varphi}) p_{\boldsymbol{\varphi}} d\boldsymbol{\varphi},$$

where the approximately equals sign expresses empirical approximation. Since the last line has exactly the form of the ML estimation problem (15.4), one can reinterpret this method as finding the reconstruction whose simulated projection distribution best matches that of the real data.

In principle, distribution matching should work for any reasonable choice of the distribution distance. This suggests that, by changing the distance, one should be able to avoid the challenging computations of likelihood that are inherent to the current ML methods. This observation is a major motivator for CryoGAN.

15.3.2 GANs for Distribution Matching

Recently, generative adversarial networks (GANs) have emerged as a promising technique for distribution matching [21]. These algorithms are adversarial in the sense that they comprise two parts, one that tries to generate samples from the target distribution and another that discriminates between the generated samples and samples from the real dataset (Fig. 15.2, upper panel). Each of the two entities is typically a convolutional neural network (CNN). GANs using CNNs have shown impressive ability to generate realistic samples.[1]

What is particularly attractive about GANs for distribution matching is that the step of likelihood estimation is avoided; only a reliable sample for each of the two distributions is required [22]. The samples from the distribution of real data are readily available in the form of the acquired projections, while those from the simulated distribution can be generated by a model of the cryo-EM physics (see Fig. 15.2, lower panel).

15.3.3 Mathematical Framework for CryoGAN

In this section, we describe in detail how CryoGAN makes use of adversarial learning to provide a likelihood-free distribution-matching method for cryo-EM reconstruction.

---

[1] https://thispersondoesnotexist.com/, https://thisartworkdoesnotexist.com/

15 CryoGAN: A Deep Generative Adversarial Approach to Single-Particle Cryo-EM

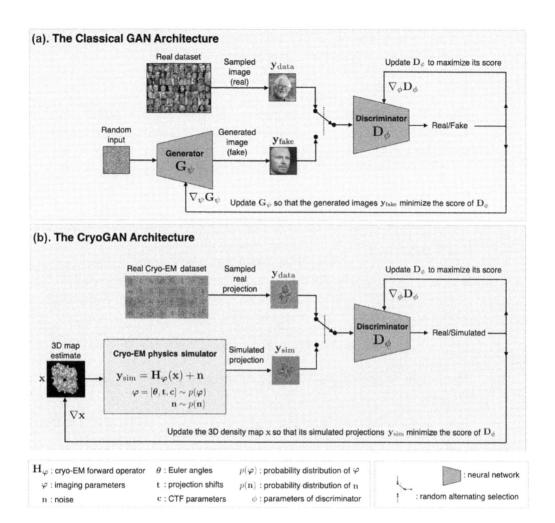

Figure 15.2 Schematic comparison between (a) a classical GAN architecture and (b) the CryoGAN architecture. Both frameworks rely on a deep adversarial learning scheme to capture the distribution of the real data. CryoGAN exploits this ability to look for the volume whose simulated projections have a distribution that matches that of the real data. This is achieved by adding a "cryo-EM physics simulator," which produces measurements following a mathematical model of the cryo-EM imaging procedure. Importantly, CryoGAN does not rely on a first low-resolution volume estimate but is initialized with a zero-valued volume. Note that, for both architectures, the updates involve backpropagation through the neural networks. © 2021 IEEE. Reprinted, with permission, from [1]

We begin by assuming that the measured projections $\{\mathbf{y}^n\}_{n=1}^N$ are i.i.d. samples from a distribution specified by the cryo-EM forward model (15.1) along with known distributions for the imaging parameters and noise. We refer to the probability density function for this distribution as $p(\cdot|\mathbf{x}_{\text{true}})$.

Theorem 1 from [1] states that, under idealized conditions,

$$p(\cdot|\mathbf{x}_{\text{true}}) = p(\cdot|\mathbf{x}) \leftrightarrow \mathbf{x}_{\text{true}} \cong \mathbf{x}, \tag{15.7}$$

where $\cong$ denotes equality up to rotation and reflection. This theorem allows us to formulate the reconstruction task as the minimization problem

$$\arg\min_{\mathbf{x}} D\big(p(\mathbf{y}|\mathbf{x}), p(\mathbf{y}|\mathbf{x}_{\text{true}})\big), \tag{15.8}$$

where D is some distance between the two distributions. In other words, Eq. (15.8) states that when two projection distributions match, the underlying 3D structures also match. For brevity, we introduce the notation $p(\mathbf{y}|\mathbf{x}) = p_{\mathbf{x}}(\mathbf{y})$.

For the distance D in Eq. (15.8), we choose the Wasserstein distance, defined as

$$D(p_1, p_2) = \inf_{\gamma \in \Pi(p_1, p_2)} \mathbb{E}_{(\mathbf{y}_1, \mathbf{y}_2) \sim \gamma} [\|\mathbf{y}_1 - \mathbf{y}_2\|], \tag{15.9}$$

where $\Pi(p_1, p_2)$ is the set of all the joint distributions $\gamma(\mathbf{y}_1, \mathbf{y}_2)$ whose marginals are p_1 and p_2. Our choice is driven by work demonstrating that the Wasserstein distance is more amenable to minimization than other popular distances (e.g., the total variation or the Kullback–Leibler divergence) due to its nondegeneracy and the fact that it provides a nonzero gradient even when distributions have nonoverlapping support [23]. Using Eq. (15.9), the minimization problem (15.8) expands as

$$\mathbf{x}_{\text{rec}} = \arg\min_{\mathbf{x}} \inf_{\gamma \in \Pi(p_{\mathbf{x}}, p_{\text{data}})} \mathbb{E}_{(\mathbf{y}_1, \mathbf{y}_2) \sim \gamma} [\|\mathbf{y}_1 - \mathbf{y}_2\|]. \tag{15.10}$$

By using the formalism of [23–25], this minimization problem can also be stated in the dual form

$$\mathbf{x}_{\text{rec}} = \arg\min_{\mathbf{x}} \max_{f: \|f\|_L < 1} \Big(\mathbb{E}_{\mathbf{y} \sim p_{\text{data}}}[f(\mathbf{y})] - \mathbb{E}_{\mathbf{y} \sim p_{\mathbf{x}}}[f(\mathbf{y})]\Big), \tag{15.11}$$

where $\|f\|_L$ denotes the Lipschitz constant of the function $f : \mathbb{R}^M \to \mathbb{R}$.

One approach to solving Eq. (15.11) is to use a Wasserstein GAN (WGAN) [23]. In the standard WGAN, the function f is a neural network $\mathbf{D}_\phi$ with parameters ϕ and is called the discriminator. The task of this discriminator is to learn to differentiate between real samples (typically coming from an experimental dataset) and "fake" samples. The latter are produced by another neural network, called the generator, which aims at producing samples that are realistic enough to fool the discriminator. This adversarial learning scheme progressively drives the generator to accurately sample the distribution of the experimental data.

In CryoGAN, we use an adversarial schemes similar to a WGAN to learn the volume $\mathbf{x}$ whose simulated projections follow the captured real-data distribution. To do so, we use a cryo-EM physics simulator, whose role is to produce projections of a volume estimate $\mathbf{x}$ using (15.1). These simulated projections then follow a distribution $\mathbf{y} \sim p_{\mathbf{x}}$. Hence, (15.11) translates into

$$\arg\min_{\mathbf{x}} \max_{\phi: \|\mathbf{D}_\phi\|_L \leq 1} \Big(\mathbb{E}_{\mathbf{y} \sim p_{\text{data}}}[\mathbf{D}_\phi(\mathbf{y})] - \mathbb{E}_{\mathbf{y} \sim p_{\mathbf{x}}}[\mathbf{D}_\phi(\mathbf{y})]\Big). \tag{15.12}$$

As proposed in [26], the Lipschitz constraint $\|\mathbf{D}_\phi\|_L \leq 1$ can be enforced by penalizing the norm of the gradient of $\mathbf{D}_\phi$ with respect to its input. This gives the final formulation of our reconstruction problem as

$$\arg\min_{\mathbf{x}} \max_{\phi} \left(\mathbb{E}_{\mathbf{y} \sim p_{\text{data}}}[\mathbf{D}_\phi(\mathbf{y})] - \mathbb{E}_{\mathbf{y} \sim p_{\mathbf{x}}}[\mathbf{D}_\phi(\mathbf{y})] - \lambda \cdot \mathbb{E}_{\mathbf{y} \sim p_{\text{int}}}[(\|\nabla_{\mathbf{y}} \mathbf{D}_\phi(\mathbf{y})\| - 1)^2] \right), \tag{15.13}$$

where p_{int} denotes the uniform distribution along the straight line between points sampled from p_{data} and $p_{\mathbf{x}}$ and $\lambda \in \mathbb{R}_+$ is an appropriate penalty coefficient (see [26], Section 4).

15.4 The CryoGAN Algorithm

We now discuss our implementation of an optimization scheme for Eq. (15.13), along with specifications for its constituent parts: the cryo-EM simulation that creates the simulated measurements $\mathbf{y}_{\text{sim}}^n$ and the discriminator network $\mathbf{D}_\phi$.

Problem (15.13) involves a min–max optimization. By replacing the expected values with their empirical counterparts (sums) [26], we can reformulate it as the minimization of

$$L_S(\mathbf{x}, \phi) = \sum_{m=1}^{M} \mathbf{D}_\phi(\mathbf{y}^{n_m}) - \sum_{m=1}^{M} \mathbf{D}_\phi(\mathbf{y}_{\text{sim}}^m) - \lambda \sum_{m=1}^{M} (\|\nabla_{\mathbf{y}} \mathbf{D}_\phi(\mathbf{y}_{\text{int}}^m)\| - 1)^2, \tag{15.14}$$

where $\{n_m\}_{m=1}^{M}$ is a set of random indices labeling the dataset of measured projection figures $\{\mathbf{y}^n\}_{n=1}^{N}$, M is the number of samples in the empirical estimates, $\{\mathbf{y}_{\text{sim}}^m\}_{m=1}^{M}$ is a set of projections from the current estimate $\mathbf{x}$ generated by the cryo-EM physics simulator, and $\mathbf{y}_{\text{int}}^m = \alpha_m \mathbf{y}^{n_m} + (1 - \alpha_m) \mathbf{y}_{\text{sim}}^m$, where α_m is sampled from a uniform distribution between 0 and 1.

In practice, we minimized Eq. (15.14) through stochastic gradient descent (SGD) using batches and the adaptive moment estimation (ADAM) optimizer [27]. The discriminator $\mathbf{D}_\phi$ (for n_{discr} iterations) and the volume $\mathbf{x}$ (for one iteration) are alternately updated. Our implementation is written using the PyTorch package [28] in Python. A schematic view of the CryoGAN algorithm and its pseudocode is given in Fig. 15.2(b) and Algorithm 15.1, respectively. We provide further details of the CryoGAN physics simulator and the discriminator network in Sections 15.4.1 and 15.4.2. For the full implementation details and links to the source code, see [1].

15.4.1 The Cryo-EM Physics Simulator

The role of the physics simulator is to sample $\mathbf{y}_{\text{sim}} \sim p_{\mathbf{x}}(\mathbf{y})$. This sampling is done in three steps. We start by sampling the imaging parameters φ from the distribution p_φ: $\varphi \sim p_\varphi$. We then generate noiseless CTF-modulated and shifted projections based on the current volume estimate with $\mathbf{H}_\varphi(\mathbf{x})$. Finally, we sample the noise model to simulate noisy projections $\mathbf{y} = \mathbf{H}_\varphi(\mathbf{x}) + \mathbf{n}$, where $\mathbf{n} \sim p_{\mathbf{n}}$. The pseudocode of this cryo-EM physics simulator is given in Algorithm 15.2, and we now detail these steps.

The set of imaging parameters is given by $\varphi = (\theta_1, \theta_2, \theta_3, t_1, t_2, d_1, d_2, \alpha_{\text{ast}})$. We first sample the Euler angles $\boldsymbol{\theta} = (\theta_1, \theta_2, \theta_3)$ from a distribution $p_{\boldsymbol{\theta}}$ decided *a priori*

Algorithm 15.1 Pseudocode for CryoGAN

Parameters: number n_{train} of training iterations; number n_{discr} of iterations of the discriminator per training iteration; size N of the batches used for SGD; penalty parameter λ

1: **for** n_{train} **do**
2:    **for** n_{discr} **do**
3:       sample real projections: $\{\mathbf{y}_{\text{batch}}^1, \ldots, \mathbf{y}_{\text{batch}}^M\}$
4:       simulate projections from current $\mathbf{x}$: $\{\mathbf{y}_{\text{sim}}^1, \ldots, \mathbf{y}_{\text{sim}}^M\}$ ▷ using Alg. 14.1
5:       sample $\{\alpha_1, \ldots, \alpha_m\} \sim U[0, 1]$
6:       for $m \in \{1, \ldots, M\}$, compute $\mathbf{y}_{\text{int}}^m = \alpha_m \mathbf{y}_{\text{batch}}^m + (1 - \alpha_m) \cdot \mathbf{y}_{\text{sim}}^m$
7:       take a gradient step using $\nabla_\phi L_S(\mathbf{x}, \phi)$
8:    **end for**
9:    take a gradient step using $\nabla_\mathbf{x} L_S(\mathbf{x}, \phi)$
10: **end for**
11: **return** reconstruction $\mathbf{x}$ and discriminator parameters ϕ

Algorithm 15.2 Pseudocode for cryo-EM physics simulator

Parameters: current volume estimate $\mathbf{x}$

1: sample the Euler angles $\boldsymbol{\theta} = (\theta_1, \theta_2, \theta_3) \sim p_{\boldsymbol{\theta}}$
2: sample the 2D shifts $\mathbf{t} = (t_1, t_2) \sim p_{\mathbf{t}}$
3: sample the CTF parameters $\mathbf{c} = (d_1, d_2, \alpha_{\text{ast}}) \sim p_{\mathbf{c}}$
4: generate a synthetic projection $\mathbf{y}_{\text{noiseless}} = \mathbf{H}_\varphi \mathbf{x}$ ▷ using Eq. (15.2)
5: sample the noise $\mathbf{n} \sim p_{\mathbf{n}}$; add to the projection as $\mathbf{y}_{\text{sim}} = \mathbf{y}_{\text{noiseless}} + \mathbf{n}$
6: **return** $\mathbf{y}_{\text{sim}}$

---

and based on the acquired dataset. In a similar fashion, the projection shifts $\mathbf{t} = (t_1, t_2)$ are sampled from the prior distribution $p_{\mathbf{t}}$. The CTF parameters $\mathbf{c} = (d_1, d_2, \alpha_{\text{ast}})$ are sampled from the prior distribution $p_{\mathbf{c}}$. In practice, we exploit the fact that the CTF parameters can often be efficiently estimated for all micrographs. We then uniformly sample from the whole set of extracted CTF parameters.

We generate noiseless projections $\mathbf{y}_{\text{noiseless}}$ by applying $\mathbf{H}_\varphi$ to the current volume estimate $\mathbf{x}$. The projection operator $\mathbf{P}_{\boldsymbol{\theta}}$ in Eq. (15.2) is implemented using the ASTRA toolbox [29]. To facilitate gradient computations through the ASTRA projector, we wrap it in a PyTorch module, relying on the fact that the gradient of a linear operator with respect to its inputs can be computed using its adjoint, which ASTRA provides.

The precise modeling of the noise is a particularly challenging feat in single-particle cryo-EM. To produce noise realizations that are as realistic as possible, we extract random background patches from the micrographs themselves at locations where particles do not appear. For consistency, the noise patch added to a given noiseless projection originates from the same micrograph that was used to estimate the CTF parameters previously applied to that projection.

15.4.2 The CryoGAN Discriminator Network

The role of the discriminator is to distinguish the real projections in the experimental dataset from the simulated projections generated by the cryo-EM physics simulator. In other words, it is an image classifier that takes an image as input and outputs a scalar value indicating how realistic it appears. The gradients obtained from the discriminator implicitly contain information about the differences between the simulated and real projections. Those gradients are used to update the current estimate **x** in order to make its simulated projections more realistic.

The discriminator network in our CryoGAN implementation (Fig. 15.3) is a standard image classifier, consisting of several convolutional layers followed by max pooling and ReLU units. As the max pooling reduces the spatial dimensions the convolutions add more channel dimensions, but the total size of the data still decreases with each subsequent layer. After several such layers, there are two final fully connected layers that result in a scalar output.

15.4.3 Reconstruction from a Realistic Synthetic Dataset

To evaluate the performance of CryoGAN, we deployed it on a synthetic dataset that consisted of 41 000 β-galactosidase projections. This dataset was designed to mimic the real EMPIAR-10061 data [30] in terms of CTF parameters and noise level. We sampled projection poses from a uniform distribution over $SO(3)$, which is the group of 3D rotations around the origin in $\mathbb{R}^3$. In order to apply random CTFs and noise,

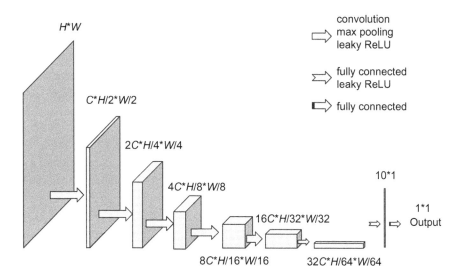

Figure 15.3 Example of a CryoGAN discriminator architecture. The asterisks indicate scalar multiplication; C indicates a channel. It outputs a scalar by iteratively filtering and downsampling the input image of size $H \times W$. © 2021 IEEE. Reprinted, with permission, from [1]

we randomly picked micrographs in the EMPIAR-10061 dataset, extracted their CTF parameters and multiple background (noise) patches, and applied them to the synthetic projections. (Note that hereafter we refer to these synthetic projections as the "real" projections, in contrast with the projections generated by the CryoGAN machinery, which we term "simulated.")

We then divided the dataset randomly into two halves and fed each half separately into the CryoGAN algorithm, to obtain two half-maps. The comparison of half-maps is a standard validation approach in the cryo-EM field. It allows one to assess whether fine details are coming from the data (in which case they will match between the half-maps) or from noise (in which case they might be different in each half-map).

The prior information given to the cryo-EM physics simulator of CryoGAN before the reconstruction includes the distribution of the imaging parameters (poses, shifts, CTFs), which is set to be identical to the distribution used to generate the dataset. The noise is handled by randomly extracting random background patches from the EMPIAR-10061 micrographs (see also Section 15.4.1). The simulator is initialized with a volume of zeros and the discriminator is initialized with random weights.

After running for 400 minutes on an NVIDIA V100 GPU, CryoGAN produces a reconstruction with a resolution of 8.64 Å, as quantified by the Fourier shell correlation (FSC) [31, 32]. The evolution of the reconstruction over time (Fig. 15.4(a)) reveals that CryoGAN progressively updates the 3D structure, quickly finding a low-resolution approximation and refining the details slowly. The evolution of the synthetic projections (Fig. 15.4(b)) tells a similar story, with the early projections looking (by eye) quite realistic. At the end of its run, the volume learned by CryoGAN has simulated projections (Fig. 15.4(c), rows 1–3) that are similar to the real projections (Fig. 15.4(c), row 4) in a *distributional* sense. The evolution of the FSC between the reconstructed half-maps (Fig. 15.4(d)) testifies to the progressive increase in resolution that derives from this adversarial learning scheme.

Hence, this experiment demonstrates the ability of CryoGAN to resolve a structure in such a way that the distribution of its simulated projections approaches that of the experimental particles. These results validate the CryoGAN paradigm and the viability of its current implementation. Indeed, without any prior training and starting from a zero-valued volume, the algorithm is able to autonomously capture the relevant statistical information from the dataset of noise-corrupted, CTF-modulated projections, and to progressively learn the volume that best explains these statistics.

Additional experiments were described in [1], including some with real experimental data.

15.4.4 Next Steps

The CryoGAN algorithm has been validated on synthetic and real data, yet significant efforts remain to obtain a high-resolution structure from a real dataset. Nonetheless, although CryoGAN reconstructions on real data are not yet competitive with state-of-the-art methods in terms of spatial resolution, the algorithm already allows one to obtain a valid first structure solely using the projection measurements and CTF

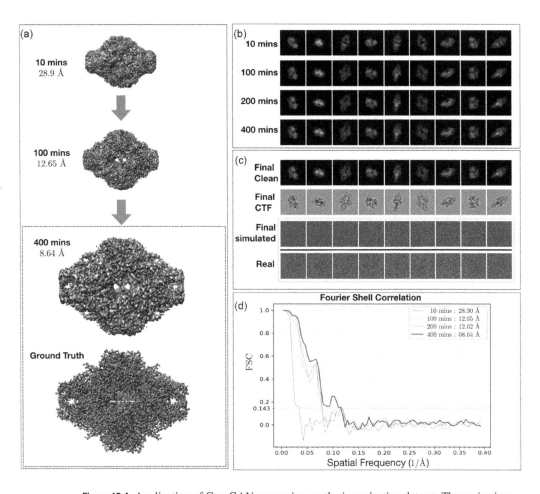

Figure 15.4 Application of CryoGAN on a noisy synthetic projection dataset. The projections were generated using a 2.5Å β-galactosidase ground-truth structure. These synthetic projections are referred as "real," and are in contrast with the "simulated" projections obtained from the cryo-EM physics simulator in CryoGAN. (a) The structure is initialized with zeros and is updated progressively in such a way that the distribution of its simulated projections matches that of the real projections. (b) Evolution of simulated clean projections during the reconstruction. These were obtained from the cryo-EM physics simulator before corruption by CTF and noise. (c) Row 1: Simulated clean projections (before corruption by CTF and noise) obtained at the end of reconstruction. Row 2: Simulated CTF-modulated projections (before corruption by noise) obtained at the end of reconstruction. Row 4: Real projections obtained from the synthetic dataset, for comparison. (d) Evolution of the FSC (between the two reconstructed half-maps) during reconstruction. © 2021 IEEE. Reprinted, with permission, from [1]

estimations. We believe that further developments will permit CryoGAN to more routinely achieve high-resolution reconstruction in single-particle cryo-EM.

An interesting line of research would be to use a coarse-to-fine strategy within CryoGAN to progressively refine the reconstruction. The motivation is that increased robustness during the low-resolution regime may improve the convergence at higher resolution. Several GAN architectures rely on such a multiscale approach [33, 34], and the benefits for CryoGAN could be considerable as well, given the extremely challenging imaging conditions faced in single-particle cryo-EM.

The performance of the CryoGAN algorithm should also increase as our ability to precisely model the physics behind single-particle cryo-EM progresses. For example, other noise models (e.g., Poisson [4, 35]) could be considered in the forward model, provided that backpropagation through the noise distribution is feasible. The cryo-EM physics simulator could also be extended to directly simulate nonaligned micrographs rather than the individual projections. This would bypass pre-processing tasks such as particle picking.

Finally, as previously discussed, the CryoGAN algorithm needs knowledge of the distribution of poses (as do likelihood-based methods). While the assumption of a uniform distribution is a good starting point, we know that, in practice, the angular distribution deviates significantly. A possible improvement to the CryoGAN framework could thus be to learn this distribution during the reconstruction procedure. This could be done by parameterizing it with learnable variables.

15.5 Conclusion

In this chapter, we have described CryoGAN, which uses ideas from deep generative adversarial learning to perform single-particle cryo-EM reconstruction; we now end by discussing our long-term perspective on this approach and the lessons learned during the project.

It is natural to wonder which other biomedical imaging problems might be amenable to reconstruction with an adversarial method like CryoGAN; we believe there are not many. Adversarial learning is a good fit for cryo-EM because there is missing information (the projection directions) that is low-dimensional and reasonably modeled as coming from a known random distribution. Moreover, there are many samples with which to train a large CNN without fear of overfitting. When there is no missing information, e.g., in standard X-ray CT or MRI, ML methods or supervised deep learning are probably better approaches. Adversarial learning is very challenging because the training is sensitive to the selection of hyperparameters – which can be numerous – relating to network architecture, training parameters, and more. Despite this warning, we will be excited to see whether the computational imaging community identifies other modalities where similar ideas can work.

We note a few key differences between CryoGAN and modern ML-based approaches. The ML approaches are currently superior in terms of resolution; whether this is the case because of an inherent limitation of GAN-based reconstruction or because CryoGAN is a less mature method is unclear. We have found the generative approach in CryoGAN to be very flexible compared with our experience with ML formulations. For example, using background patches as noise is straightforward in CryoGAN, but it is hard to imagine doing something equivalent in the ML framework.

We think that this flexibility may allow approaches like CryoGAN to eventually surpass (or at least usefully augment) ML reconstructions.

One place in particular where we think generative modeling may be advantageous is for molecules with continuous conformational changes. A major challenge is that many proteins are actually dynamic entities: they perform functions in cells by changing their configuration upon receiving stimuli. The range of configurations adopted by a protein, called its conformational landscape, means that micrographs may actually contain a mix of projections from distinct 3D structures [36]. If left unaddressed, this can lead to a blurred reconstruction. Indeed, reconstructing the dynamics of structures that exhibit a continuum of conformational states has been called the greatest technical challenge ahead for single-particle cryo-EM [36]. In the CryoGAN framework, this problem could be addressed by using a CNN to parameterize the reconstruction volume itself, transforming a low-dimensional latent random variable into a 3D volume with continuously varying conformations. We have already begun explorations in this direction [37] and see great promise in this approach.

Acknowledgement

This work was carried out with support from the European Research Council (ERC) under the European Union's Horizon 2020 research and innovation program, Grant Agreement No. 692726, GlobalBioIm: Global integrative framework for computational bio-imaging.

References

[1] H. Gupta, M. T. McCann, L. Donati, and M. Unser, "CryoGAN: A new reconstruction paradigm for single-particle cryo-EM via deep adversarial learning," *IEEE Transactions on Computational Imaging*, vol. 7, pp. 759–774, 2021.

[2] J. Frank, *Three-Dimensional Electron Microscopy of Macromolecular Assemblies: Visualization of Biological Molecules in Their Native State*. Oxford University Press, 2006.

[3] ——, *Electron Tomography: Methods for Three-Dimensional Visualization of Structures in the Cell*. Springer, 2008.

[4] H. Rullgård, L.-G. Öfverstedt, S. Masich, B. Daneholt, and O. Öktem, "Simulation of transmission electron microscope images of biological specimens," *Journal of Microscopy*, vol. 243, no. 3, pp. 234–256, 2011.

[5] M. Vulović, R. B. Ravelli, L. J. van Vliet, A. J. Koster, I. Lazić, U. Lücken, H. Rullgård, O. Öktem, and B. Rieger, "Image formation modeling in cryo-electron microscopy," *Journal of Structural Biology*, vol. 183, no. 1, pp. 19–32, 2013.

[6] F. Natterer, *The Mathematics of Computerized Tomography*. SIAM, 2001.

[7] H. Shigematsu and F. Sigworth, "Noise models and cryo-EM drift correction with a direct-electron camera," *Ultramicroscopy*, vol. 131, pp. 61–69, 2013.

[8] Y. Cheng, N. Grigorieff, P. A. Penczek, and T. Walz, "A primer to single-particle cryo-electron microscopy," *Cell*, vol. 161, no. 3, pp. 438–449, 2015.

[9] C. Sorzano, L. De La Fraga, R. Clackdoyle, and J. Carazo, "Normalizing projection images: A study of image normalizing procedures for single particle three-dimensional electron microscopy," *Ultramicroscopy*, vol. 101, nos. 2–4, pp. 129–138, 2004.

[10] R. Henderson, "Avoiding the pitfalls of single particle cryo-electron microscopy: Einstein from noise," *Proceedings of the National Academy of Sciences*, vol. 110, no. 45, pp. 18 037–18 041, 2013.

[11] P. A. Penczek, R. A. Grassucci, and J. Frank, "The ribosome at improved resolution: New techniques for merging and orientation refinement in 3D cryo-electron microscopy of biological particles," *Ultramicroscopy*, vol. 53, no. 3, pp. 251–270, 1994.

[12] T. Baker and R. Cheng, "A model-based approach for determining orientations of biological macromolecules imaged by cryoelectron microscopy," *Journal of Structural Biology*, vol. 116, no. 1, pp. 120–130, 1996.

[13] J. Carazo, C. Sorzano, J. Otón, R. Marabini, and J. Vargas, "Three-dimensional reconstruction methods in single particle analysis from transmission electron microscopy data," *Archives of Biochemistry and Biophysics*, vol. 581, pp. 39–48, 2015.

[14] A. Singer and F. J. Sigworth, "Computational methods for single-particle electron cryomicroscopy," *Annual Review of Biomedical Data Science*, vol. 3, pp. 163–190, 2020.

[15] F. J. Sigworth, "A maximum-likelihood approach to single-particle image refinement," *Journal of Structural Biology*, vol. 122, no. 3, pp. 328–339, 1998.

[16] S. H. Scheres, "RELION: Implementation of a Bayesian approach to cryo-EM structure determination," *Journal of Structural Biology*, vol. 180, no. 3, pp. 519–530, 2012.

[17] A. Punjani, J. L. Rubinstein, D. J. Fleet, and M. A. Brubaker, "cryoSPARC: algorithms for rapid unsupervised cryo-EM structure determination," *Nature Methods*, vol. 14, no. 3, pp. 290–296, 2017.

[18] N. Sharon, J. Kileel, Y. Khoo, B. Landa, and A. Singer, "Method of moments for 3D single particle ab initio modeling with non-uniform distribution of viewing angles," *Inverse Problems*, vol. 36, no. 4, p. 044 003, 2020.

[19] Z. Kam, "The reconstruction of structure from electron micrographs of randomly oriented particles," *Journal of Theoretical Biology*, vol. 82, no. 1, pp. 15–39, 1980.

[20] V. M. Panaretos, "On random tomography with unobservable projection angles," *Annals of Statistics*, vol. 37, no. 6A, pp. 3272–3306, 2009.

[21] I. Goodfellow, J. Pouget-Abadie, M. Mirza, B. Xu, D. Warde-Farley, S. Ozair, A. Courville, and Y. Bengio, "Generative adversarial nets," in *Advances in Neural Information Processing Systems*, 2014, pp. 2672–2680.

[22] S. Mohamed and B. Lakshminarayanan, "Learning in implicit generative models," *arXiv:1610.03483*, 2016.

[23] M. Arjovsky, S. Chintala, and L. Bottou, "Wasserstein generative adversarial networks," in *Proc. International Conference on Machine Learning*, 2017, pp. 214–223.

[24] C. Villani, *Optimal Transport: Old and New*. Springer, 2008, vol. 338.

[25] G. Peyré, M. Cuturi et al., "Computational optimal transport," *Foundations and Trends® in Machine Learning*, vol. 11, nos. 5–6, pp. 355–607, 2019.

[26] I. Gulrajani, F. Ahmed, M. Arjovsky, V. Dumoulin, and A. C. Courville, "Improved training of Wasserstein GANs," in *Advances in Neural Information Processing Systems*, 2017, pp. 5767–5777.

[27] D. Kingma and J. Ba, "Adam: A method for stochastic optimization," *arXiv:1412.6980*, 2014.

[28] A. Paszke, S. Gross, F. Massa, A. Lerer, J. Bradbury, G. Chanan, T. Killeen, Z. Lin, N. Gimelshein, L. Antiga *et al.*, "Pytorch: An imperative style, high-performance deep learning library," in *Advances in Neural Information Processing Systems*, 2019, pp. 8024–8035.

[29] W. van Aarle, W. J. Palenstijn, J. De Beenhouwer, T. Altantzis, S. Bals, K. J. Batenburg, and J. Sijbers, "The ASTRA toolbox: A platform for advanced algorithm development in electron tomography," *Ultramicroscopy*, vol. 157, pp. 35–47, 2015.

[30] A. Bartesaghi, A. Merk, S. Banerjee, D. Matthies, X. Wu, J. L. S. Milne, and S. Subramaniam, "2.2 Å resolution cryo-EM structure of β-galactosidase in complex with a cell-permeant inhibitor," *Science*, vol. 348, no. 6239, pp. 1147–1151, 2015.

[31] N. Biyani, R. D. Righetto, R. McLeod, D. Caujolle-Bert, D. Castano-Diez, K. N. Goldie, and H. Stahlberg, "Focus: The interface between data collection and data processing in cryo-EM," *Journal of Structural Biology*, vol. 198, no. 2, pp. 124–133, 2017.

[32] R. D. Righetto, N. Biyani, J. Kowal, M. Chami, and H. Stahlberg, "Retrieving high-resolution information from disordered 2D crystals by single-particle cryo-EM," *Nature Communications*, vol. 10, no. 1, pp. 1–10, 2019.

[33] T. Karras, T. Aila, S. Laine, and J. Lehtinen, "Progressive growing of GANs for improved quality, stability, and variation," *arXiv:1710.10196*, 2017.

[34] T. Karras, S. Laine, and T. Aila, "A style-based generator architecture for generative adversarial networks," in *Proc. IEEE Conference on Computer Vision and Pattern Recognition*, 2019, pp. 4401–4410.

[35] M. Vulović, R. B. Ravelli, L. J. van Vliet, A. J. Koster, I. Lazić, U. Lücken, H. Rullgård, O. Öktem, and B. Rieger, "Image formation modeling in cryo-electron microscopy," *Journal of Structural Biology*, vol. 183, no. 1, pp. 19–32, 2013.

[36] J. Frank, "New opportunities created by single-particle cryo-EM: the mapping of conformational space," *Agricultural Science and Technology*, vol. 57, no. 2, p. 888, 2018.

[37] H. Gupta, T. H. Phan, J. Yoo, and M. Unser, "Multi-CryoGAN: Reconstruction of continuous conformations in cryo-EM using generative adversarial networks," in *Proc. Workshops on Computer Vision – ECCV 2020*, A. Bartoli and A. Fusiello, eds. Springer, 2020, pp. 429–444.